W9-BKJ-116

HOW THINGS WORK

Weighing objects with a spring balance

Discovering the strength of different types of bridge

Making a miniature fire extinguisher

Recording your voice with a working microphone

Showing the strength of a paper barrel roof

Testing the strength of a miniature skyscraper's frame

HOW THINGS WORK

Exploring the way in which an aircraft's autopilot works

Neil Ardley

Playing your own home-made guitar

Seeing how a camera focuses an image

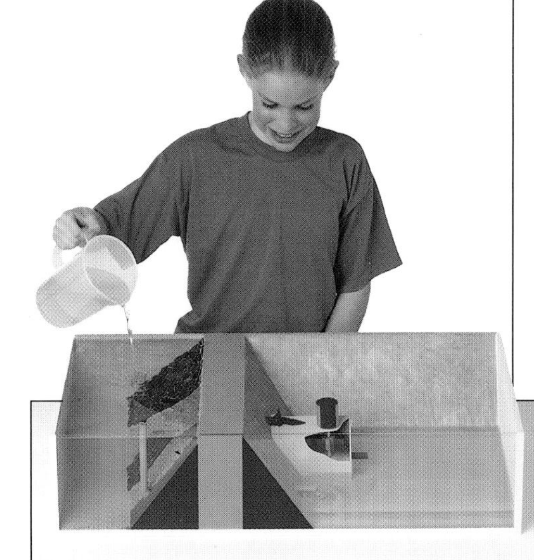

Finding out how dams generate electric power from water

The Reader's Digest Association, Inc.
Pleasantville, New York • Montreal

A READER'S DIGEST BOOK
Designed and edited by
Dorling Kindersley Limited, London

Project Editor Paul Docherty
Senior Art Editors Bryn Walls, Geoff Manders
Designers Philip Ormerod, Christina Betts
Editors Mukul Patel, Katriona John
Production Hilary Stephens
Managing Editor Ruth Midgley
Managing Art Editor Philip Gilderdale

Educational Consultant Tim Kirk
Photography Andy Crawford

The credits and acknowledgments that appear on page 192 are hereby made a part of this copyright page.

Library of Congress Cataloging in Publication Data

Ardley, Neil
How things work / Neil Ardley
p. cm
Includes index.
ISBN 0–89577–694–4
1. Machinery—Juvenile literature. [1. Machinery.
2. Machinery—Experiments. 3. Experiments.] I. Title.
TJ147.A72 1995
621.8—dc20 94–46531

READER'S DIGEST and the Pegasus logo are registered trademarks of The Reader's Digest Association, Inc.

Printed in Singapore

Contents

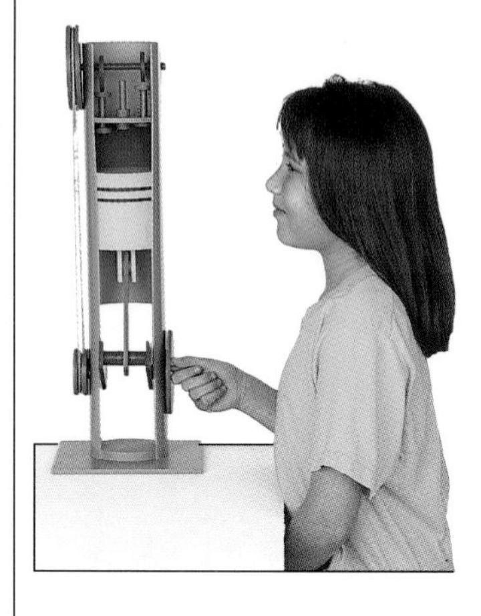

The Basics of machines

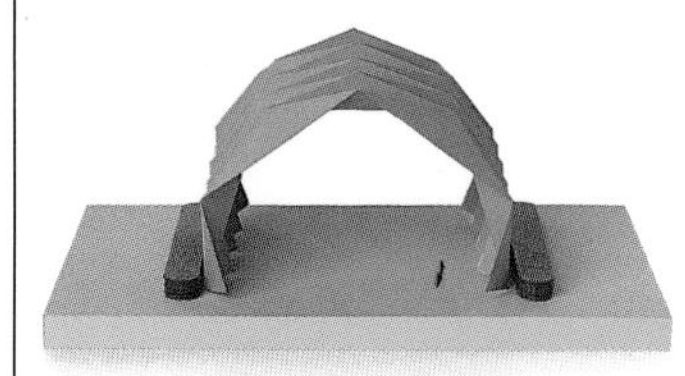

Construction and buildings

Household machines

Transport

Leisure

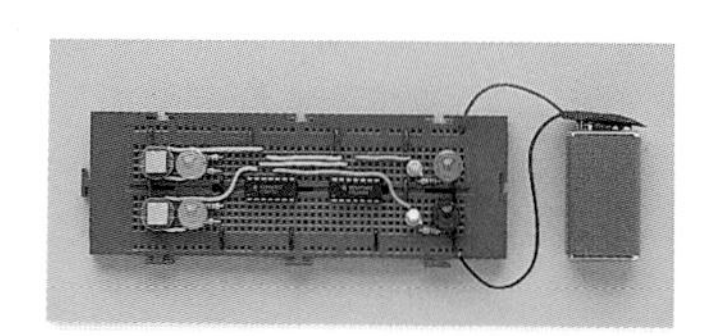

Information technology

INTRODUCTION

RUB A BALLOON on a woolen glove, then make it stick to the wall. Did you know that you are using the same scientific principle that makes a photocopier work? In the 1930s, an inventor realized that static electricity, which makes the balloon cling to the wall, could also be used to make instant copies of documents. Today, photocopiers are common in offices and schools.

Many such machines and devices perform useful tasks for us, and they all apply scientific principles to everyday life. Although the basic idea behind a machine is usually quite simple, like the clinging balloon, it is often difficult to see how machines work in practice. Like photocopiers, many have covers that hide their inner workings. And even if you remove the covers, you are likely to discover only a maze of components that gives few clues to the way in which the machine carries out a task.

This book guides you through the hearts of machines by showing you how to carry out experiments that explain how things work, using step-by-step instructions, clear photographs, and diagrams. You can build real machines and other devices that actually work and which you can use. These include a battery to power an alarm clock; a microphone that you can use to record

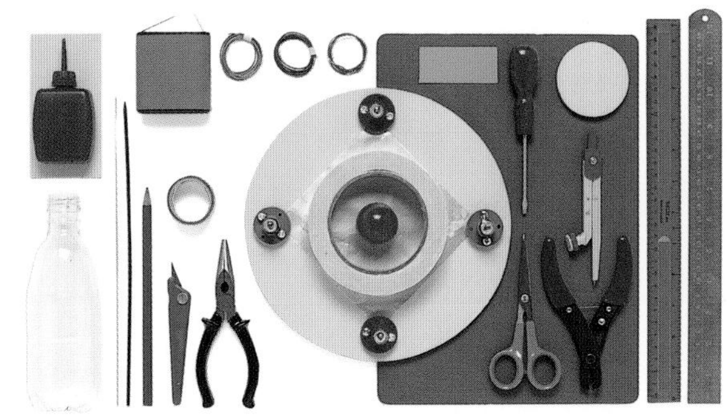

your voice on tape and which also operates as a loudspeaker; a one-way balloon valve; a metal detector; a box with a combination lock; a burglar alarm; an electric motor; a simple computer; musical instruments that play in tune; and a kite that flies.

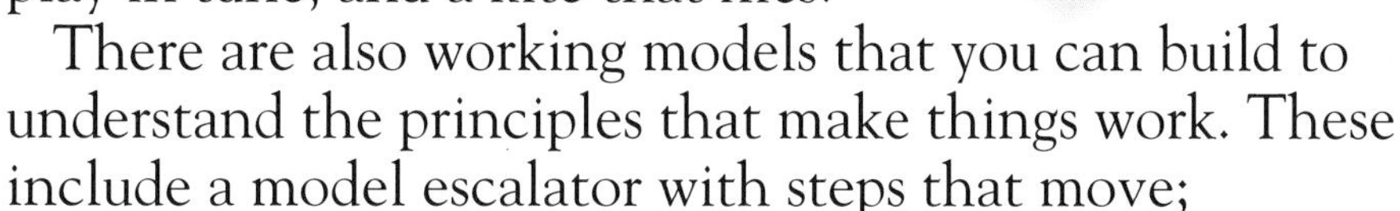

There are also working models that you can build to understand the principles that make things work. These include a model escalator with steps that move; a model autopilot that can sense its direction of movement; a bar-code reader that reads codes in the same way as a real one; a model aqualung that supplies air in the same way as a real aqualung used by divers to breathe underwater; a model rocket that takes off; and a model disk drive that stores information like the disk drive in a computer.

These are only a few of the many experiments in this book. Some are simple enough for you to do on your own, while others are more complex and may require help. Carry them out carefully and you will find out about all kinds of things that you use or see every day—from cars to computers, from toasters to television, and from simple levers to mighty skyscrapers.

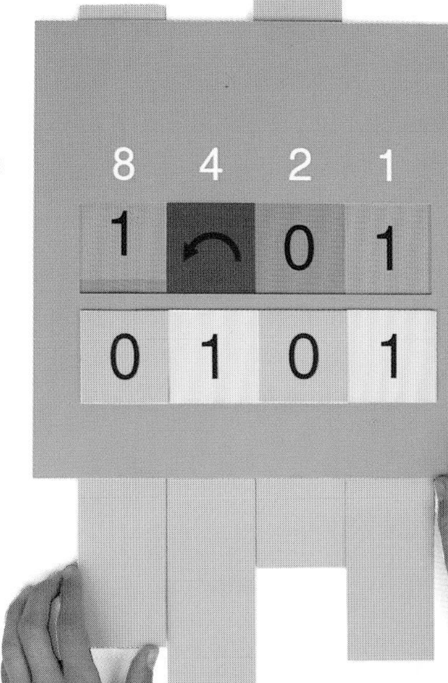

The home laboratory

YOU CAN CARRY OUT MOST of the experiments in this book with simple things that you will find at home, or which you can easily buy at little expense. On these two pages you can see the kinds of equipment you will need. If you do not have the exact items, you can probably find something similar that will do. Follow the instructions and you will find out how all kinds of things work. But always take great care, especially when using electricity, hot objects, or sharp implements.

Tools

You will need tools to cut things, make holes, and hold items together. A vise may help to grip an object as you work on it. In some experiments goggles are essential to protect your eyes. Use wire strippers to cut and strip wires, not scissors. A hand drill is easier and much safer to use than an electric drill. A compass can be used to make small holes as well as to draw perfect circles.

Weights and measures

All weights and measures in this book are given in two systems: English, which uses units such as inches (in), and metric, which uses units such as centimeters (cm). Make measurements in only one of these systems. Do not use both systems in the same experiment. Most measurements are specific, but where approximate amounts are indicated the numbers are rounded to the nearest full unit.

Containers

Plastic or glass containers of several sizes are needed: plastic cups, pots, and food containers often work well. Some containers must be watertight or airtight, so check for cracks and badly-fitting lids or tops.

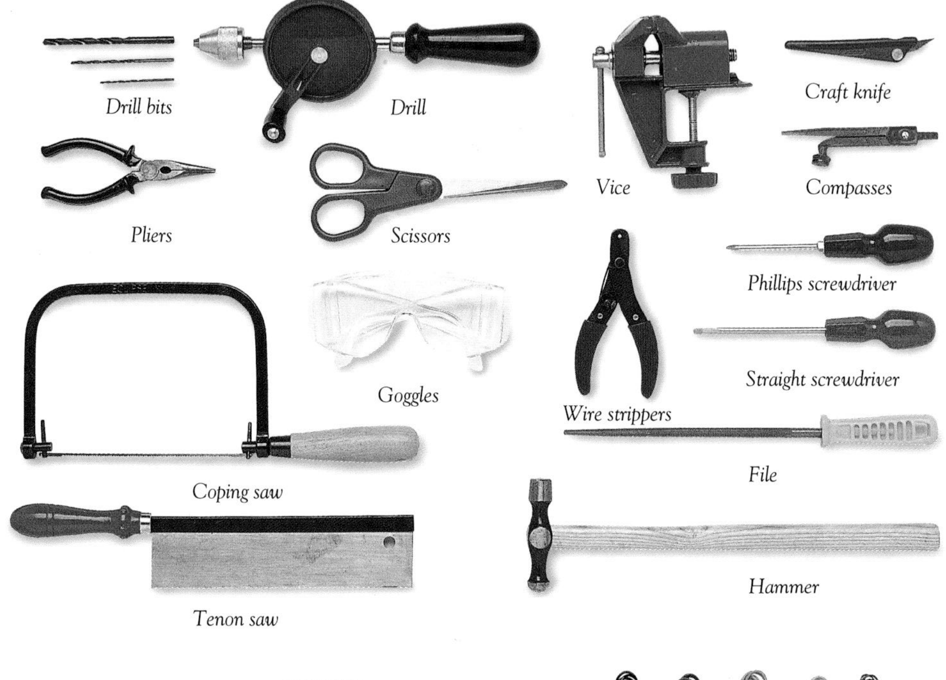

Drill bits *Drill* *Pliers* *Scissors* *Coping saw* *Tenon saw* *Goggles* *Vice* *Craft knife* *Compasses* *Phillips screwdriver* *Straight screwdriver* *Wire strippers* *File* *Hammer*

Plastic bottles *Funnel* *Bowl* *Jug* *Plastic containers* *Glass containers* *Wire* *Bulb and holder* *Buzzer* *Crocodile clips* *4.5V battery* *9V battery*

Electrical equipment

Many experiments need electricity in order to work. Connect wires very carefully. Remember that an experiment may not work, or may stop working, because just one wire has come loose—or because the battery has died!

Materials

Many experiments require you to build or make things. Cut materials on a cutting mat to avoid damage to the table top. When you glue parts together, always give the glue time to set. Smooth rough edges with sandpaper. Foamcore is a useful material available in art supply shops. It is light and strong, easy to cut and glue, and does not splinter.

Useful items

Your home will contain lots of other useful items, many of which are shown below. A steel ruler, for example, will help you to cut in a straight line when using a craft knife. Drinking straws that bend are the best type to use. Some models in this book have been colored with paints or pens; you may like to do the same.

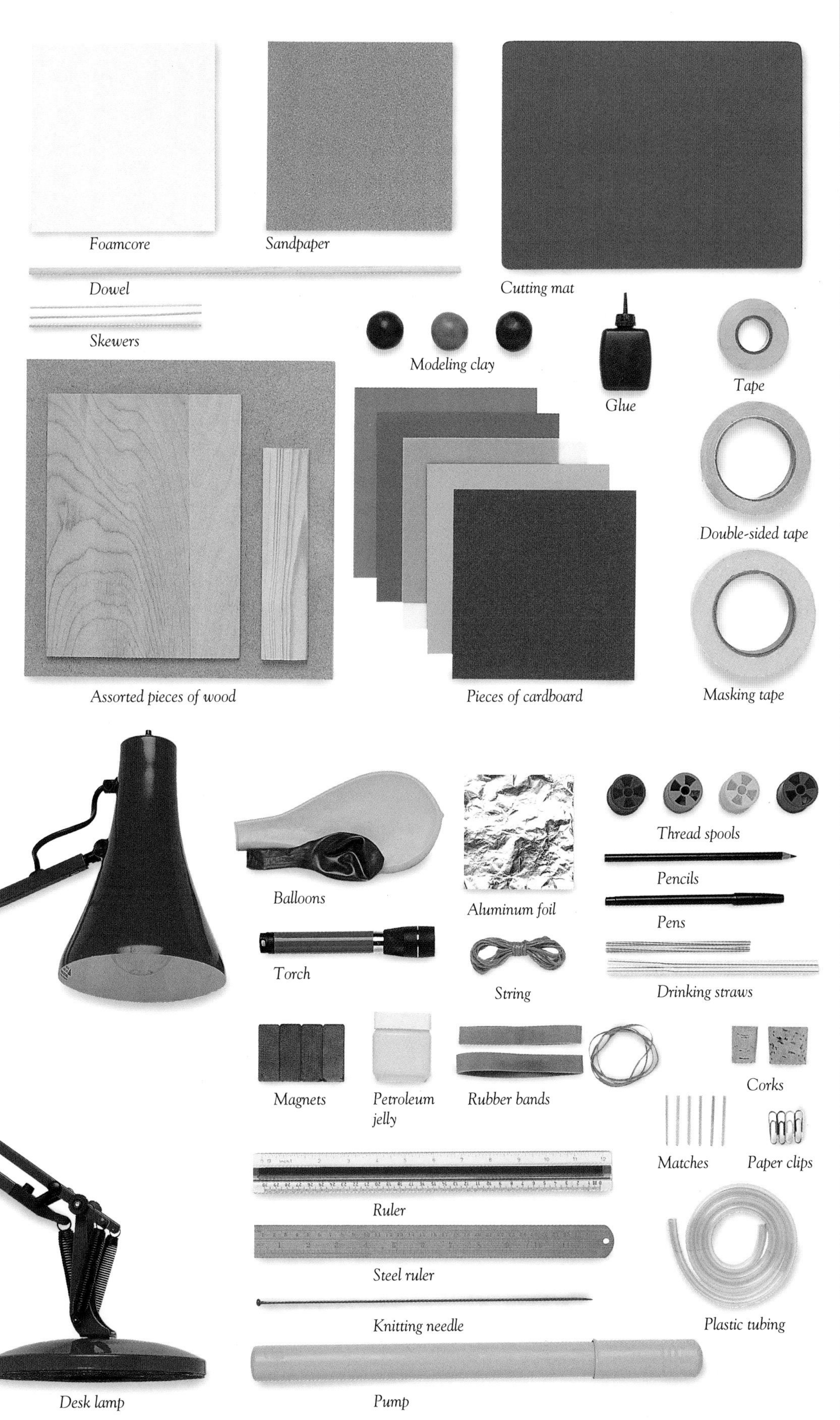

Building circuits

ELECTRONIC COMPONENTS like transistors and microchips are used in circuits that are at the heart of many machines. This book has experiments that require you to build a circuit. Do this by assembling components directly, without soldering, on a breadboard (prototyping board). Each experiment has a diagram and wiring list that clearly show the connections.

Breadboard
A breadboard has a grid of holes into which the components and wires fit. The holes are 0.1 in (approx. 2.5 mm) apart and have spring-loaded contacts. Use a board with at least 47 holes per row, two groups of five rows (marked B to K), and two power supply rows (A and L).

Top of breadboard — *Rows and columns of holes*

Bottom of breadboard — *Metal strips connect the rows*

Wires
Use insulated solid-core wire that fits the breadboard holes (usually 22-gauge wire). Position the wire to find the length needed, bending it to fit neatly. Cut it and strip ¼ in (0.5 cm) of insulation from each end. Bend the ends at 90° and insert into the breadboard.

Insulated wire — *Wire ends are stripped and bent*

Wires

A7–B7	A14–B14	A33–B33
A45–B45	C39–I19	D19–D30
D37–J18	E8–E18	E20–G23
E37–E38	E39–E40	F7–G7
F30–G30	F45–I40	H38–H40
J1–J9	K8–K17	K20–L20
K25–L25	K31–L31	

Demonstration circuit

This breadboard has been fitted with the components and wires shown in the diagram on the opposite page. It is for demonstration purposes only, and does not actually do anything useful.

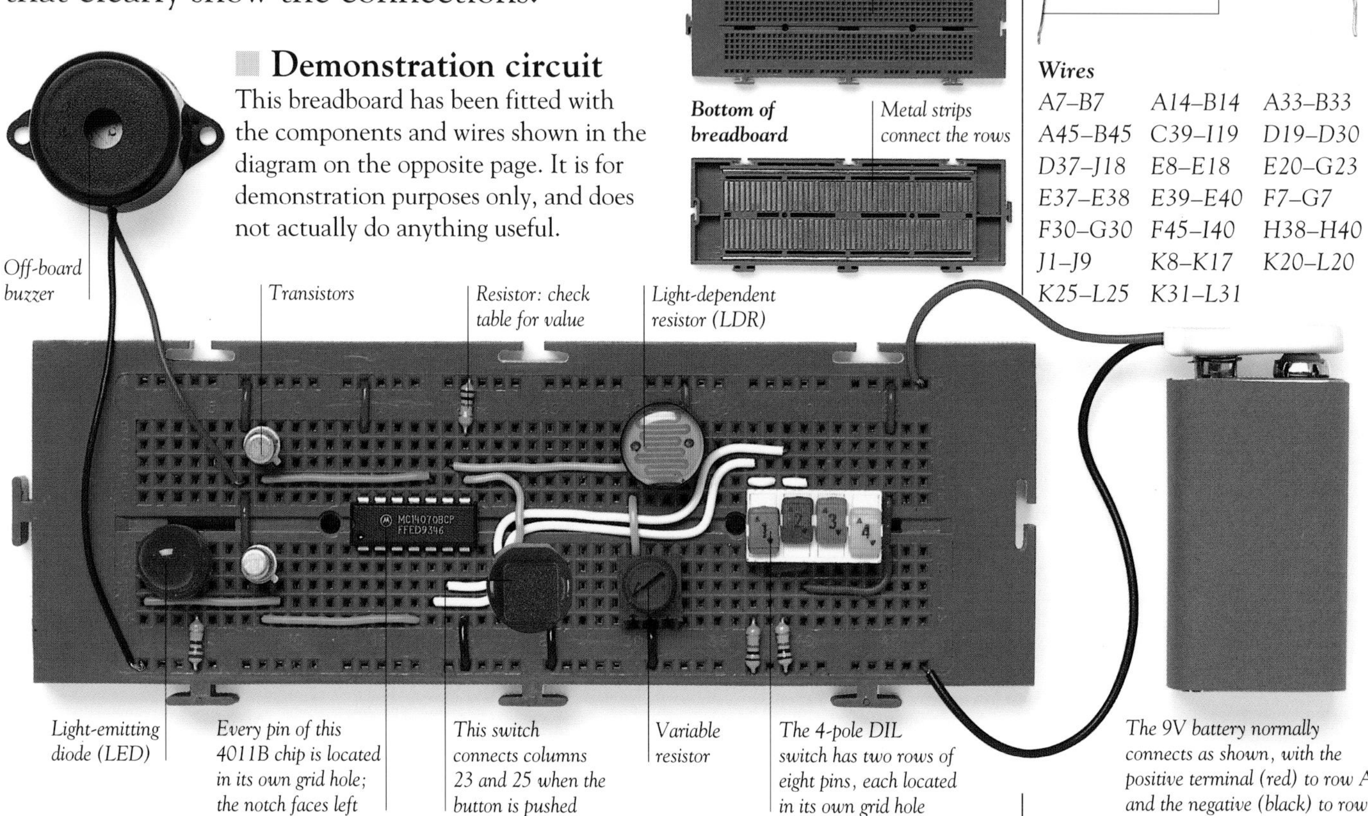

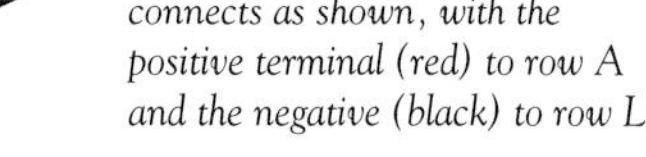

The 9V battery normally connects as shown, with the positive terminal (red) to row A and the negative (black) to row L

Wiring circuits

Use the circuit diagram when building a circuit. Each component has a symbol with small circles bearing numbers. The row in which a specified pin sits is shown by the position of the circle on the diagram; the number inside indicates the column.

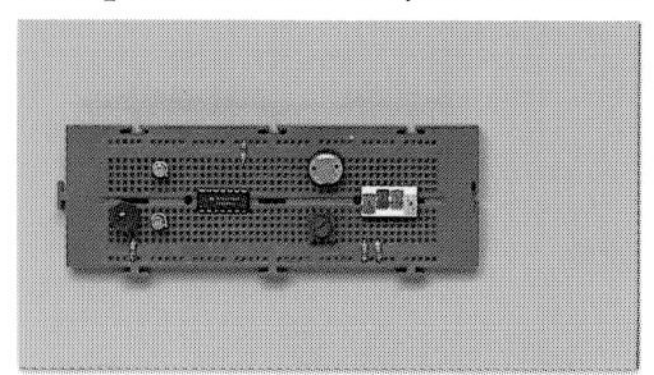

1 PLUG THE component pins into the board locations shown in the diagram. Check that you have the correct components, and that the pins are positioned correctly.

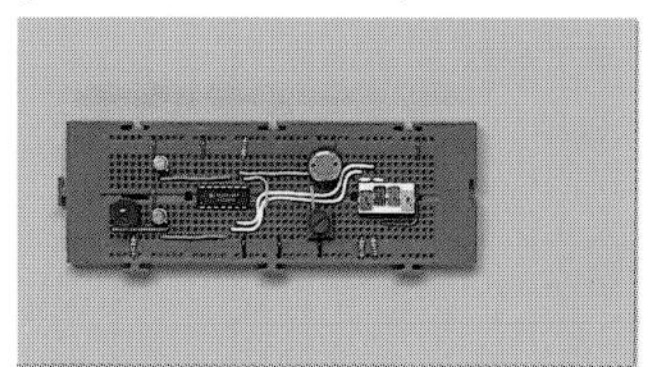

2 FIT WIRES into the board around the components. The wiring list locates a pair of holes for each wire. The stripped ends of each wire fit into these holes.

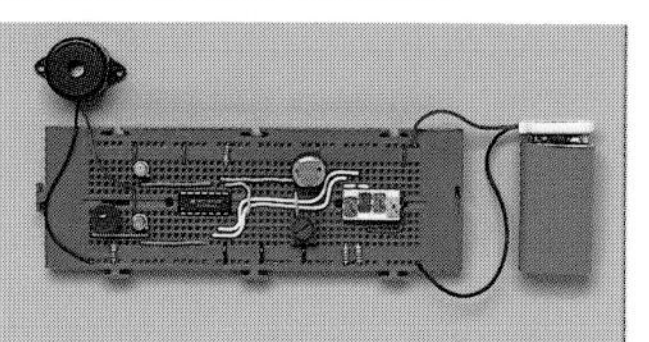

3 FINISH BY adding the off-board components and battery. Check the board to ensure that bare wires do not touch each other or any component pins.

Off-board components
Some components connect indirectly to the breadboard with wires. Buzzers and magnetic switches usually come with wires attached, but LED's and LDR's must have wires connected to their pins before being used off-board.

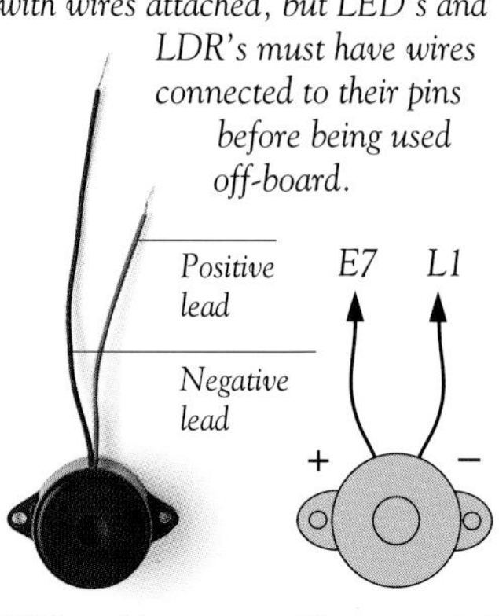

Off-board buzzer **Buzzer symbol**

Resistors
Use 0.25W resistors of the following values: 220R, 1K, 10K. They have two leads and can be inserted into the breadboard in either direction. Trim the leads to length so that they fit neatly into the specified holes; bend each end before inserting. The resistor list gives a pair of locations for each resistor. Be sure to use the correct value of resistor at each location.

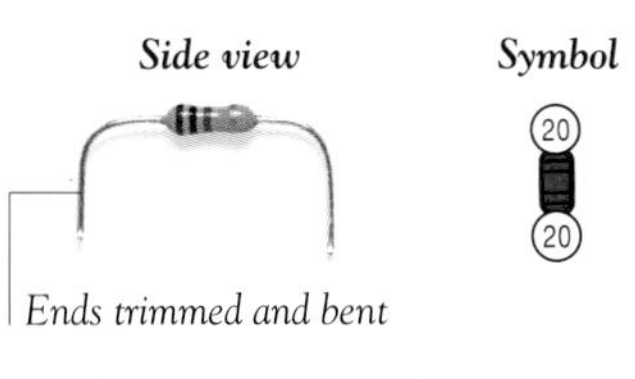

10K resistors	**220R resistors**
A20–B20	K4–L4
K37–L37	
K39–L39	

Variable resistors
These resistors are also known as potentiometers or trimmers. Their resistance can be varied by turning a dial. Use 5K variable resistors suitable for 0.1 in (approx. 2.5 mm) pitch grid. This type has three pins: a middle (wiper) pin and one pin on each side. The pins must be inserted a particular way around; the symbol on the circuit diagram provides this information.

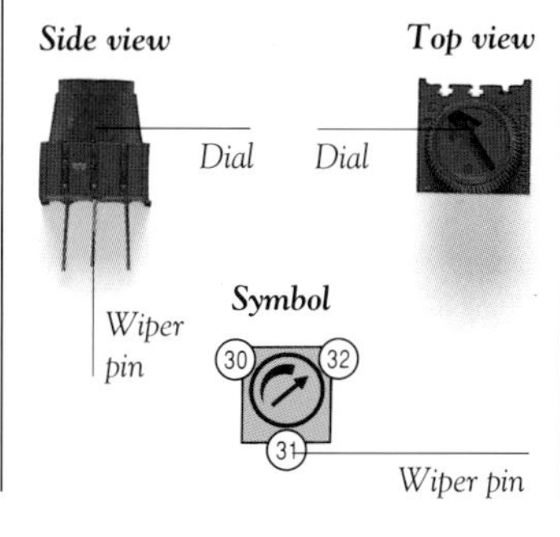

Light-dependent resistors
Light-dependent resistors (LDR's) are two-pin devices that may be inserted in either direction. Their resistance varies according to the amount of light falling on them. They have higher resistance in the dark and lower resistance in bright light. The type used throughout experiments that use breadboards in this book is NORP–12.

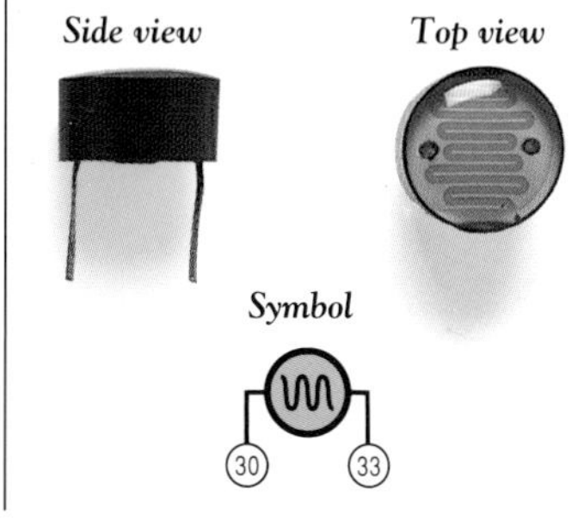

Transistors
These three-pin devices amplify or switch currents. Use NPN transistors, of types BC108, BC441, or equivalent. Use the diagram on the package and the tab on the body of the transistor to help you identify the pins. The three pins are the collector (C), base (B), and emitter (E). They must always be connected exactly as indicated by the circuit diagram.

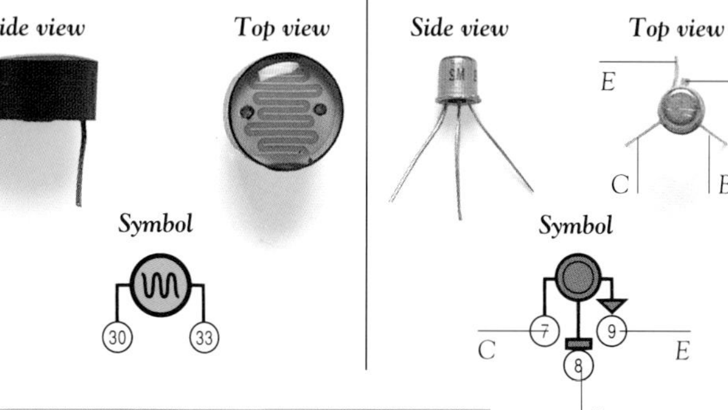

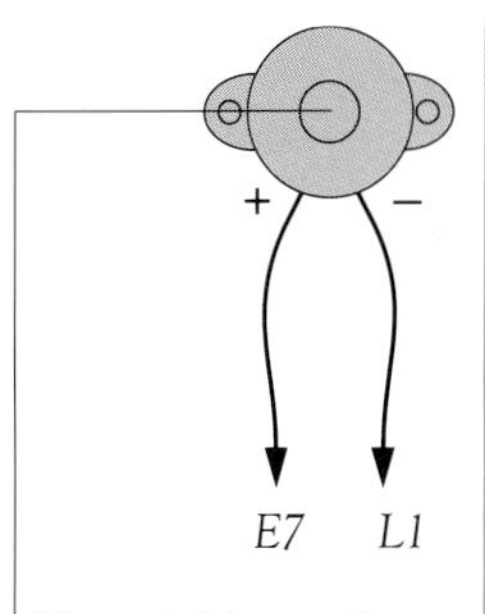

The symbol for an off-board component shows board coordinates for the wires, which may be identified by + and – signs

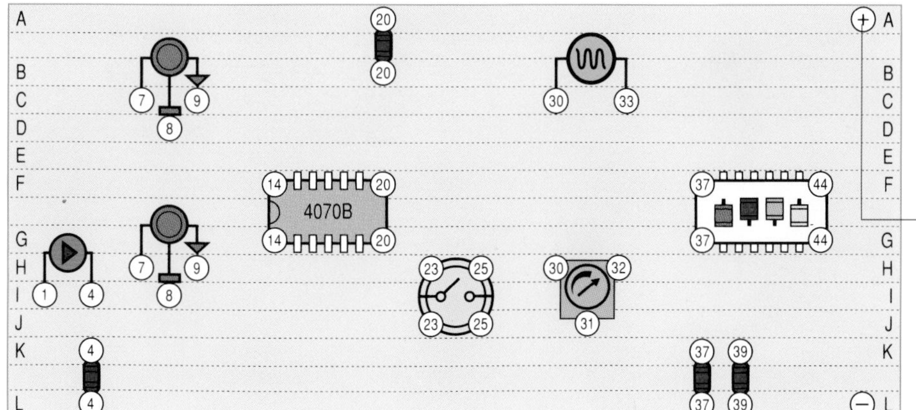

The positive or negative sign inside each circle shows the board location of a battery lead

Switches
Use normally open momentary SPST switches suitable for 0.1 in (approx. 2.5 mm) pitch grid. These may have two or four pins. Check the package to find out which pins are connected when the button is pressed. In the type shown here, two pairs of pins are connected.

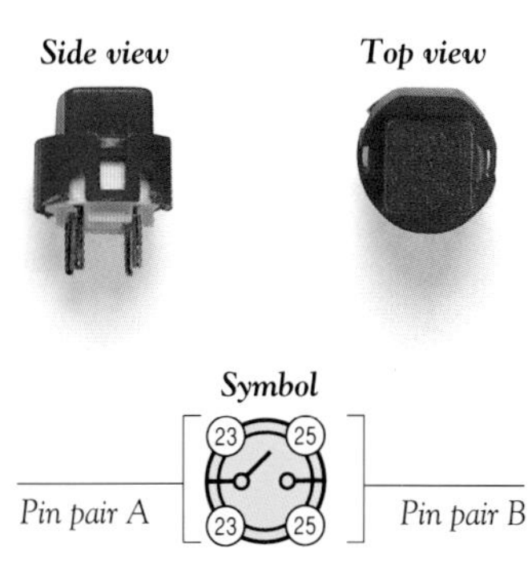

Dual in-line switches
Dual in-line (DIL) switches plug into the breadboard like chips. A single-pole changeover DIL switch has one sliding button and four pins. A 4-pole changeover DIL switch has four buttons and 16 pins in two rows. The symbol gives the locations of the four corner pins.

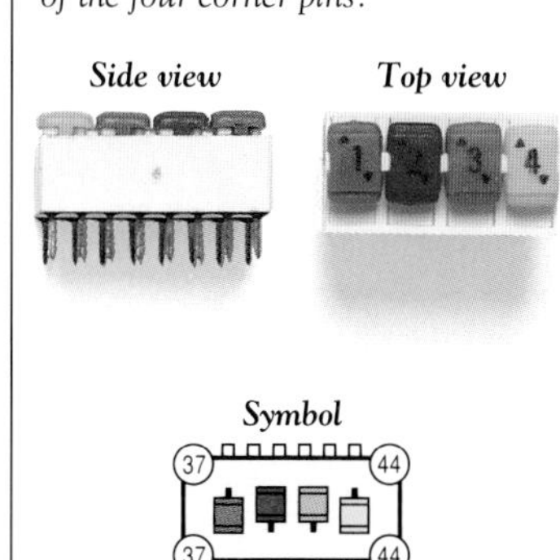

Light-emitting diodes
Light-emitting diodes (LED's) have two pins (the anode and cathode) that must be connected in the right directions. The cathode is usually the shorter pin, but check the package to identify it. The LED symbol has an arrow pointing to the cathode. LED's come in various colors.

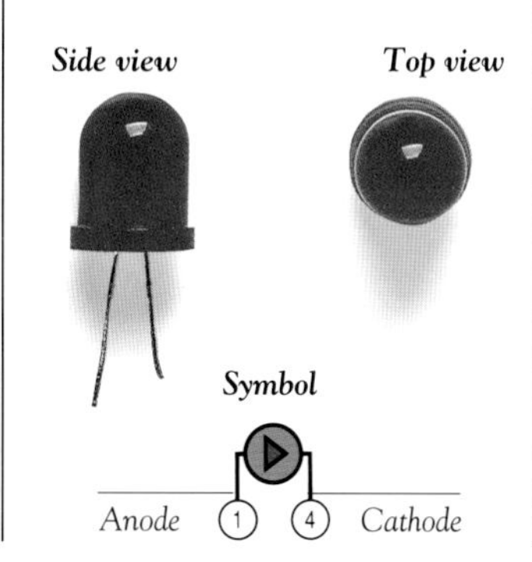

Microchips
In this book we use both TL071 and CMOS chips, which may have 8, 14, 16, or 24 pins in two rows. The pins fit into a breadboard on each side of the central horizontal gap. The symbol gives locations of the four corner pins. Ensure that the notched end of the chip points left.

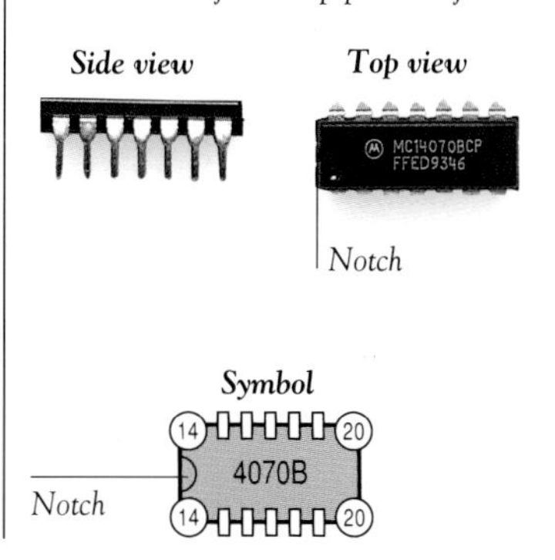

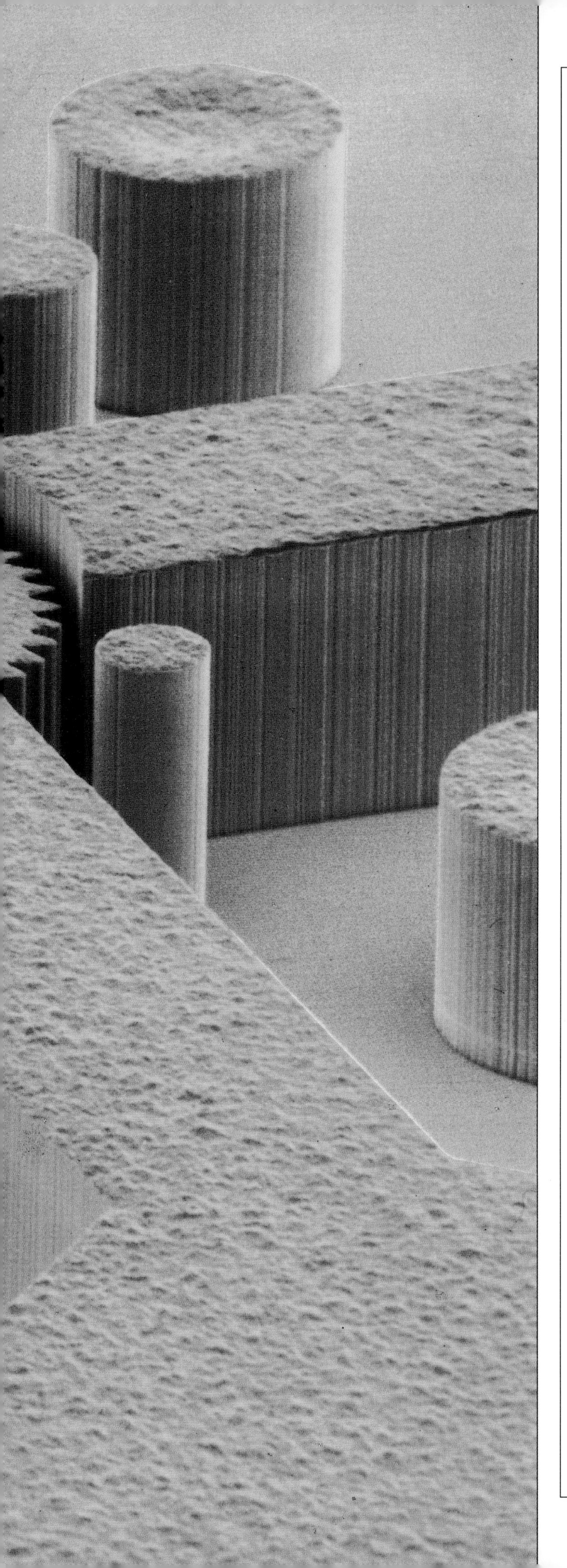

The BASICS of MACHINES

Energetic applications
All machines need an energy source in order to work, and they apply this energy while they perform a task. A car engine (above) burns gasoline to produce heat energy, which it then turns into motion. A miniature set of gears (left), each no wider than a human hair, transfers motion in a tiny sensing device that can work accurately under extremely high acceleration because it is so small.

A BEWILDERING ARRAY OF machines may surround you in your everyday life—in your home, at work or school, in your leisure time, and as you travel from place to place. They can perform all kinds of useful tasks for you. Because they have different purposes, machines come in a huge variety of shapes and sizes. Yet whatever they do and whatever they look like, all of them work on the same basic principles. Understanding these principles will greatly help you to find out how things work.

WHAT DRIVES MACHINES?

MACHINES, EVEN THE MOST FULLY AUTOMATED, do not start working or continue to operate totally unaided. To produce any action, a machine needs a source of energy. That energy, in the form of movement or electricity, may come from muscles, engines, batteries, or power stations. Machines apply this energy to perform useful tasks. They may use the energy to perform a physical movement, such as turning a wheel, or they may convert the energy into another form—for example, microphones are used to convert sound energy into electric signals that can be changed back into sound energy by loudspeakers.

Wind turbines are modern versions *of the ancient windmill. Both capture wind energy to provide power to drive machines—the windmill directly, and the turbine by generating electricity.*

The first source of energy used to drive machines was muscle power, both human and animal. Even today, it is still commonly used to drive machines in many parts of the world. And even where electricity and other sources of energy are readily available, it is often easier to work devices such as can openers and corkscrews by hand. However, people and beasts soon tire and cannot provide a constant flow of energy. About 2,000 years ago, the flow of water in rivers was put to use to drive waterwheels that powered machines. And later, windmills were used to capture the energy of moving air. Wind and water still drive machines that produce much of our electricity.

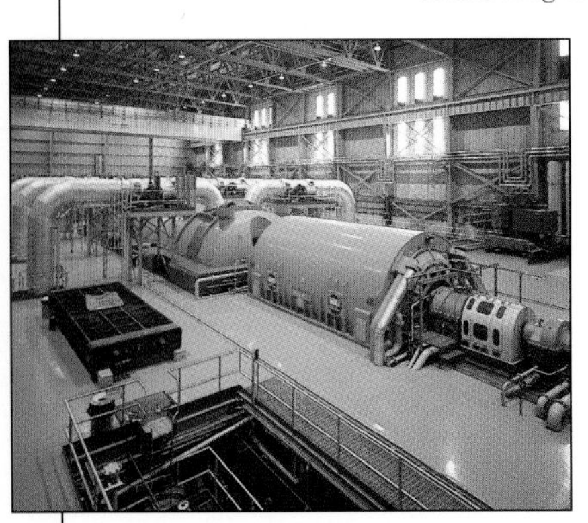

At power stations, turbines *drive big electrical generators. Such generators produce most of the electrical energy needed to power homes, schools, offices, and factories.*

New power

Although rivers and winds can be powerful, they are not always constant or convenient energy sources. Few locations are blessed with swift rivers, and winds often drop without warning. To be of real service to us, machines need compact and reliable sources of energy that are either part of the machine or can be brought to it. The eighteenth century saw the development of the first such energy source: the steam engine. Steam power is now mainly used to drive generators in power stations, having been displaced in the last century by gasoline and diesel engines. The most convenient energy source of all, electricity, was first harnessed in the nineteenth century. Electricity is either generated in power stations and carried along power lines, or supplied on the spot by batteries. Electric motors now power many kinds of machines, and electricity is essential for computers and communications equipment.

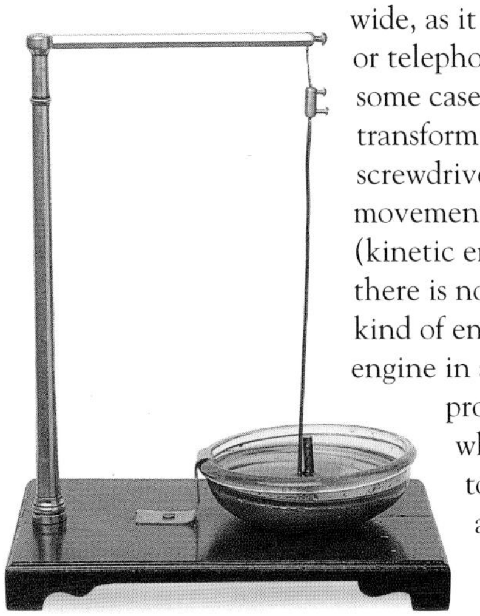

The first electric motor was *invented by the British scientist Michael Faraday in 1821. In this machine, a wire carrying an electric current rotated around a magnet.*

Energy transfers

Every machine has one basic purpose: to take energy from a source and use it to perform a task. In doing so, the machine transfers the energy from the source to the place where it is required. Sometimes, the energy travels only a short distance—a screwdriver transfers energy from your hand along the screwdriver to the head of the screw. But energy may also travel world-wide, as it does when a radio or telephone is used. In some cases, a machine may transform energy. When a screwdriver transfers the movement of muscles (kinetic energy) to a screw, there is no change in the kind of energy. But the engine in a car burns fuel to produce heat energy, which is changed to kinetic energy as the engine turns the wheels. And in a hydroelectric power station, the kinetic energy of moving water drives a turbine that powers a generator, which turns the kinetic energy into electricity.

Mechanical marvels

An efficient machine gets the right amount of energy to the right place. Many machines do this using mechanical parts, such as levers, gears, or pulleys. These machines often produce more force or speed than the energy source that powers them. For example, a bicycle uses a chain and gears to turn the rear wheel faster than your feet turn the pedals. In such ways, machines can do things that are beyond our physical powers. It might seem that a machine that increases force or speed also adds to the energy that you put into it, but this is not the case: you cannot get more energy out of a machine than you put in. When any object is set in motion, the amount of energy involved depends on the amount of force acting on the object and the distance that the object moves. If a machine magnifies the force acting on an object, then the distance that the object moves decreases, and vice versa. Therefore, the total quantity of energy transferred does not change.

A cylinder of a car engine *shows how heat energy from burning fuel can be turned into motive power to drive a machine.*

Fluids and currents

Many machines use fluids (liquids and gases) or electricity, rather than solid components, to transfer energy from one part to another. For example, excavators have pipes filled with liquid that transfers energy from a motor to a digging bucket, causing the bucket to move slowly but with great force. And a dentist's drill is driven by air to turn with great speed rather than great force. While mechanical machines are powered by kinetic, chemical, or heat energy, electric and electronic machines are powered by electric energy. All these machines, however, apply the same basic principles to the transfer and transformation of energy. In an electrical washing machine, for example, electric energy from the mains supply is delivered to a motor then transformed into kinetic energy; and in a light bulb, electric energy is transformed into light energy. Electronic machines transfer and transform energy using few, if any, moving parts. Instead, they use tiny components such as transistors and microchips that carry and process electric signals. Such components are not only powered by electricity, but also use it as a medium for carrying data (information) such as sounds or images. Electronic components may also transform electric energy into radio waves or light to transmit data along wires or through air to telephones, radio and television sets, or computers.

A train of four different kinds of gears shows *how mechanical parts transfer energy in machines and put it to use. The parts often change the force or speed with which the machine operates.*

A waste of energy

Some of the energy that allows a machine to do its work is diverted in the process. The energy does not disappear; it is converted to a form that is not useful to the machine. All machines with moving parts, or which move over the ground or through air or water, are subject to friction as their moving parts rub against each other or against other substances. Friction changes some of their energy into heat and sound, so that parts get hot and make noise. When friction wears parts down, a machine starts to perform badly. Electronic machines may have no moving parts, but electricity flowing through their circuits makes them warm up, so the machines still lose some energy in the form of heat. A well-designed machine keeps energy losses to a minimum to work as efficiently as possible.

Controls

Complex machines can be difficult to operate. Fully automated machines, however, can operate unaided. For example, a manual camera requires many delicate adjustments to take a good picture, but the user of an automatic camera need only press a button and let the camera do the rest. Automatic control is important in many machines. Some, such as vending machines and washing machines, follow a set routine of operations once started. Others are totally independent of humans and use detectors to sense their own environment or performance. For example, automatic doors open to allow entry when they sense an approaching person. An autopilot continually checks the altitude and direction of a plane and acts to keep it on course. The ultimate automatic machine is the robot.

High-pressure fluids *such as oil and air drive many machines. Valves like this one-way balloon valve are required to control their flow.*

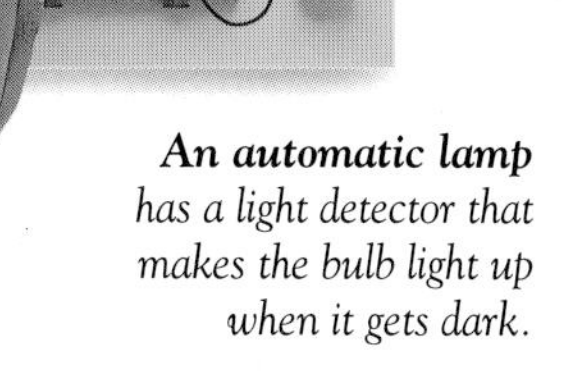

An automatic lamp *has a light detector that makes the bulb light up when it gets dark.*

Turbines

TURBINES ARE IMPORTANT sources of power that drive many kinds of machine. A turbine is basically a shaft with a circle of blades attached. A stream of gas, such as air, or water strikes the blades and causes the shaft to rotate. The turning shaft then powers the machine. Turbines are the main source of power for electricity generators in power stations. Some power stations burn fuels such as coal, oil, or gas to heat water; steam from the water drives turbines, which then power electricity generators. In hydroelectric power stations, turbines are driven by falling water. In wind farms, turbines that look like windmills are driven by wind to generate electricity. Turbines are also a source of power in many smaller machines. Aircraft jet engines (pp.20–21) contain turbines driven by hot gases. Powerful road-vehicle engines may contain turbochargers, in which a turbine driven by exhaust gases helps suck more air into the engine to generate more power. A dentist's drill contains an air-powered turbine like a miniature windmill.

EXPERIMENT

Water turbine

Adult help is advised for this experiment

Build a simple water turbine driven by falling water like that used in a hydroelectric power station. The speed of the turbine depends on the distance that the water falls before it strikes the turbine blades.

YOU WILL NEED

• 12 in (30 cm) dowel • round file • scissors • masking tape • pen • modeling clay • ruler • funnel • craft knife • adhesive tape • pitcher of water • tray • cardboard disk about 5 in (12 cm) in diameter with pattern drawn on • thread spool that fits on dowel • glue • cork with a lengthways dowel-sized hole drilled • two 3 qt (3 l) plastic bottles

Hydroelectric power station

Dams built across rivers form big reservoirs of water behind them. A hydroelectric power station can be built inside or alongside a dam to generate electricity from this water. A wide pipe carries water through the dam and down to turbines in the base of the power station. The falling water spins the blades of the turbines, which drive an electricity generator mounted above. Some hydroelectric power stations also pump water back up to the reservoir from the river or lake below. Electricity from other power stations drives a pump beside the turbine, sending water back up the pipe to the reservoir above. This is done during off-peak periods when demand for electricity is low. The extra water gives the power station more generating capacity during peak periods. This method of power generation is called pumped storage.

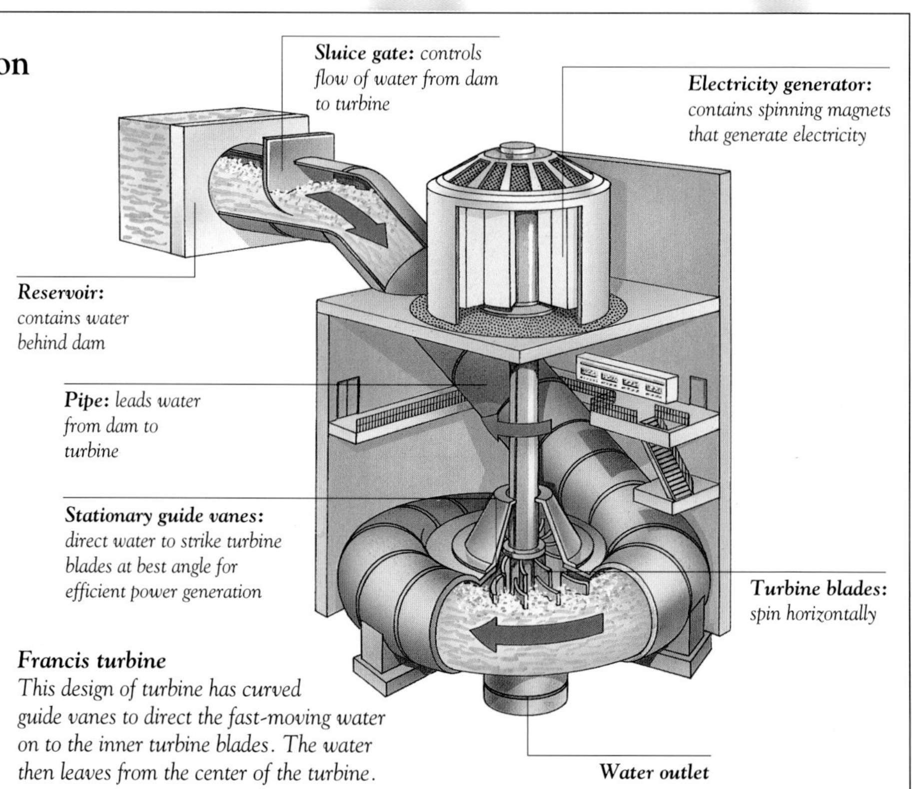

Francis turbine
This design of turbine has curved guide vanes to direct the fast-moving water on to the inner turbine blades. The water then leaves from the center of the turbine.

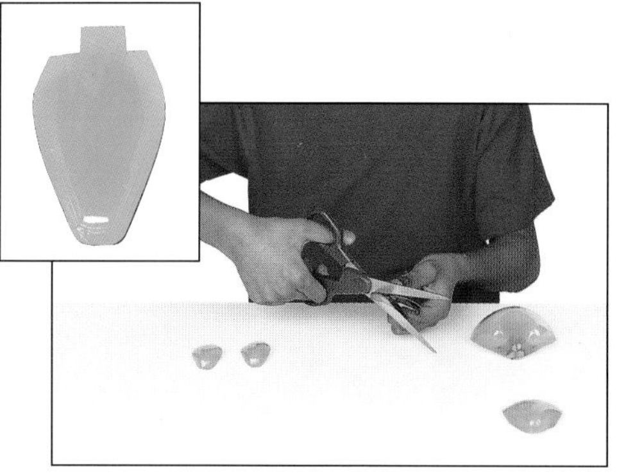

1 CUT FOUR sections, each with a scoop and a flat edge, from the base of a plastic bottle to form four turbine blades (inset, above).

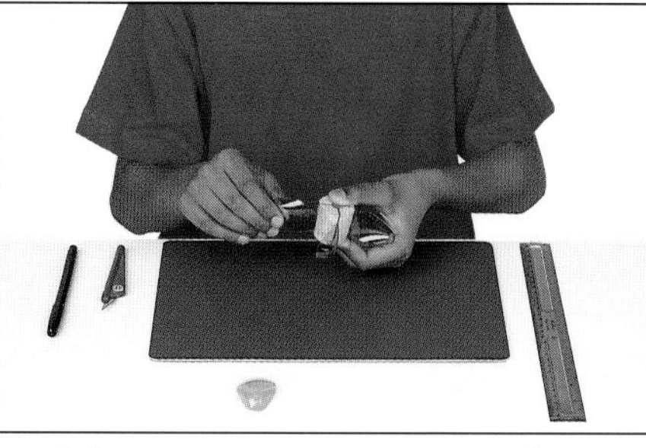

2 MARK A CORK into quarters along its length. Cut a slit along each mark to fit the flat edge of a blade. Insert a blade in each slit to make a turbine.

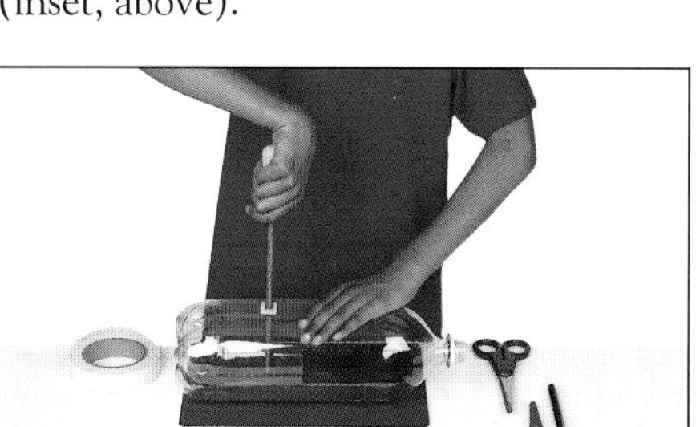

3 CUT A TALL hole 6 x 3 in (15 x 8 cm) in the top half of the second bottle. Hold the turbine in the bottle so one blade scoop directly faces the neck. Cut two small slits in the bottle aligned with the hole in the cork, and a third 4 in (10 cm) below these. Enlarge the slits into dowel-sized holes with a file.

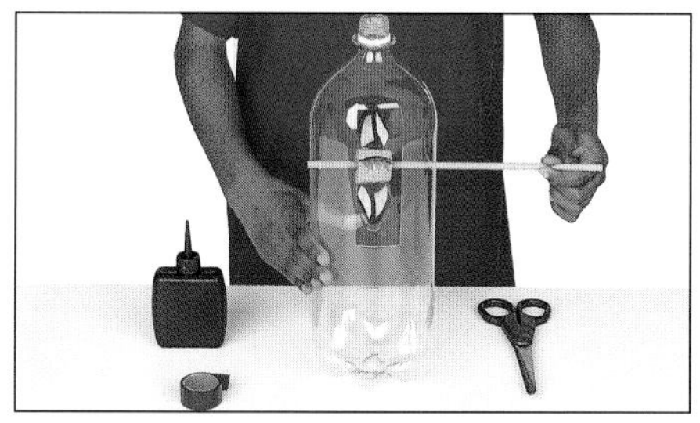

4 PUT GLUE in the hole in the cork. Hold the turbine inside the bottle and align the cork between the pair of upper holes. Insert a dowel through these holes and through the cork. Wrap tape around the dowel outside the bottle to keep it from moving sideways.

5 PUT GLUE IN the central shaft of the thread spool, then place the spool over one end of the dowel. Make a dowel-sized central hole in the patterned cardboard disk. Push the disk on to the dowel and glue it to the cotton spool.

6 PLACE THE APPARATUS in a tray so that any water flowing from the lowest small hole falls into the tray. Use modeling clay to secure the funnel in the neck of the bottle. Pour water through the funnel so that it strikes the scooped side of the turbine blades and makes the turbine spin.

■ DISCOVERY ■

Power production

The turbine was the first step towards a world in which machines work for us and relieve both people and animals of physical labor. The Roman architect Vitruvius first described a waterwheel in the second century BC, and the Romans used waterwheels to power grindstones in flour mills. Although sails were common in ancient times, it was not until the invention of the windmill in the seventh century AD that sails captured the power of the wind to drive machines as well as ships.

Roman waterwheels in Syria

Combustion engines

BURNING FUEL produces a lot of heat energy, which combustion engines can transform into motion (kinetic energy). The heat from the burning fuel creates hot gases or steam, and these expand and drive the engine parts. External combustion engines burn fuel outside the engine; a steam engine has a separate boiler to make steam that goes to the engine turbines (p.187). Internal combustion engines burn fuel inside the engine. Gasoline or diesel engines, for example, burn fuel inside the engine cylinders. These compact, powerful engines drive many machines, including cars, speedboats, and chainsaws.

Four-cycle engine

Below is a model of one cylinder of a typical four-cycle gasoline engine. Fuel burns in the cylinder to produce hot, expanding gases that push the piston down. In a four-cycle engine the piston makes four linear (up-and-down) movements for each power stroke. The piston is joined to a connecting rod and crankshaft that turn its linear motion into rotary motion, which is more useful to machines. A diesel engine is similar to a gasoline engine, but has no spark plugs. Instead, compression of the diesel air-fuel mixture causes it to heat up and ignite.

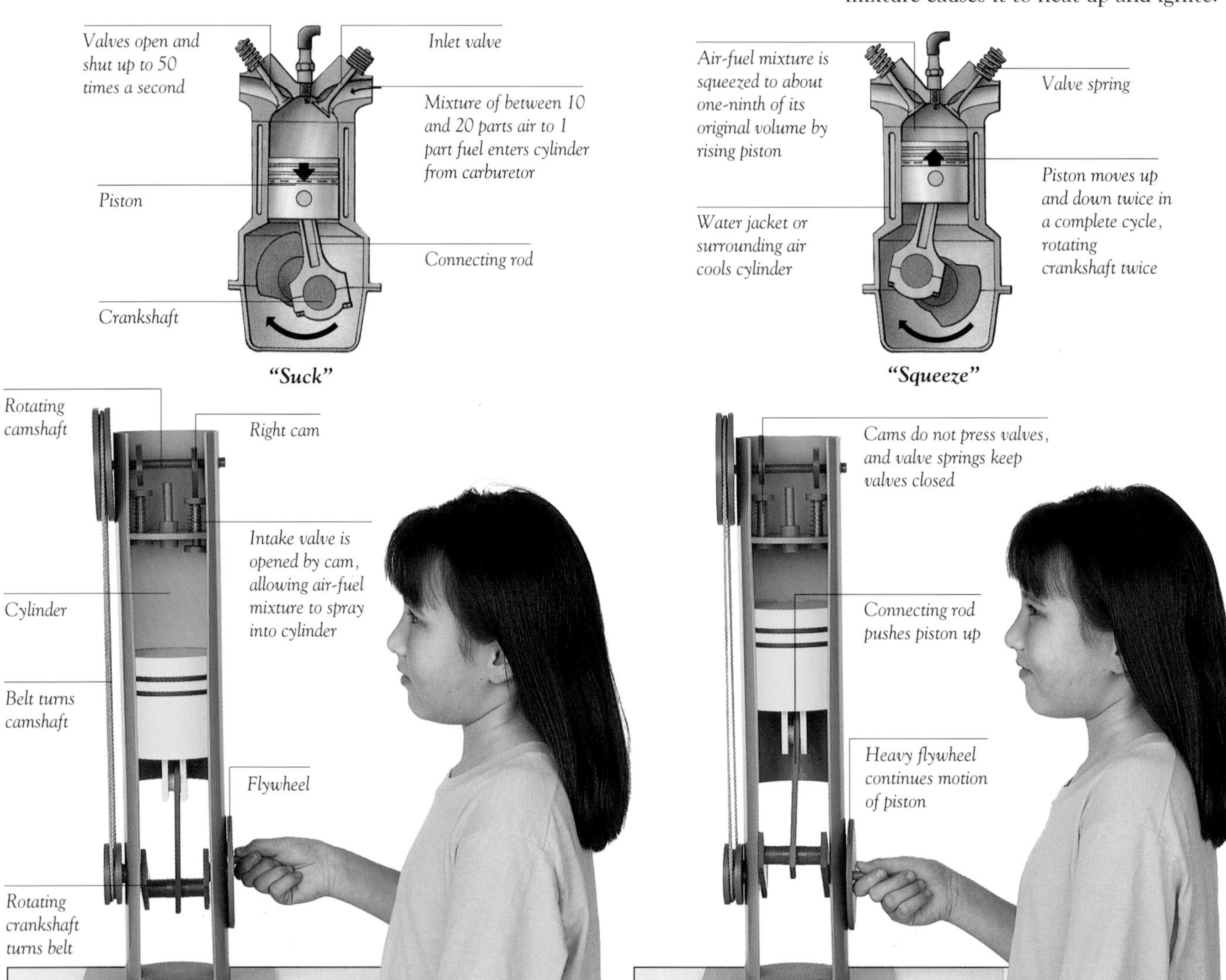

First stroke (induction stroke)
The flywheel, which is turning as a result of the last power stroke, turns the camshaft via a crankshaft and belt. The right cam depresses the intake valve to allow the air-fuel mixture into the cylinder.

Second stroke (compression stroke)
The heavy flywheel continues to rotate. The piston, connected to the flywheel via the connecting rod and crankshaft, rises in the cylinder. Both valves are closed, so the air-fuel mixture is compressed by the rising piston.

Smooth power

The energy released during the power stroke of each cycle moves the piston, crankshaft, and other parts of the engine with great force. In an engine with only one cylinder, this force makes the engine vibrate with each power stroke. Most car engines therefore have four or more cylinders that fire at different times. Each cylinder produces less vibration than one large cylinder, and the movements are balanced so that the engine runs smoothly.

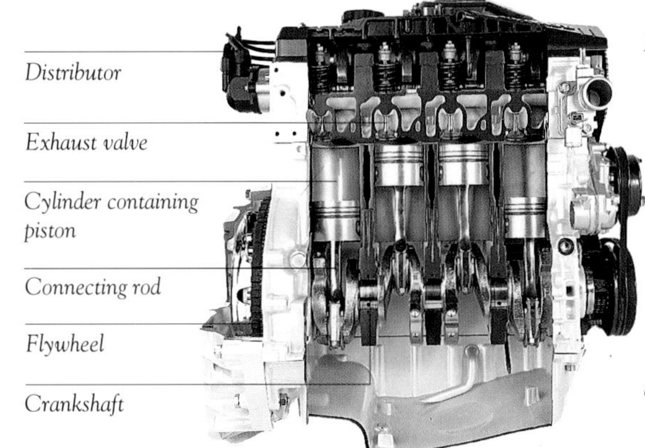

Four-cylinder car engine
Each of these cylinders is at a different stage of the power cycle. The first (left) cylinder is about to begin the compression stroke. The second is on the power stroke, the third is on the induction stroke, and the fourth is on the exhaust stroke.

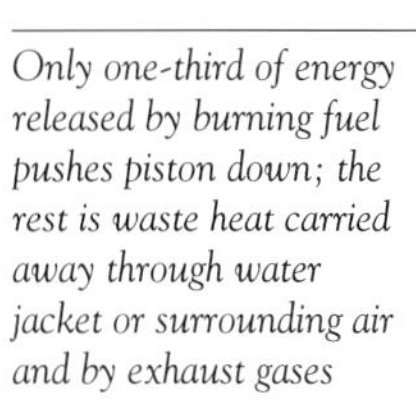
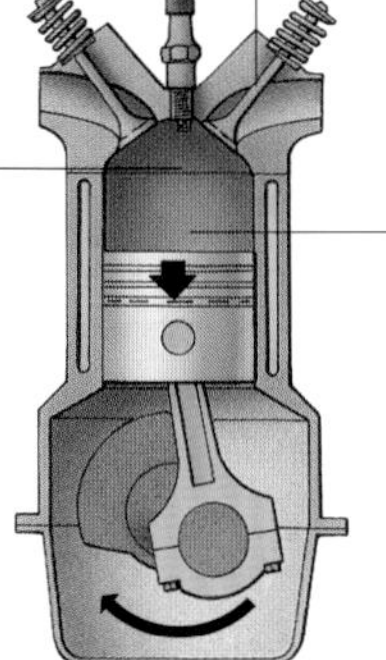

"Bang"

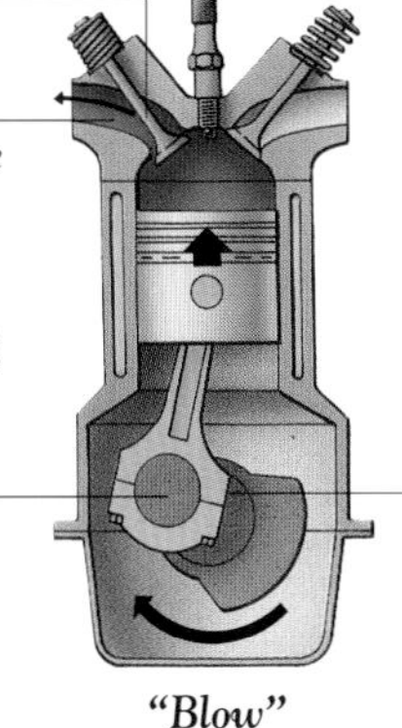

"Blow"

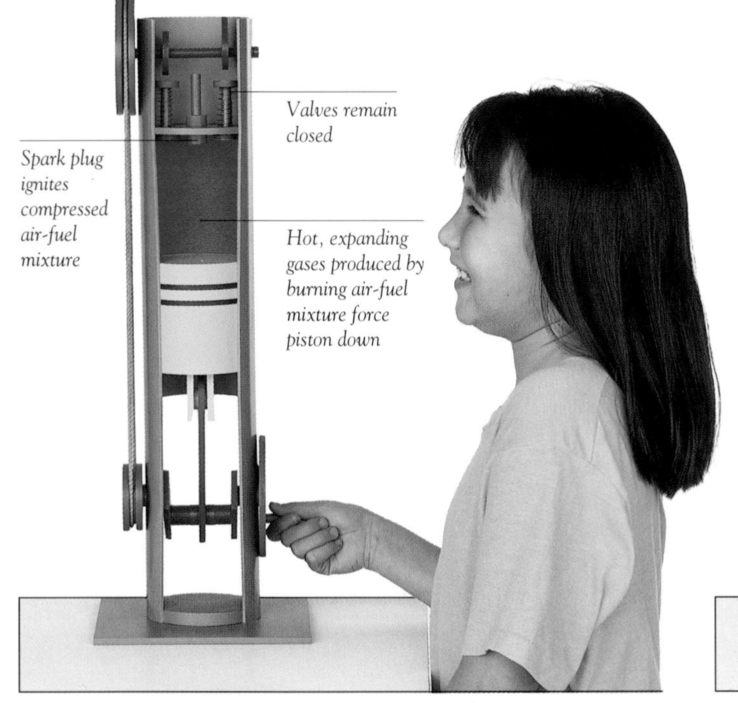

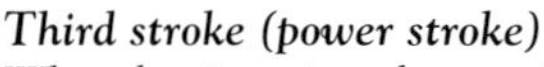

Third stroke (power stroke)
When the piston is at the top of the cylinder and the air-fuel mixture is most compressed, the spark plug ignites the mixture. The mixture explodes, forcing the piston down, which gives renewed momentum to the flywheel.

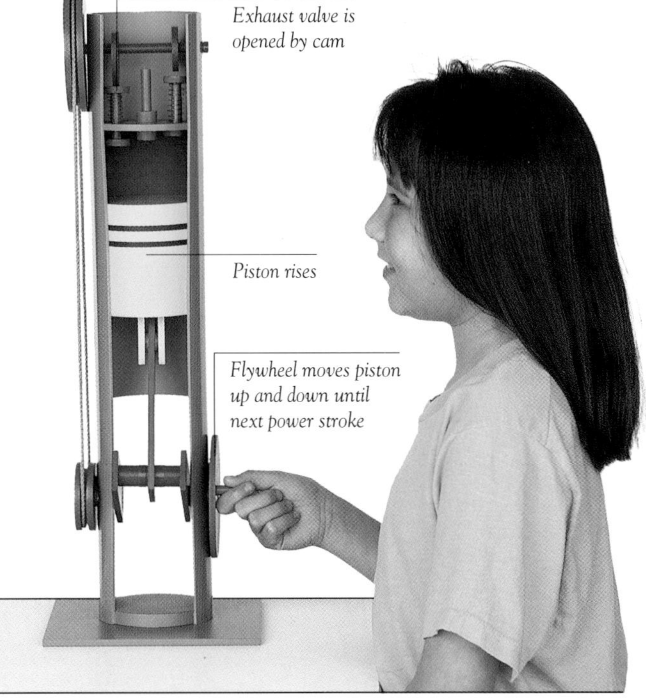

Fourth stroke (exhaust stroke)
The rotating flywheel forces the piston up again. The left cam opens the exhaust valve so that the rising piston can push out the exhaust gases. The four-stroke cycle then begins again.

Propellers and jet engines

SHIPS AND AIRPLANES are powered by combustion engines, which burn fuel to produce the huge amounts of power needed to move them. Jet engines, or propellers driven by combustion engines, thrust the vehicles through the air or water. The rotating blades of a propeller, and the exhaust pipe of a jet engine, both produce a backward stream of water or air that drives the ship or aircraft forward. Ships use large diesel engines or steam turbines (p.187) to turn propellers immersed in the water. Light aircraft may have piston engines to power their propellers. Most modern aircraft, however, are powered by jet engines, and some fast boats have engines that expel a strong jet of water.

Three-bladed ship's propeller

How a propeller works

The blades of a propeller slope at an angle and are curved. They may also taper at the tip, which moves faster than the rest of the blade. As the propeller whirls, either in water or in air, each blade pushes some water or air backward. The curved surface of the blade also causes the water or air to move faster as it flows over the surface. The blade acts in the same way as a wing (p.108), reducing water or air pressure in front of the blade. This pressure reduction, and the backward flow behind the propeller, combine to thrust the propeller forward.

1920s aircraft propeller

EXPERIMENT

Air power

Adult help is advised for this experiment

A portable fan blows cool air over your face in hot weather. Turn it into an aircraft engine, and show how a backward-moving stream of air can drive an aircraft through the sky.

Caution: avoid touching the blades as you switch the fan on and off.

YOU WILL NEED

- *2 jar lids, one slightly smaller than the other*
- *marbles*
- *hand-held fan*
- *ruler*
- *safety goggles*
- *scissors*
- *double-sided adhesive tape*
- *about 1/2 lb (250 g) modeling clay*
- *supporting block*

1 PUT A SINGLE layer of marbles in the small lid and stick it on top of the block with double-sided tape. Attach the ruler to the top of the large lid with modeling clay.

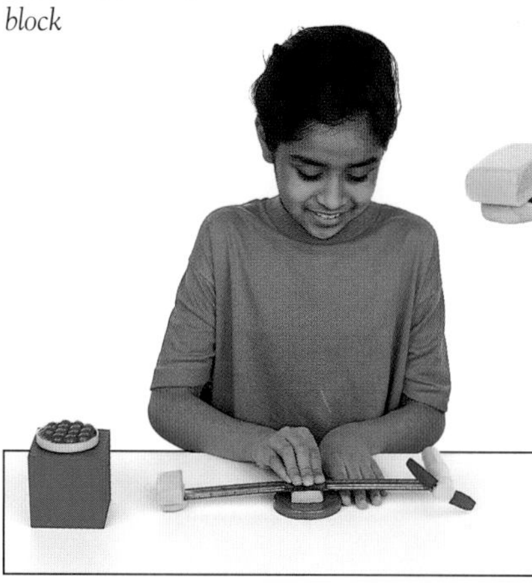

2 ATTACH THE FAN to one end of the ruler with modeling clay. Weight the other end of the ruler with clay until it is evenly balanced. Place the large lid on top of the marbles in the small lid.

3 HOLD THE FAN firmly, switch it on, then let go. As the blades of the fan push air in one direction, the body of the fan and the ruler rotate in the opposite direction.

■ DISCOVERY ■

A great leap forward

The first operational jet engine powered a German fighter plane in 1939, at the start of World War II. Two years later, a British fighter took to the air powered by a jet engine invented by Frank Whittle (born 1907). This engine developed into the jet engines used in aviation today. Whittle was not a scientist, but an aircraft engineer. He realized in 1929 that high-speed flight would be possible with an engine expelling a powerful backward jet of air. He received little encouragement at first because many officials believed that a jet engine would be too heavy to fly. Nevertheless Whittle persisted, and helped bring about an age of rapid and safe worldwide travel as jet planes easily outpaced their propeller-driven forebears.

How a jet engine works

This jet engine is a turbofan, like the very powerful engines that propel most airliners. At the front, a large fan spins to suck air into the engine. Part of this air enters the compressors, which increase the pressure of the air. The high-pressure air then enters the combustion chambers, where it mixes with kerosene and is ignited. The mixture burns, producing hot, high-pressure gases. These gases next pass through turbines, which spin to drive both the fan and the compressors. The hot gases then rush from the exhaust of the engine in a powerful jet. The thrust produced by the gas jet is augmented by the rest of the air drawn in by the fan, which bypasses the engine. This air, mixed with the exhaust gases, speeds from the turbofan, driving the aircraft forward with tremendous force.

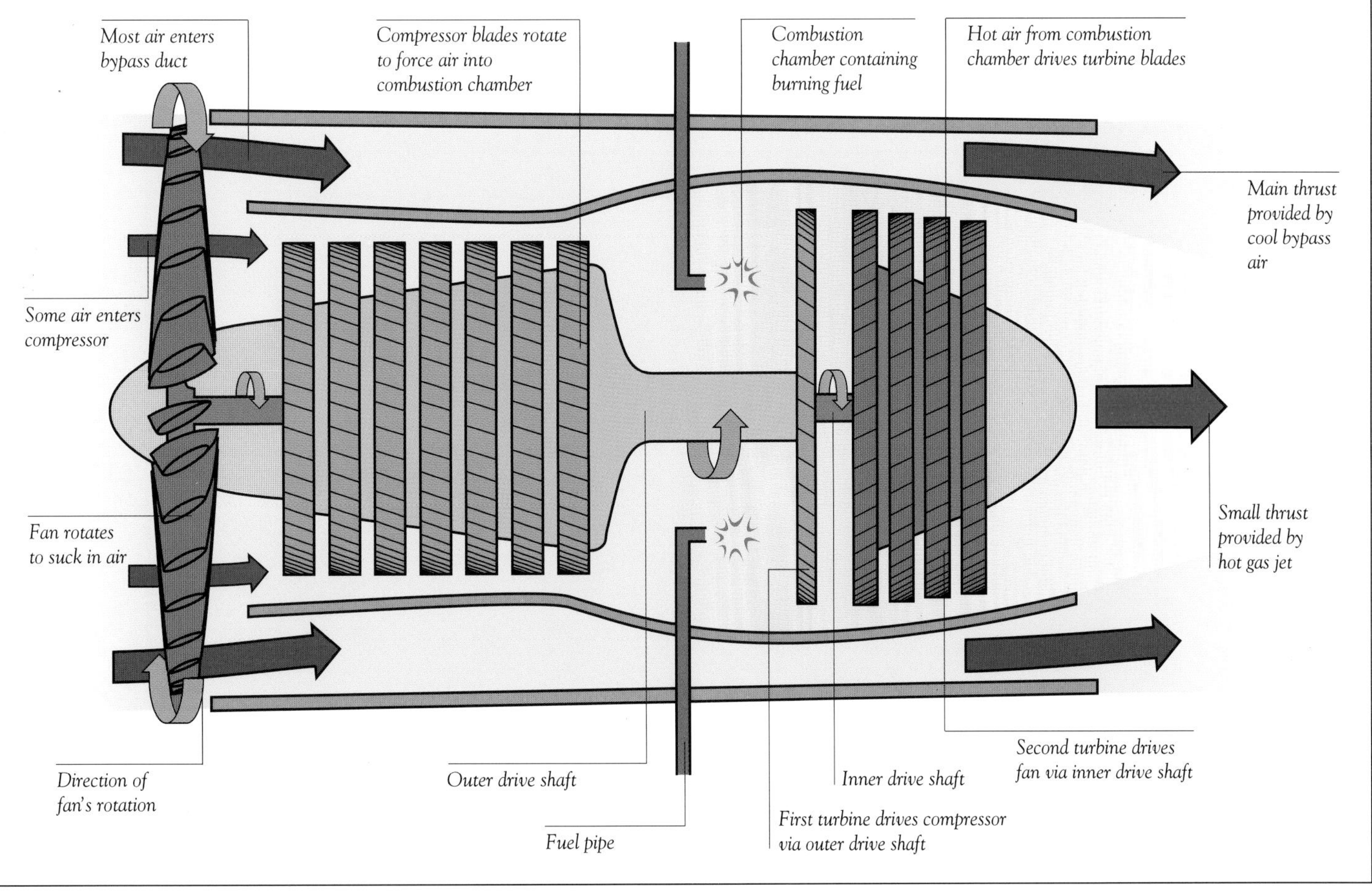

Rockets

ORDINARY FIREWORK ROCKETS, which speed into the sky to burst into showers of stars, work in basically the same way as space rockets that carry satellites and crews into orbit around the Earth or send probes to the planets. All rockets contain a fuel that burns without air, which is why rockets can travel in airless space. The burning fuel produces gases that rush from the exhaust of the rocket to drive it forward. There are two main kinds of rocket engine: solid fuel and liquid fuel. Solid-fuel engines contain a powder that burns rapidly. Firework rockets, and the booster rockets of the space shuttle use solid fuel. Once ignited, solid-fuel rockets continue to burn until their fuel runs out. The engines of liquid-fuel rockets are fed with liquid fuel and liquid oxidizer, which burn together. Unlike solid-fuel engines, these can be turned on and off.

■ DISCOVERY ■

Robert Goddard

The first rockets were solid-fuel fireworks, which were invented in China at least 800 years ago. Liquid-fuel rockets, however, date back to only 1926, when the American engineer Robert Goddard (1882–1945) fired a small prototype rocket over a remote farm in Massachusetts. His rocket was powered by gasoline and liquid oxygen. It reached a speed of about 60 mph (about 100 kph) and a height of 41 feet (12.5 meters) during a flight lasting only 2.5 seconds. The first space rockets used liquid-fuel engines like the one in Goddard's rocket.

EXPERIMENT

Bottle rocket

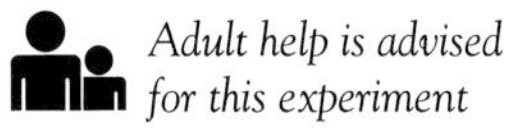

Adult help is advised for this experiment

Build a safe, simple rocket that takes off at great speed. Instead of burning fuel, it pressurizes water in order to produce a spray that drives it upward. This has the same effect as the exhaust gases expelled by a real rocket.

YOU WILL NEED

- *bicycle pump and connector* • *tape* • *plastic bottle* • *cork trimmed to fit tightly in bottle neck* • *string* • *petroleum jelly* • *drinking straw* • *protective goggles* • *compass* • *scissors* • *inflating needle* • *pitcher of water*

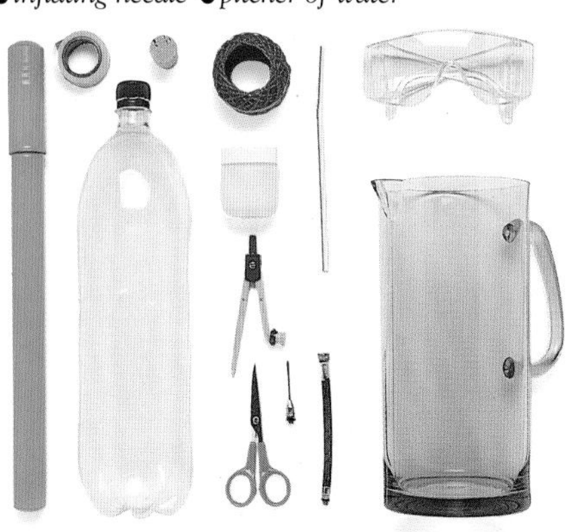

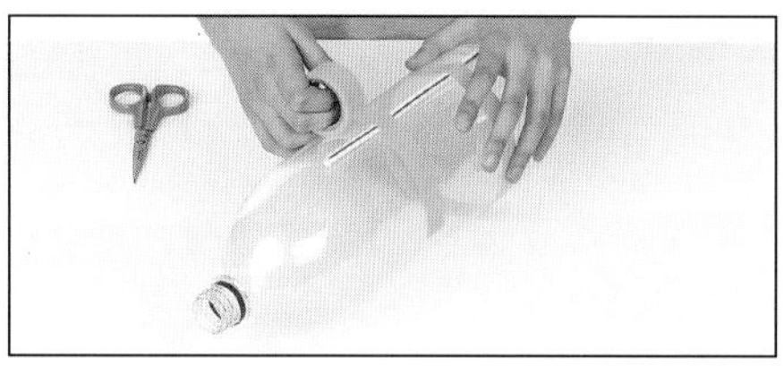

1 HOLD THE STRAW lengthwise against the side of the bottle. Secure it very tightly with tape, making sure the straw does not bend.

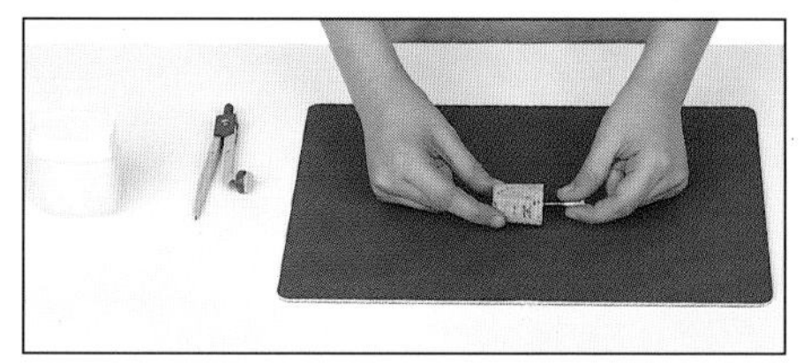

2 USING THE POINT of the compass, pierce a hole in the top of the cork. Push the inflating needle into the hole and right through the cork.

3 POUR 2 IN (5 cm) of water into the bottle. Grease the cork with petroleum jelly, and push it firmly into the bottle neck so that the needle inlet sticks out of the top.

4 TIE ONE END of the string high up on a tree or tall post outside. Thread the string through the straw, starting at the bottom of the bottle. Secure the loose end very firmly near the ground.

The space shuttle

The space shuttle is a hybrid rocket containing both solid-fuel and liquid-fuel engines. Both kinds are required to provide the enormous power needed to thrust the Shuttle into orbit. The clouds of smoke at liftoff come from the two solid-fuel booster rockets at the sides. The booster rockets are released after they burn out on the way up to space, and they parachute back to earth to be recharged with fuel and used again. The shuttle is fixed to a large external tank containing liquid hydrogen and liquid oxygen. The liquids are piped separately to three engines in the tail of the shuttle, where they mix and burn to take the shuttle into space. Once in space, the external tank is discarded and it then falls back and burns up in the atmosphere. Smaller liquid fuel engines are used to reach the desired orbit, to maneuver the shuttle as it carries out its mission, and then to leave orbit. The shuttle reenters the atmosphere and uses its wings to glide down and land. It is then made ready for its next mission in space.

Space rockets

Inside a liquid-fuel rocket, pumps send liquid fuel and oxidizer from tanks to a combustion chamber in the rocket engine. There they burn, producing gases that press against the chamber walls and rush out of the exhaust to push the rocket forward. Many rockets have several parts called stages, each with its own tanks and engines. As the rocket soars toward space, each stage drops away when its fuel is gone, and the next stage fires. Only the final stage and the payload, such as a satellite, enter space. Discarding stages saves fuel.

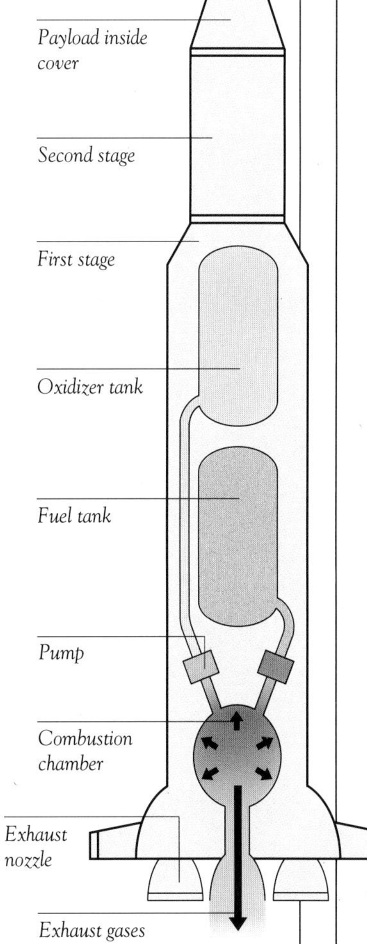

5 THE BASE of the bottle should now be pointing upward at about 45°. Attach the pump to the needle. The water should cover the neck of the bottle. If not, remove the cork, add a little water to the bottle, then replace the cork in the bottle neck. Begin pumping and be ready for the water to spray out of the rocket when it takes off.

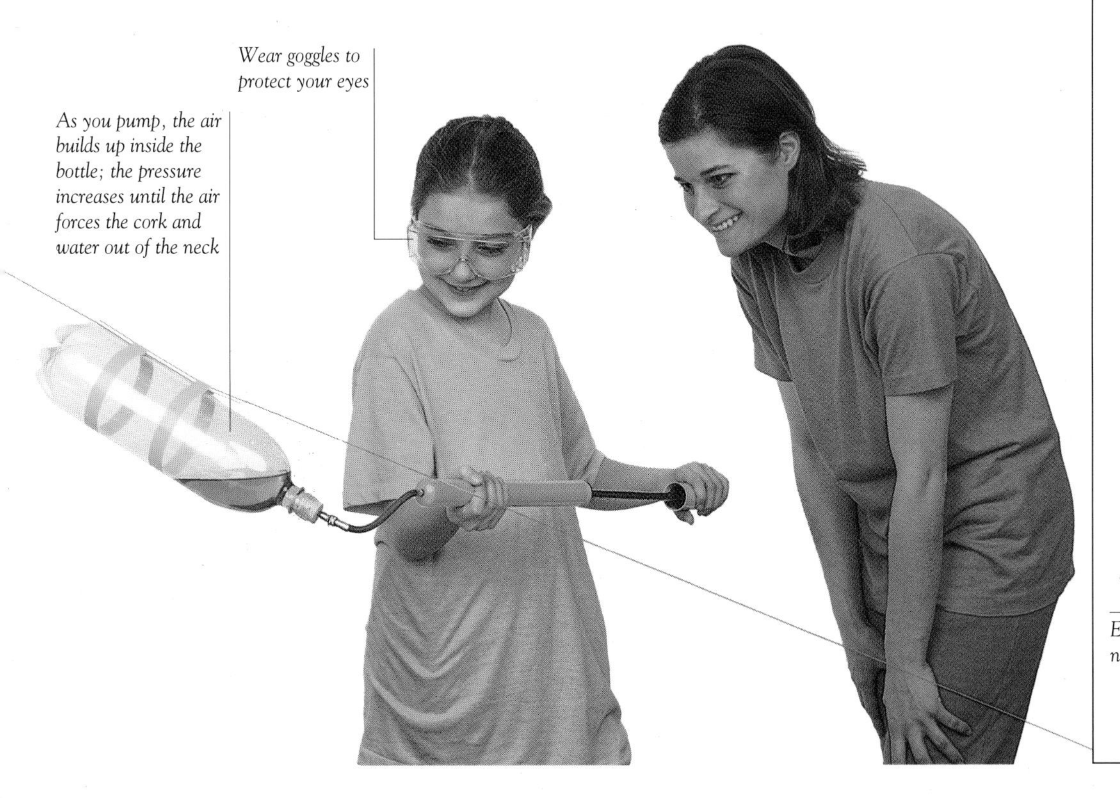

Wear goggles to protect your eyes

As you pump, the air builds up inside the bottle; the pressure increases until the air forces the cork and water out of the neck

Batteries

A BATTERY IS A VERY convenient source of energy for machines. Light, clean, and safe, it provides electricity at any time and in any place. The invention of the battery has made possible such useful devices as flashlights, hearing aids, personal stereos, and remote controllers. Batteries also provide off-duty power supplies in computers and fax machines, enabling these machines to remember the time and date when they are disconnected from a power outlet. A battery contains chemicals that react together to produce electricity when a device connected to the battery is switched on. When the chemicals are used up, the battery stops working, although some batteries, such as those used in cars, can be recharged with electricity. An electric current is fed back into the battery, and the reaction proceeds in reverse to restore the original chemicals. These can then react again to provide electricity. Depending on its voltage, a battery contains one or more electric cells. A typical cell produces about 1.5 volts. A 12-volt car battery has six 2-volt cells.

■ DISCOVERY ■

Piles of power

The Italian scientist Alessandro Volta invented the battery in 1800. He built a pile of copper and zinc disks separated by pasteboard disks soaked with salt solution or a weak acid such as vinegar. A set of three disks formed an electric cell, in which the copper and zinc reacted with the salt or acid to produce an electric current. Piling up the sets of disks connected the cells together, combining their low voltages to make a high-voltage battery. The electric current was drawn off by wires attached to the top and bottom disks of the battery, which was known as a voltaic pile. Volta's invention was based on a strange observation made a few years before by Luigi Galvani, an Italian anatomist. He noticed that the legs of a dead frog twitch when touched simultaneously by two different metals. Volta realized that the metals and salty fluid in the frog react to produce an electric current that activates the frog's muscles.

EXPERIMENT

Homemade battery

Make a battery with enough power to work an alarm clock. It consists of a set of electric cells connected together. Each cell gives about 0.6 volts. To find out how many cells you need, first remove the clock's battery and check its voltage. If it is 1.5 volts, you will need to make a battery with two or three cells to power the clock. A 3-volt clock will need a battery with four or five cells. Make sure that all the pieces of wire in the battery are firmly connected to the foil squares and to the battery terminals in the alarm clock. Just one tiny gap will stop the battery from working.

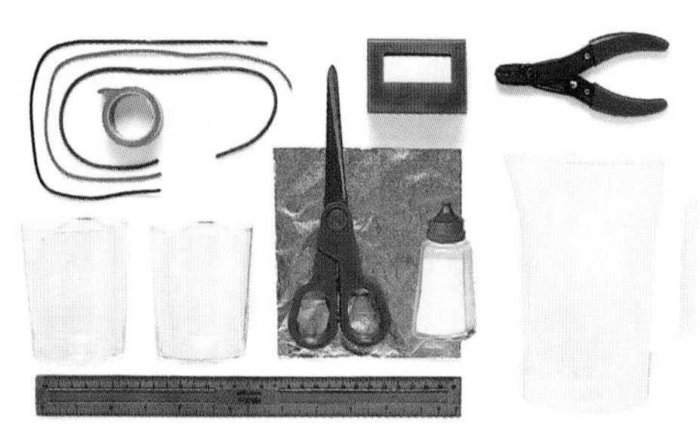

YOU WILL NEED

• three 18 in (45 cm) lengths of insulated copper wire • adhesive tape • 2 beakers • scissors • two 3 in (8 cm) aluminum foil squares • pocket alarm clock with battery removed • wire strippers • ruler • salt • pitcher

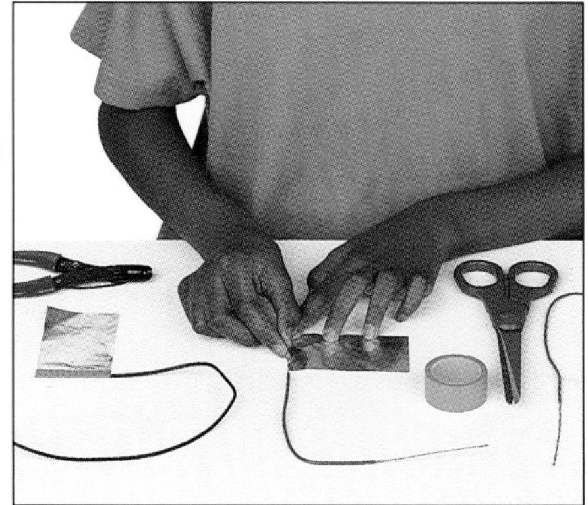

1 USING THE WIRE STRIPPERS, strip both ends of each wire. Attach a foil square to one end of each wire by folding the foil over the wire and securing with tape. Press the foil firmly around the wire to make a good contact.

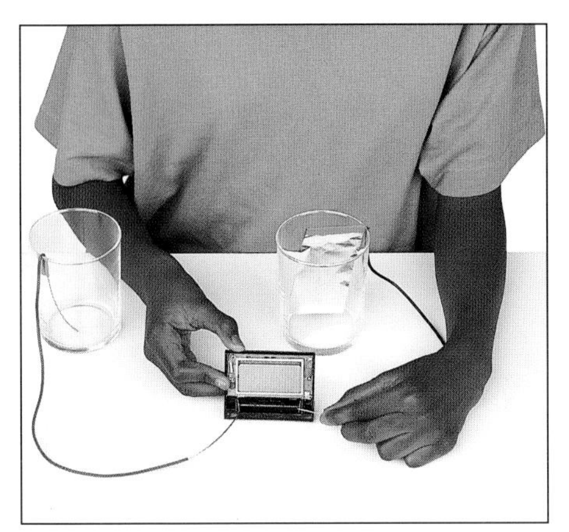

2 ATTACH THE WIRE without foil and one wire with foil to the positive and negative clock terminals as shown in the diagram below right. Tape the other ends of these wires inside the beakers.

Long-life battery

A long-life battery has a cell made of a steel case with a central metal rod. Between the case and rod are electrodes of powdered zinc, and manganese oxide mixed with carbon. This type of cell is called an alkaline cell because potassium hydroxide, an alkaline substance, is mixed with both electrodes. It enables the zinc and manganese in the electrodes to produce an electric current. The rod and inner steel case lead the current from the electrodes to a positive terminal at the top of the battery and a negative terminal at the base.

Alkaline manganese power cell
Compacted electrodes give this type of cell a longer life.

Outer steel case

Inner steel case

Manganese oxide and carbon electrode

Metal rod

Separator between electrodes

Powdered zinc electrode

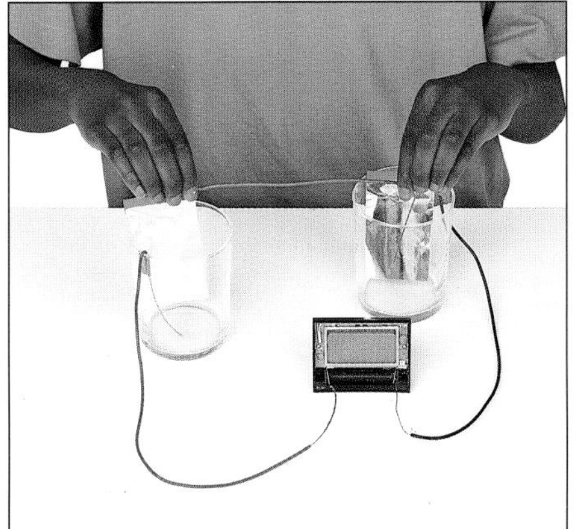

3 TAPE THE THIRD WIRE between the beakers or cells as shown in the diagram above right. Each cell should now have one foil contact and one wire contact.

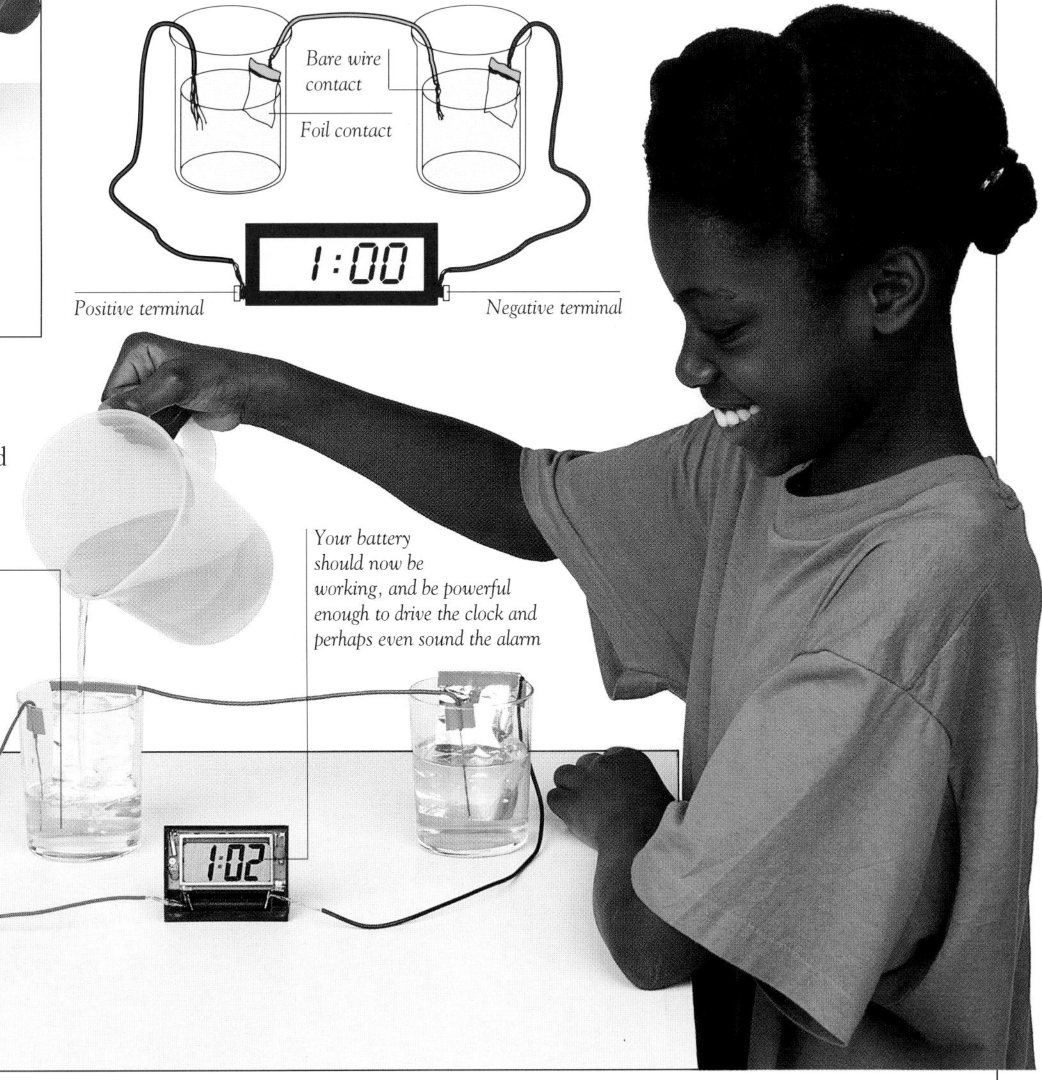

Salt water reacts with copper in the wire and aluminum in the foil to allow an electric current to flow through the circuit

Your battery should now be working, and be powerful enough to drive the clock and perhaps even sound the alarm

4 DISSOLVE two teaspoons of salt in warm water, then pour it into both beakers. Make sure that the water reaches all four contacts but does not touch the wire attached to the foil.

Electric motors and generators

ELECTRICITY, which is convenient and clean, is the main source of energy for the machines we use in the home as well as many machines in industry. Food processors, personal stereos, elevators, and electric trains all contain motors that turn electricity into motion and drive the moving parts of these machines. The starter motor in a car is an electric motor. Most electrical machines in the home and in industry use electricity provided by utility companies. The electricity is produced by large electricity generators in power stations and carried to consumers by power lines. A generator is driven by a combustion engine or water-powered turbine, and is like an electric motor in reverse. It converts motion into electricity.

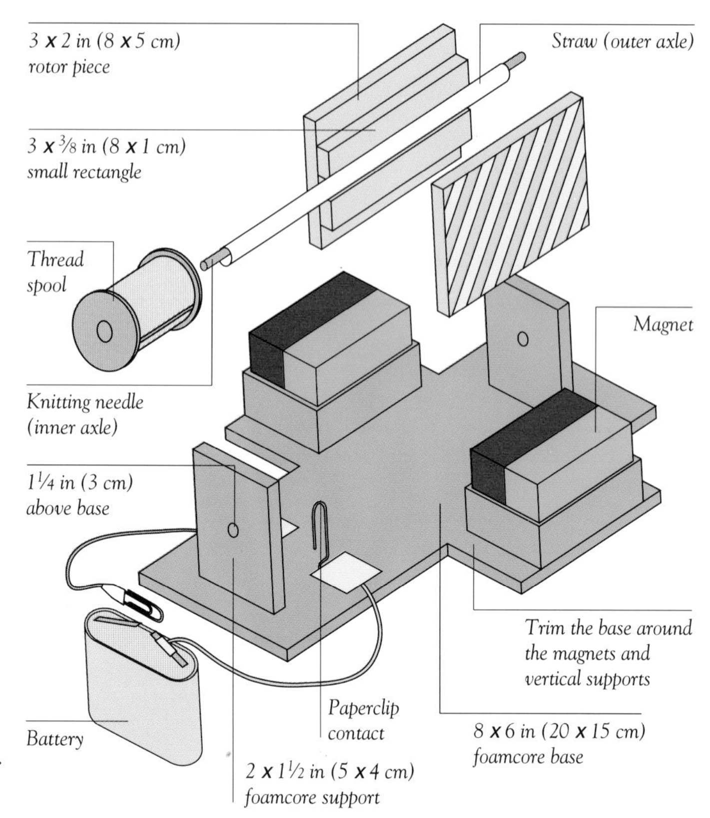

EXPERIMENT

Make a motor

 *Adult help is advised for this experiment*

Build a real electric motor using simple materials. The motor is battery-powered and has a coil of wire supported between two magnets. Electricity fed into the coil from the battery makes the coil spin. All electric motors basically work in this way.

YOU WILL NEED

• 3/16 in (5 mm) thick foamcore • thin insulated wire • 2 strong magnets (you can make these by joining several small ones together) • aluminum foil • 4.5V battery • thread spool • 2 metal paperclips • ruler • drinking straw • knitting needle or skewer to fit loosely inside straw • adhesive tape • double-sided adhesive tape • scissors • glue • craft knife • cutting mat • steel ruler • pliers • wire strippers

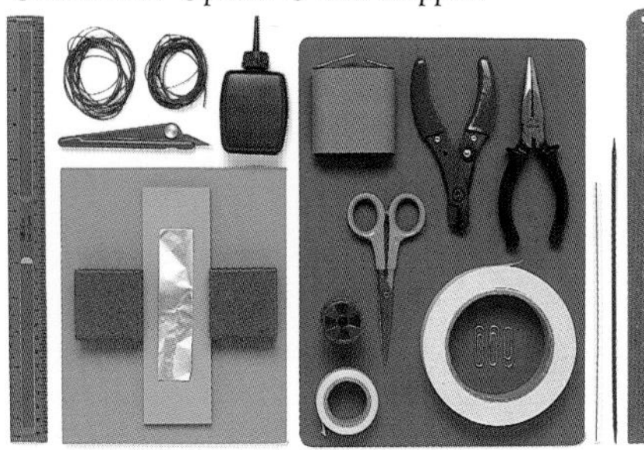

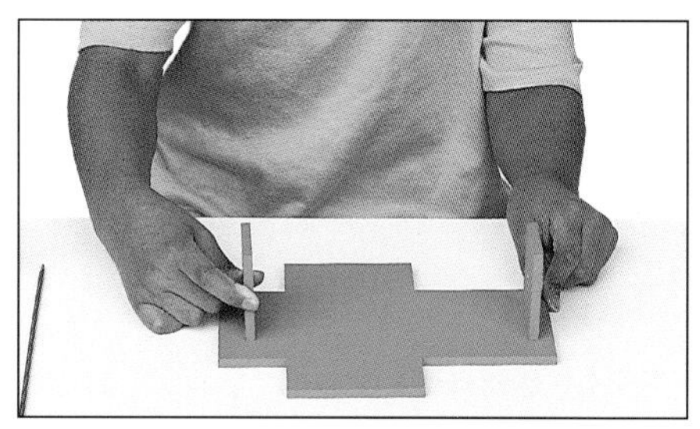

1 CUT THE BASE and the vertical supports from foamcore, following the diagram above. Make holes in the supports using the knitting needle, in the positions shown. Glue the supports to the base 6¼ in (15.5 cm) apart.

2 TRIM THE STRAW to 6 in (15 cm) long to make the outer axle. Glue the straw and two smaller rectangles between the two rotor pieces. The straw should lie along the middle of the rotor, protruding 2 in (5 cm) from one end.

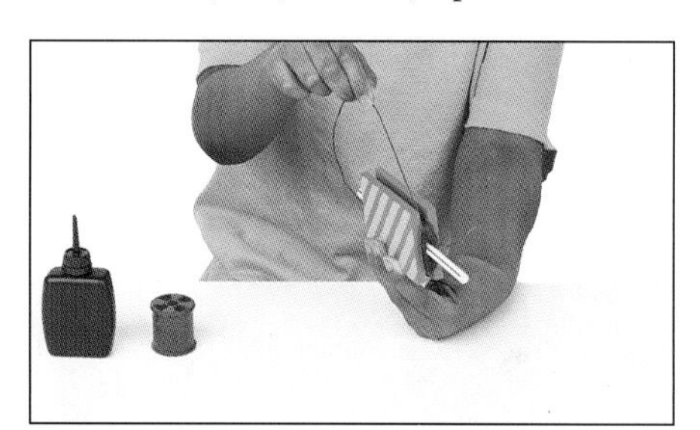

3 WIND THE WIRE about 30 times around the slot in the rotor. Leave 2 in (5 cm) of wire free at each end. Wind the coil on both sides of the outer axle to balance the rotor. Glue the long end of the axle inside the thread spool.

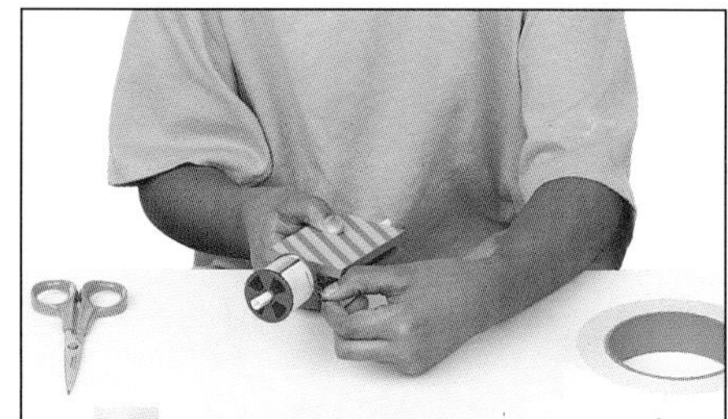

4 STICK TWO patches of double-sided tape to the spool. Separate the patches by two tiny gaps (at top and bottom when the rotor is horizontal). Stick a stripped end of wire from the coil to each patch, then cover with foil.

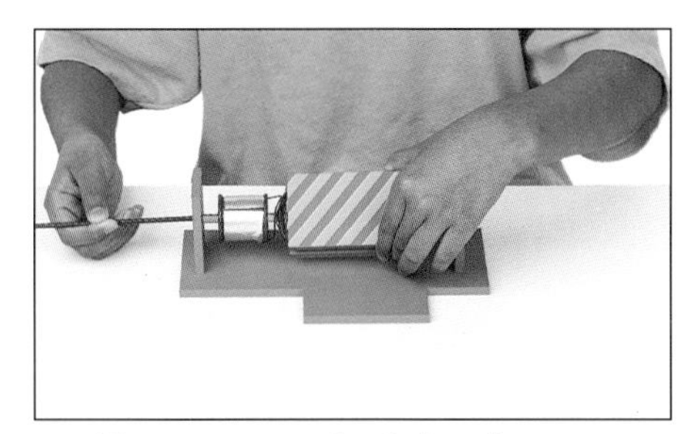

5 ENSURE THAT the foil and wire are in contact on each half of the spool, but not across the gaps. Assemble the rotor on the base, pushing the knitting needle (inner axle) through the vertical supports and outer axle.

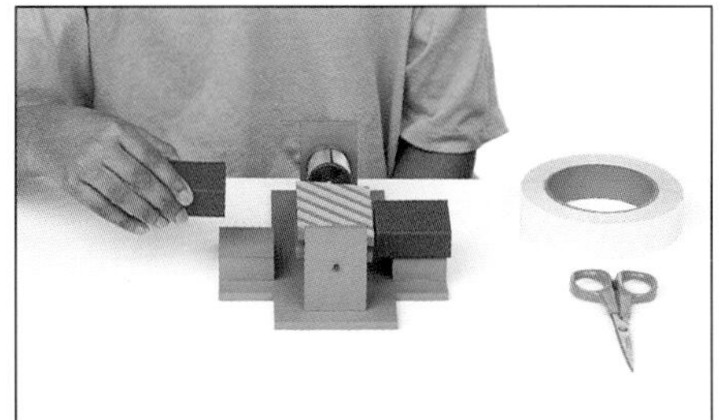

6 MOUNT THE MAGNETS on foamcore blocks so that they align with the axles, with unlike poles facing each other. Position the magnets as close as possible to the rotor, while ensuring that it can still rotate freely.

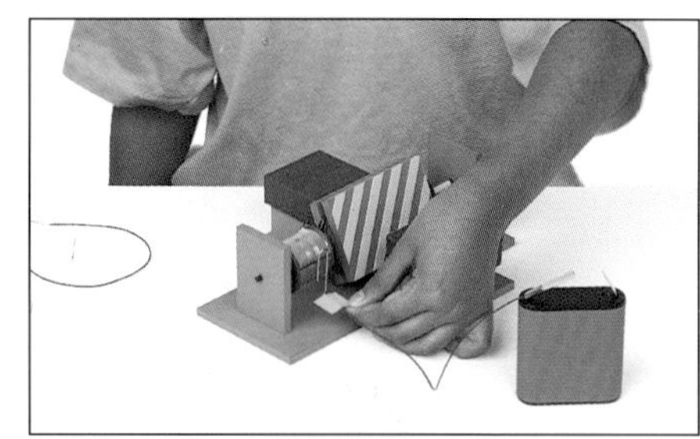

7 BEND TWO paperclips into L-shapes. Attach an 8 in (20 cm) length of wire to each. Attach the free end of one wire to a battery terminal. Stick the clips to the base on either side of the spool so that they just touch the foil surfaces.

How motors and generators work

When an electric current flows through contacts into the coil of wire in a motor, a magnetic field forms around the coil. Magnets around the coil make their own magnetic field. The two fields push or pull on each other, just as two magnets attract or repel one another, and cause the coil to spin. The spinning coil drives the shaft of the motor, which powers a machine.

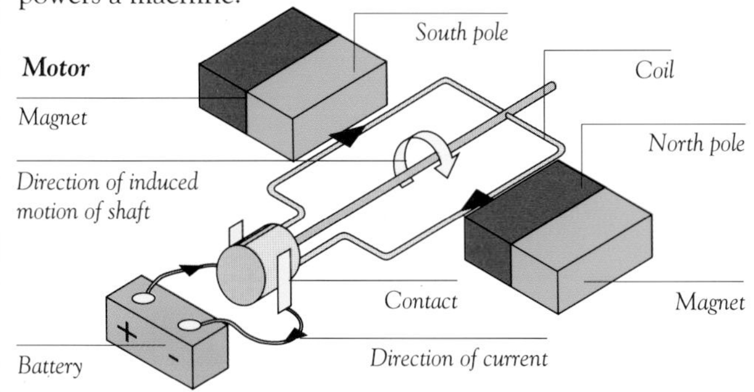

An electric generator is like a motor, but the coil is made to spin by an outside engine or turbine. As the wire in the coil cuts through the magnetic field of the magnets, an electric current is set up. This current flows out through contacts to power light bulbs or machines.

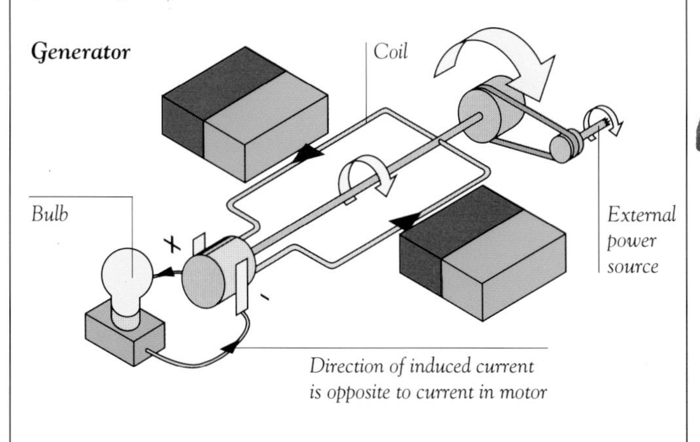

8 CHECK THAT the rotor still rotates freely. Connect the remaining end of wire to a third paperclip. Keep the rotor horizontal and slide this clip onto the second battery terminal to start the motor. You may need to spin the rotor to start it.

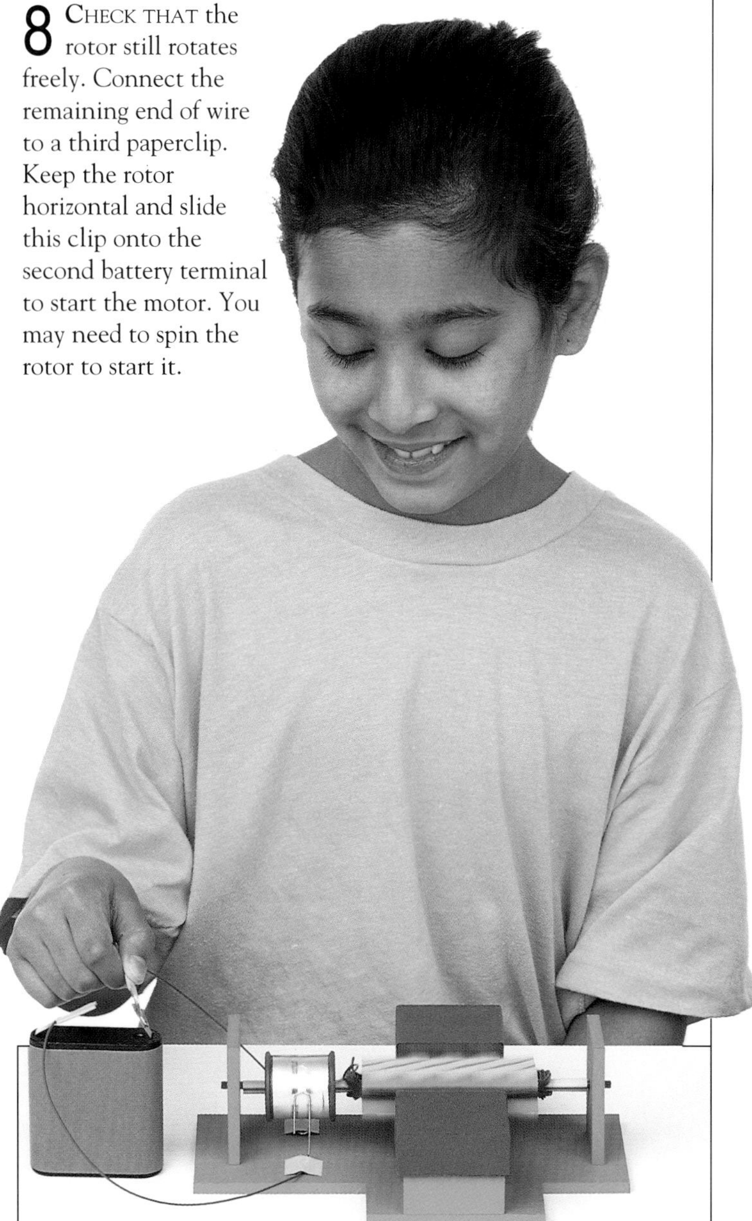

Levers

IT IS IMPOSSIBLE TO OPEN a can with your bare hands, but a can opener makes the job easy. The can opener has two levers (the parts that you squeeze) that magnify the force applied to the opener by your hands. A lever is a rigid rod or bar that pivots around a fixed point called a fulcrum. Many machines use levers to increase force. When a force (the effort) is applied to one part of a lever, the lever pivots on the fulcrum and another part of the lever moves the load. If the load-bearing part of the lever moves a shorter distance than the part that receives the effort, the load is moved with a correspondingly greater force.

Practical pairs

These everyday devices are examples of the three classes of levers that come in pairs. Scissors and nutcrackers magnify the force of your fingers to cut material or crack a shell. Tweezers don't magnify force but instead decrease it; they are useful for handling delicate objects such as rare stamps.

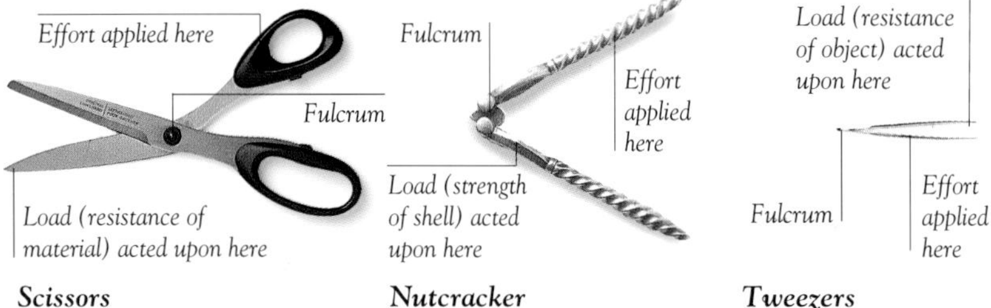

Scissors
A pair of first-class levers

Nutcracker
A pair of second-class levers

Tweezers
A pair of third-class levers

EXPERIMENT

Large lever

The force that drives a lever is called effort. It makes the lever pivot on its fulcrum, to raise a weight or overcome a resistance (both of which are called the load). There are three classes of lever. These differ only in the position of the effort, fulcrum, and load. Build a big lever and see how the three kinds work. Some levers magnify the force you apply to them and others magnify the distance you move them.

YOU WILL NEED

- *string*
- *long broom handle or wooden pole*
- *strong net bag*
- *oranges or other fruit for weights*
- *wooden rolling pin*
- *tape*
- *tall plastic bin or container*

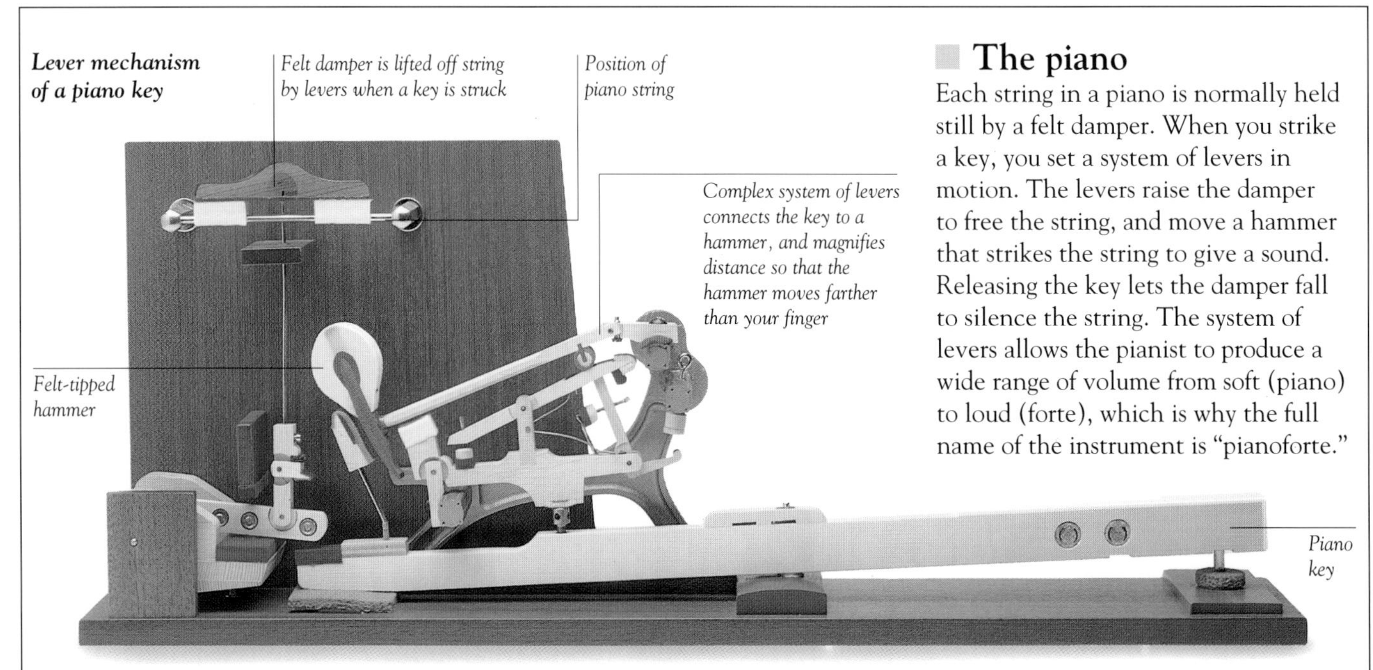

Lever mechanism of a piano key

The piano

Each string in a piano is normally held still by a felt damper. When you strike a key, you set a system of levers in motion. The levers raise the damper to free the string, and move a hammer that strikes the string to give a sound. Releasing the key lets the damper fall to silence the string. The system of levers allows the pianist to produce a wide range of volume from soft (piano) to loud (forte), which is why the full name of the instrument is "pianoforte."

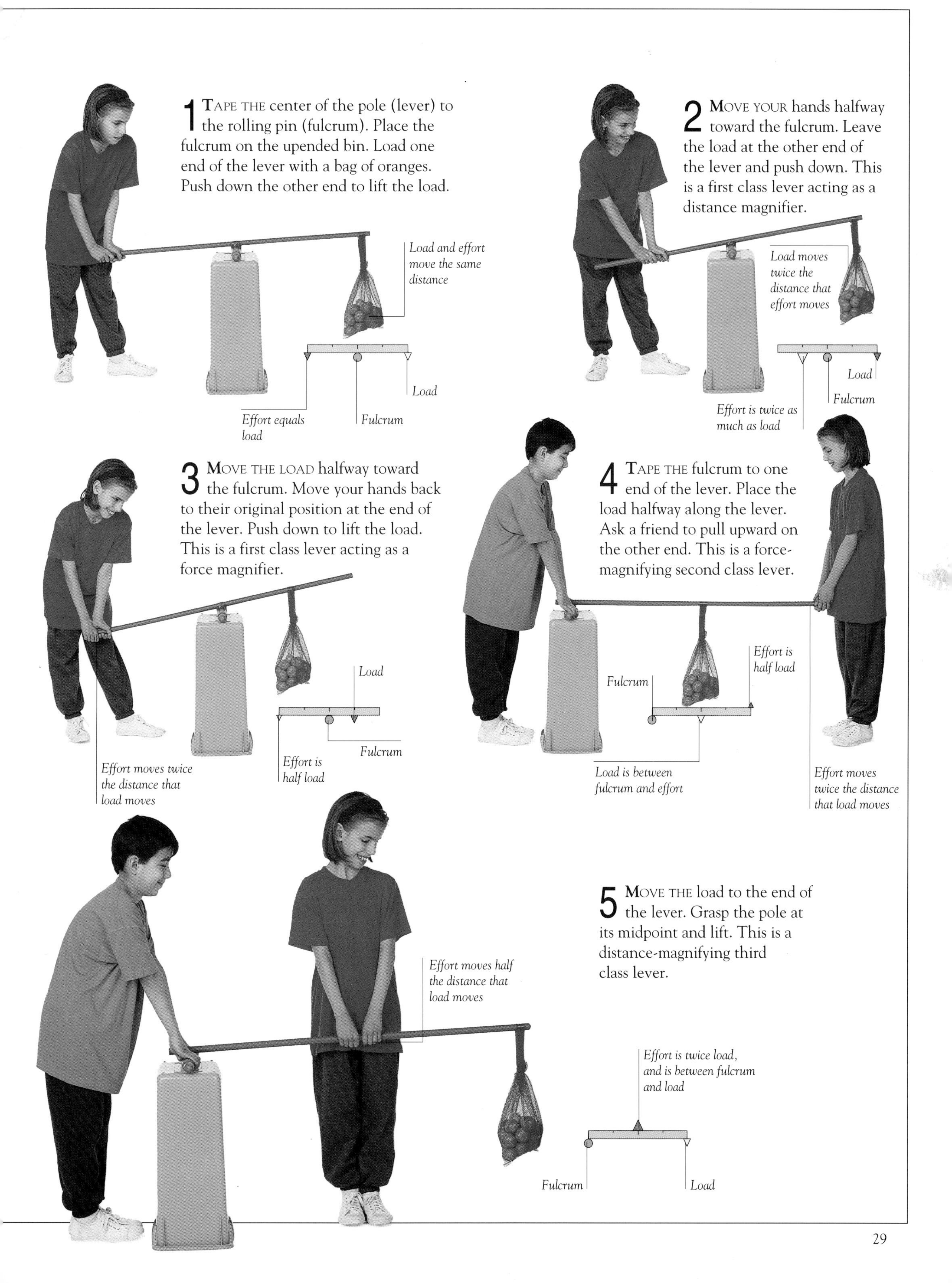

1 TAPE THE center of the pole (lever) to the rolling pin (fulcrum). Place the fulcrum on the upended bin. Load one end of the lever with a bag of oranges. Push down the other end to lift the load.

2 MOVE YOUR hands halfway toward the fulcrum. Leave the load at the other end of the lever and push down. This is a first class lever acting as a distance magnifier.

3 MOVE THE LOAD halfway toward the fulcrum. Move your hands back to their original position at the end of the lever. Push down to lift the load. This is a first class lever acting as a force magnifier.

4 TAPE THE fulcrum to one end of the lever. Place the load halfway along the lever. Ask a friend to pull upward on the other end. This is a force-magnifying second class lever.

5 MOVE THE load to the end of the lever. Grasp the pole at its midpoint and lift. This is a distance-magnifying third class lever.

Gears and screws

MANY MOTORS THAT power machines work over only a limited speed range, but the machines may have to work over a wider speed range. One way of changing speed is to use gears. Toothed gearwheels mounted on different shafts mesh together so that when one shaft is turned by a motor, the other shaft turns too. If the wheels are of different sizes, the shafts turn at different speeds and with different amounts of turning force. Gears can also change the direction of motion produced by a motor. Like gears, screws can magnify force. When a screw is turned, its thread moves the shaft of the screw forward or backward with more force than the turning effort.

Kinds of gears

A set of connected gears, like the one below, is called a gear train. The first shaft is powered, and gears link the other shafts to it. The four main types of gears shown below may change the speed, force, or direction of motion transmitted to them by the powered shaft. A spur gear has toothed wheels that mesh to connect parallel shafts. Bevel gears have toothed wheels with sloping faces that mesh at a particular angle. They are often used to join two shafts together at 90°. A worm gear has a shaft with a screw thread (the worm) that meshes with and turns a spur-gear wheel. The gearwheel shaft is at 90° to the shaft carrying the worm. A rack and pinion consists of a spur-gear wheel (the pinion) that meshes with a toothed bar (the rack).

Screw jack

A screw jack magnifies the force of the user's arm in order to lift a car. It works on the principle that a small force moving through a long distance can be turned into a greater force that moves through a shorter distance. The screw jack's handle, when turned, rotates a shaft with a screw thread. The shaft passes through a threaded bar, which can be attached to the car. When the shaft is rotated, it moves the bar up or down to raise or lower the car. The screw thread is like a sloping ramp spiraling around the shaft. If the handle is turned once, the shaft rotates and the bar moves up the vertical distance between two turns of the screw thread. The small force needed to turn the handle and rotate the screw through a long distance is converted into a larger force exerted by the bar, which lifts the car a short distance.

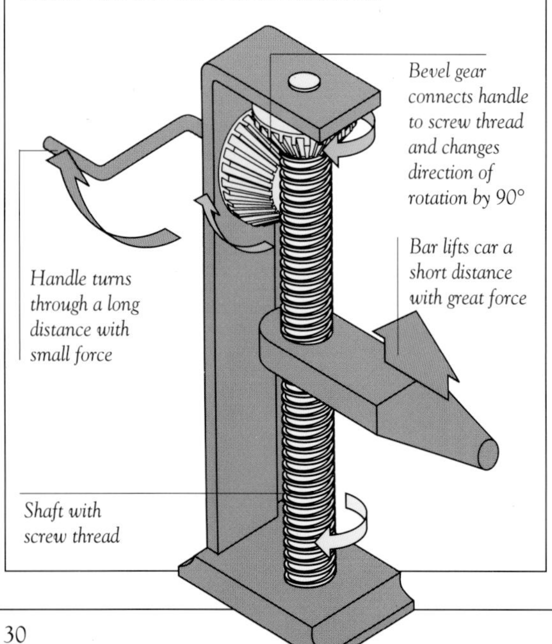

Spur gear (red)
This spur gear has two toothed wheels that connect two parallel shafts. The first shaft is powered (in this case by hand) and carries the smaller gearwheel. The second shaft carries a larger wheel and therefore rotates with less speed but greater turning force than the first shaft. Car transmissions usually contain sets of spur gears to transform the engine's speed range into a wider range of speeds for the car wheels.

EXPERIMENT

Bicycle gears

Gears allow the rear wheel to turn at different speeds to the rate at which the cyclist pedals, and with different amounts of force. They enable the rider to cycle at a wide range of speeds.

You Will Need

- *scissors* - *tape* - *gloves* - *bicycle with derailleur gears*

1 Wearing gloves, turn the bicycle upside down. Ask a friend to hold the frame steady. Turn the crank so that one crank arm points upward. Mark a point on the rear tire with a piece of tape.

2 Select a low gear, with the chain on a small front chainwheel and a large rear sprocket. Hold the wheel lightly and turn the crank once. Note how many times the wheel turns.

3 Now select a high gear, with the chain on a large front chainwheel and a small rear sprocket. Turn the crank once, and note how many times the wheel turns.

Gear ratios

Calculate a gear ratio on your bicycle. Note which rear sprocket and front chainwheel the chain passes over. Divide the number of sprocket teeth by the number of chainwheel teeth to obtain the gear ratio. If the sprocket has 10 teeth and the chainwheel has 40, the gear ratio is $\frac{10}{40}$ or $\frac{1}{4}$, written as 1:4. This is a high gear ratio, in which the rear wheel turns four times as fast as you pedal but with a quarter of the turning force. High gears are used for high-speed cruising on flat roads. In lower gears (such as 1:2), the wheel turns fewer times for each turn of the pedals but with more force. Low gears are used when cycling uphill.

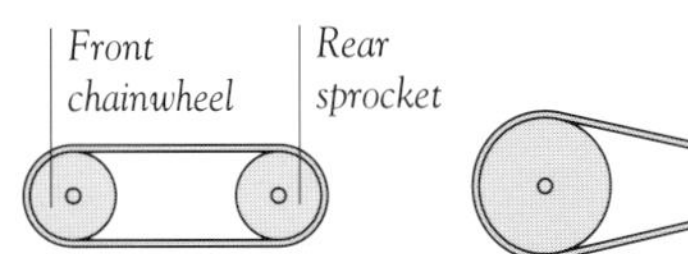

Low gear 1:1 *Chainwheel is same size as sprocket.*

High gear 1:4 *Chainwheel is four times size of sprocket.*

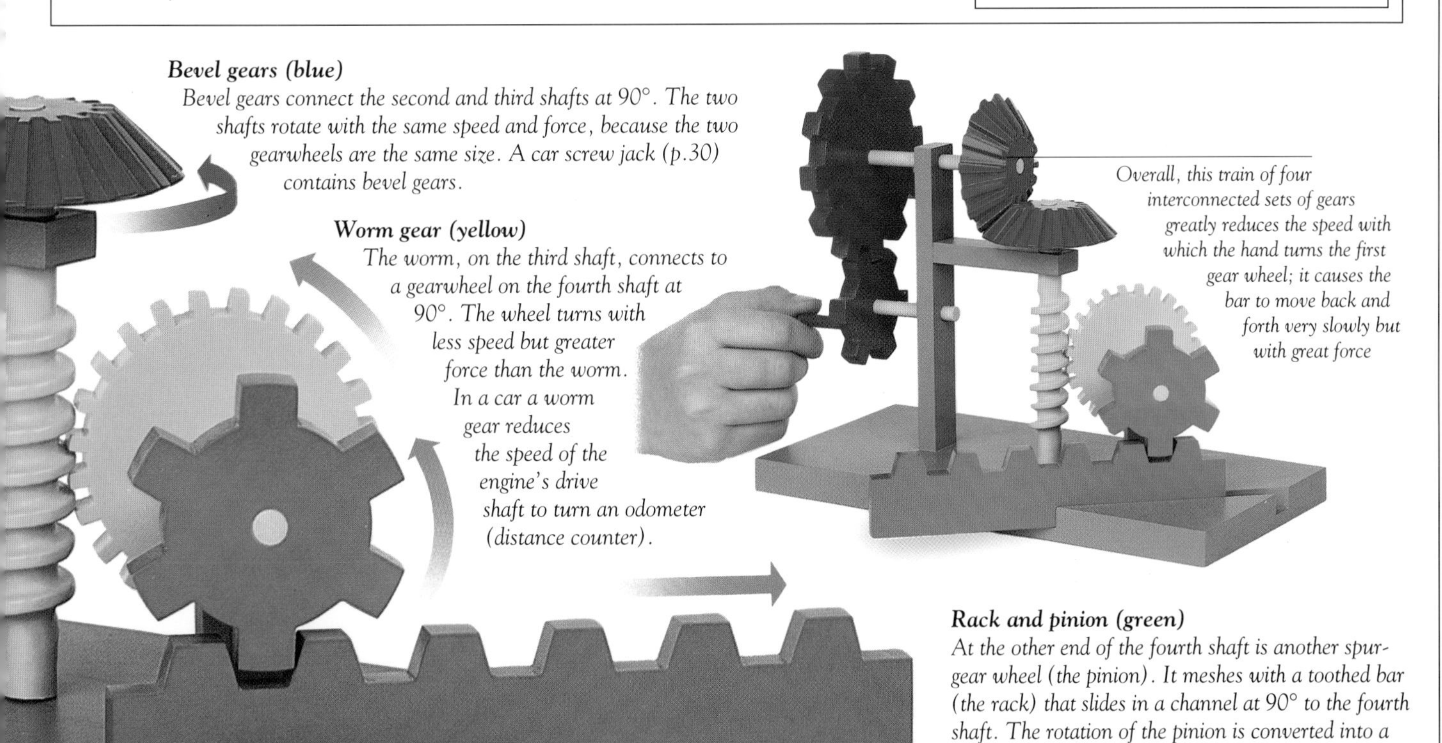

Bevel gears (blue)
Bevel gears connect the second and third shafts at 90°. The two shafts rotate with the same speed and force, because the two gearwheels are the same size. A car screw jack (p.30) contains bevel gears.

Worm gear (yellow)
The worm, on the third shaft, connects to a gearwheel on the fourth shaft at 90°. The wheel turns with less speed but greater force than the worm. In a car a worm gear reduces the speed of the engine's drive shaft to turn an odometer (distance counter).

Overall, this train of four interconnected sets of gears greatly reduces the speed with which the hand turns the first gear wheel; it causes the bar to move back and forth very slowly but with great force

Rack and pinion (green)
At the other end of the fourth shaft is another spur-gear wheel (the pinion). It meshes with a toothed bar (the rack) that slides in a channel at 90° to the fourth shaft. The rotation of the pinion is converted into a linear motion, in which the rack moves backward or forward in a straight line. The steering mechanism of a car (pp.102–103) may contain a rack and pinion.

Pulleys

LIFTING A HEAVY LOAD is much easier if you attach the load to one end of a rope and sling the rope over a wheel fixed to a high beam. You can then use your own weight to pull down the other end of the rope and raise the load. This arrangement is called a single pulley. The amount of force or effort needed to raise the load is equal to its weight. Running the rope around more pulley wheels enables less effort to be used—two wheels require half the effort, three wheels a third, and so on. Cranes have pulleys that lift heavy loads using small motors, and sailing boats have ropes with pulleys to help the crew overcome the force of the wind when trimming the sails.

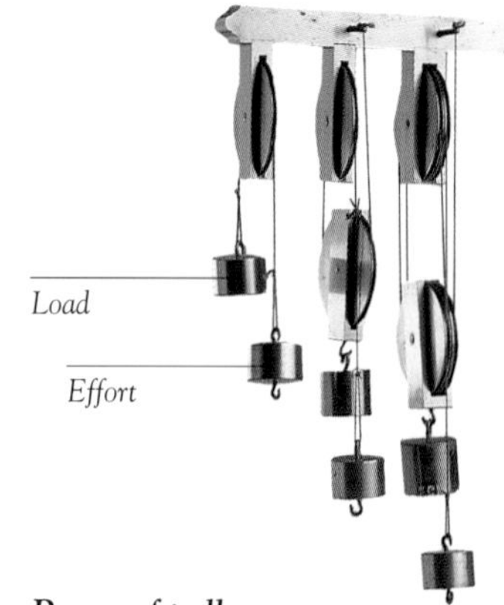

Power of pulleys
The three pulley systems above lift different loads for the same effort. The more wheels a pulley has, the more it magnifies force.

EXPERIMENT

Block and tackle

Adult help is advised for this experiment

A block and tackle has one fixed and one moving group of pulleys. A rope passes through the two groups, or blocks, of pulley wheels. The upper block is attached to a support and the lower block hangs from it by the rope. A load hangs from the lower block. A small effort applied at the free end of the rope can move a heavy load.

YOU WILL NEED

• ruler • steel ruler • 15 large marbles for weights • pencil • 30 ft (10 m) of string • C-clamp • wood strip 13 ft 8 in x 1 1/2 in x 3/8 in (4.1 m x 3.5 cm x 1 cm) • saw • drill and 3/16-in (5-mm) bit • drilling board • 12 cardboard disks of diameter 2 1/2 in (6.5 cm) • six 3/16 in (5 mm) thick foamcore disks of diameter 2 in (5 cm) • screw hook • wood glue • craft knife • cutting mat • candle • two plastic net bags • compass • wood strip 4 1/2 x 3/8 x 3/8 in (12 x 1 x 1 cm) • one 3-in (7-cm) and two 5 1/2-in (14-cm) lengths of 3/16-in (5-mm) dowel • wood base 4 1/2 x 4 1/2 x 3/8 in (12 x 12 x 1 cm) • nine 3/4 in (2 cm) square, 1/8 in (3 mm) thick foamcore spacers with central 3/16 in (5 mm) holes

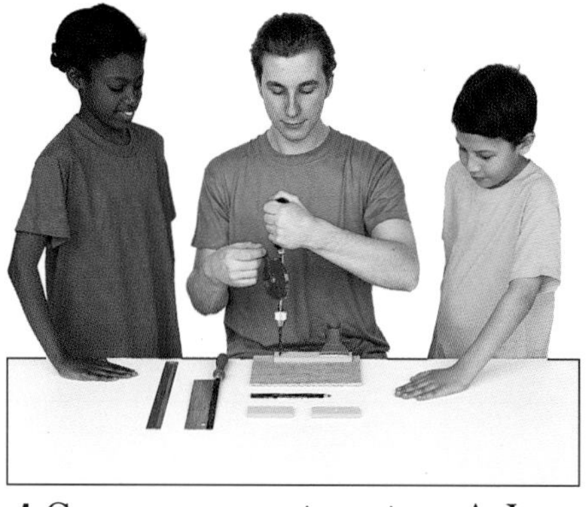

1 CUT WOOD STRIPS into pieces **A–L** as indicated in the illustration opposite. Drill a pair of holes in pieces **A** and **B** and central holes in pieces **J** and **L**, into which a dowel fits tightly.

2 USING WOOD glue, assemble pieces **A–I** together with the base to make a frame for the block and tackle. Make sure that the holes drilled in pieces **A** and **B** are aligned.

3 USE THE COMPASS TO make a hole in the center of each disk. Enlarge the holes with a pencil so the disks rotate freely on a dowel. Glue each foamcore disk between two cardboard disks to make six pulley wheels.

4 RUB THE THREE lengths of dowel with a candle to lubricate them. On one long piece of dowel put two wheels and three spacers; on the other put one wheel and two spacers. Glue the long dowel axles into the top of the frame.

5 TO MAKE the lower block, glue pieces **J**, **K**, and **L** together. Screw the hook into the middle of piece **K**. Glue one end of the short length of dowel into the hole in piece **J**.

6 SLIDE THREE pulley wheels and four foamcore spacers onto the short dowel. Glue piece **M** to the free end of the dowel and to pieces **L** and **K** to complete the lower block.

7 TIE ONE END of the string to the upper piece of dowel fitted with two pulleys. Wind the string around the pulley wheels as shown below. The loose end of the string should hang from the single pulley.

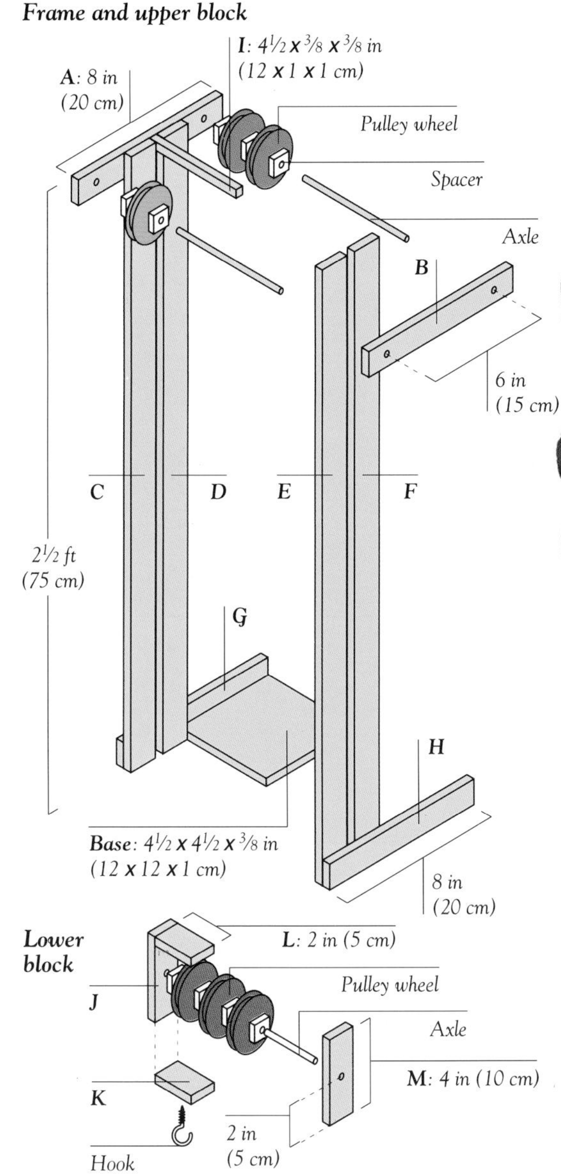

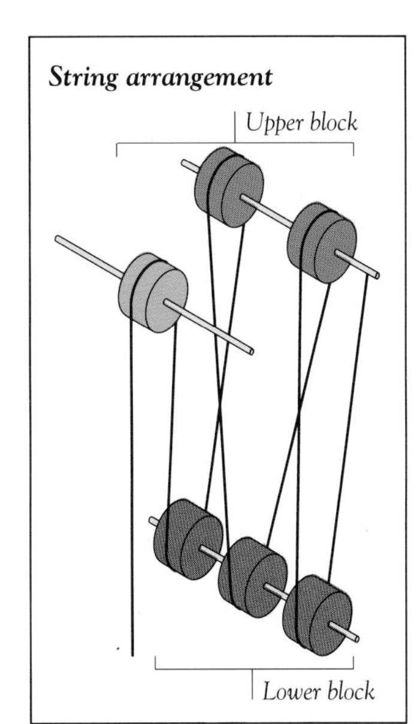

8 HANG A BAG of 12 marbles from the hook on the lower block. Make sure that the block is fully lowered. The bag and its marbles make up the load. Tie the other bag to the loose end of the string, about 6 in (15 cm) below the single pulley. Add marbles (which together represent the effort) to this bag until the block and load start to rise. Only a few marbles (i.e., a small effort) are needed to lift the block and load, but the effort moves through a greater distance than does the load.

Bearings and lubrication

A MACHINE THAT WORKS WELL wastes as little of the energy put into it as possible, but some loss of energy is inevitable. Energy loss is often caused by friction, which occurs when moving parts inside a machine rub against each other or against their supports. Some of this energy turns into heat and sound, leaving less energy to drive the machine, which slows down or loses power. Reducing friction allows a machine to run smoothly and efficiently and to use less fuel or energy. Bearings are supports that reduce friction, often by enabling a moving part to roll rather than rub on its support. Lubricants, such as oil, allow parts to slide easily over each other.

EXPERIMENT

Ball bearing

Cars, washing machines, and fans are just some of the many machines that have rotating parts. A part that turns, such as a bicycle wheel, rotates on a stationary support, in this case an axle, which connects the part to the rest of the machine. A ball bearing can be fitted between the rotating part and its support to reduce friction. The rotating part rolls over small metal balls in the bearing instead of rubbing against its support. Build a simple ball bearing, and see how it enables an object to spin easily.

EXPERIMENT

Sliding

Adult help is advised for this experiment

Ice is slippery because the surface is very smooth and has little friction. A lubricant, such as oil or grease, similarly reduces friction because it forms a smooth film over a surface. Lubricants make machines run better because the moving parts slide easily over each other. This is why you need to oil a bicycle and to check the oil level in a car engine. See how well oil lubricates by building a device that uses a scale to measure the friction between various kinds of surfaces.

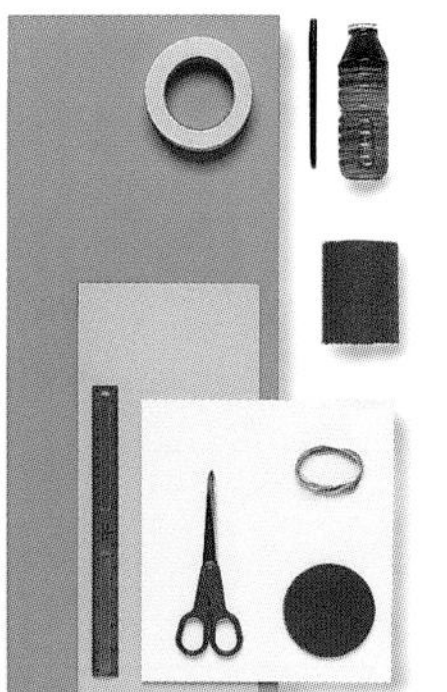

YOU WILL NEED

- *wooden board about 40 x 12 in x 1 in (100 x 30 cm x 2.5 cm)*
- *double-sided tape*
- *pen*
- *cooking oil*
- *heavy tin can*
- *cardboard strip for scale*
- *ruler*
- *self-adhesive plastic sheet*
- *scissors*
- *rubber bands*
- *flat, round wooden base for can*

1 TAPE THE cardboard strip along one long edge of the board, and mark a scale on it from 1 to 20. Tape the base to the can.

2 PLACE THE can at the zero mark on the scale. Loop a string of rubber bands around the can. Pull slowly on the string along the scale. Add rubber bands, or remove them, until the can just begins to move as your hand nears the end of the scale.

3 NOTE THE point on the scale by the free end of the string when the can just begins to move. Now cover the board with self-adhesive plastic. Place the can at zero. Pull very slowly on the end of the string as before. What does the scale read when the can begins to move?

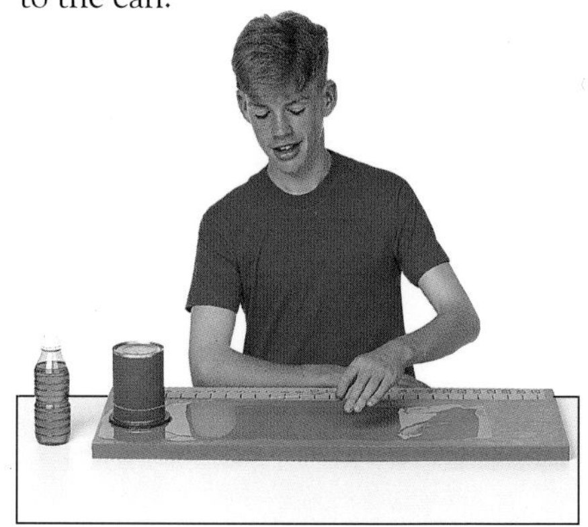

4 TO ADD a second lubricating layer, spread a thin film of oil over the plastic-coated board. Place the can at the zero mark on the scale. Pull slowly on the free end of the string as before. What does the scale read this time when the can begins to move?

YOU WILL NEED

- *2 jar lids, one fitting into the other*
- *marbles* • *pencil* • *scissors*
- *about 2 oz (50 g) modeling clay*
- *wooden block* • *double-sided tape*

1 TAPE THE smaller jar lid upside down on the wooden block. Place equal blocks of clay on each end of the pencil. Stick the middle of the pencil onto the top of the larger lid with clay.

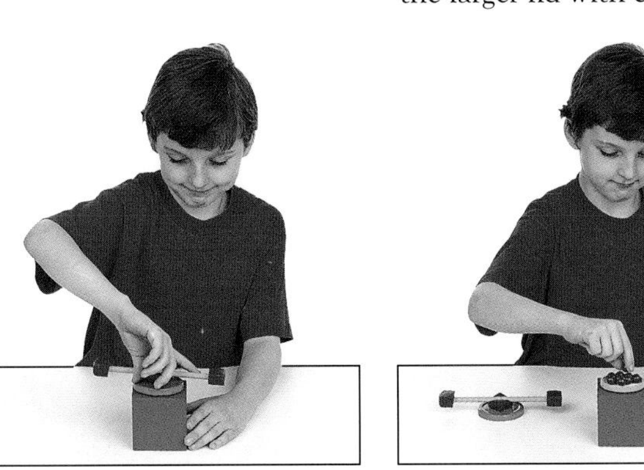

2 PUT THE larger jar lid on top of the smaller one. Hold the block steady with one hand. Hold the middle of the pencil and spin it sharply. Note how long it spins.

3 REMOVE THE larger lid. Put marbles in the smaller lid. There should be enough marbles to form a ring around the inside of the lid, while still being able to roll around.

4 REPLACE THE larger jar lid on top of the smaller lid to make a bearing. Hold the block steady with one hand. Hold the middle of the pencil and spin it sharply. Note how long it continues to spin. The pencil spins faster this time because the marbles separate the two lids and roll as the top lid turns. This greatly reduces friction between the top lid and its support.

■ DISCOVERY ■

Fighting friction

The idea of reducing friction to ease labor is very old. Before the invention of the wheel, about 5,500 years ago, very heavy loads were often moved by rolling them on logs. These simple roller bearings reduced friction and were the forerunners of modern ball bearings. Wheels, which could be permanently fixed to a vehicle, made it much easier to transport heavy loads. Lubricants, such as greases from plants or animals, were probably used to reduce friction on wheel axles. In about 100 BC, wheel bearings were invented.

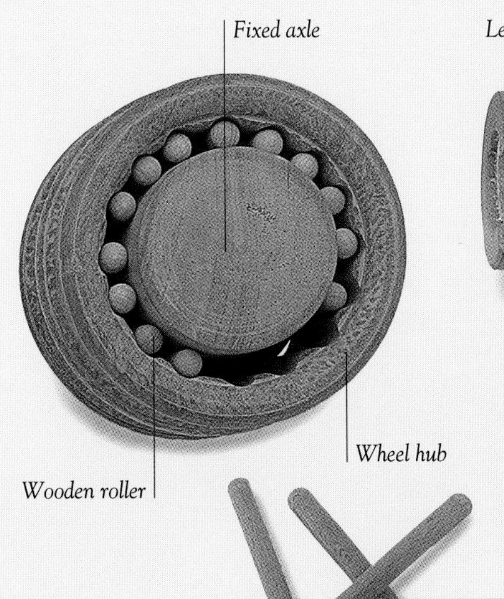

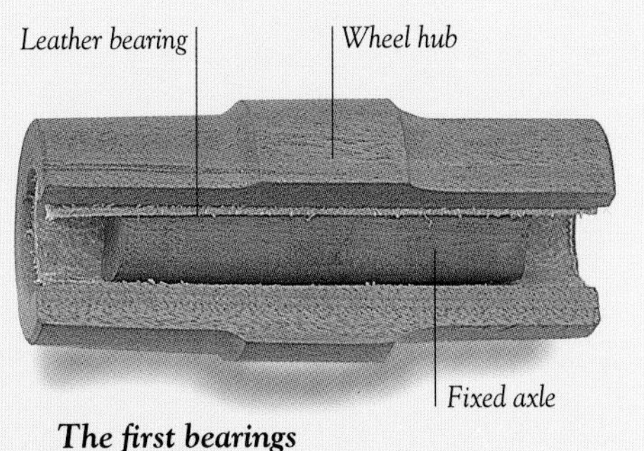

The first bearings
In about 100 BC, the Celts of France and Germany made wheel bearings (above) consisting of a leather sleeve fitted between the fixed axle and wheel hub. At the same time Danish wagon builders invented roller bearings for wheels (left), in which wooden rollers fitted between the hub and the axle.

Valves

A VALVE REGULATES the flow of a fluid (a liquid or a gas), usually in a pipe or machine. Many are one-way valves that allow the fluid to flow in one direction only. For example, a car tire valve allows air in but prevents it from getting out. A fluid often needs to be forced through a valve; this may be done by using a pump to pressurize the fluid. Valves enable many machines to work efficiently and safely. For instance, valves operated by thermostats control gas flow to keep oven temperatures constant. Safety valves in boilers prevent explosions by opening to allow steam to escape if the steam pressure reaches a dangerous level.

Car tire valve

A projection in the hose of an air pump pushes down a pin in a tire valve. This depresses the valve core, opening the valve to allow air to be pumped into the tire. When the hose is removed, a spring snaps the valve shut to prevent air from escaping. If the tire contains too much air, the driver can push down the pin to let air out.

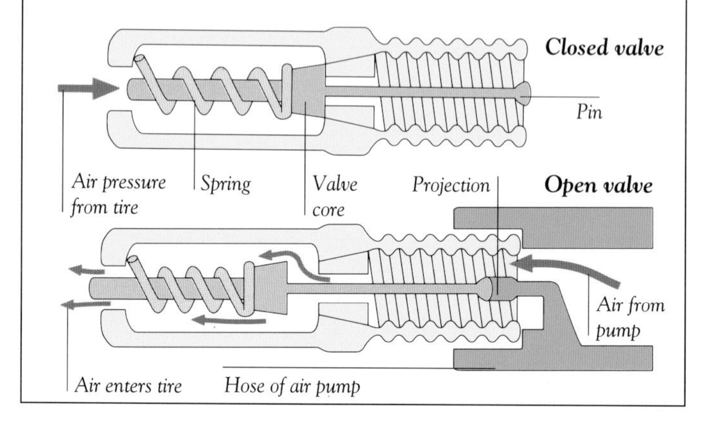

Soaring in safety

The basket of a hot-air balloon carries several cylinders of high-pressure propane fuel and a set of burners that burn this fuel. Flames shoot up from the burners to heat the air inside the balloon and send it soaring up into the sky. Several different valves allow the crew to fly the balloon safely. Safety valves on the cylinders keep the pressure of the propane inside from becoming too high. A liquid-fuel valve controls the flow of liquid propane from the cylinders to the burners. The crew opens a main valve on the burners to feed fuel into hot coils, which vaporize the fuel before it burns. The pilot light, which ignites the fuel, also has its own valve.

Hot-air balloon burners
Liquid propane first flows around the hot coils, where it vaporizes, and then out through the jet ring, where it ignites.

EXPERIMENT

One-way balloon valve

Adult help is advised for this experiment

It can take a lot of effort to blow up a balloon, whether you use your mouth or a bicycle- or balloon-pump. Air then escapes from the balloon as soon as you loosen your grip on the neck. Make a simple one-way valve that you can insert into the neck of a balloon. Each time you blow or pump air into the balloon, the air will stay in the balloon and not escape.

YOU WILL NEED

- *two 4-in (10-cm) lengths of flexible plastic tube, one narrow enough to fit tightly inside the other*
- *ball bearing that moves freely in the large tube but does not fit in the small tube*
- *tape*
- *balloon*
- *cutting mat*
- *pliers*
- *scissors*
- *petroleum jelly*
- *ruler*
- *balloon pump or bicycle pump and connecting hose*

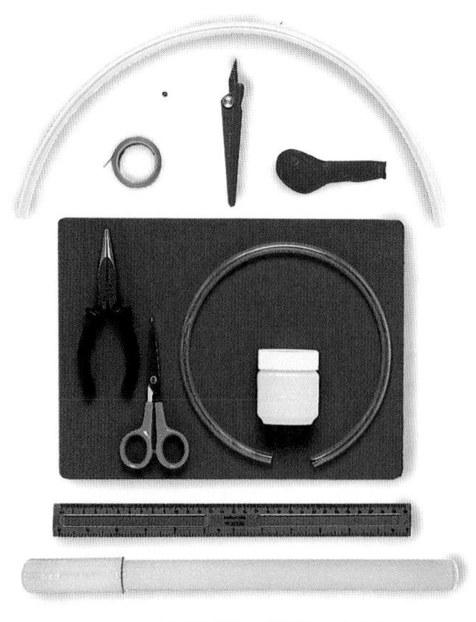

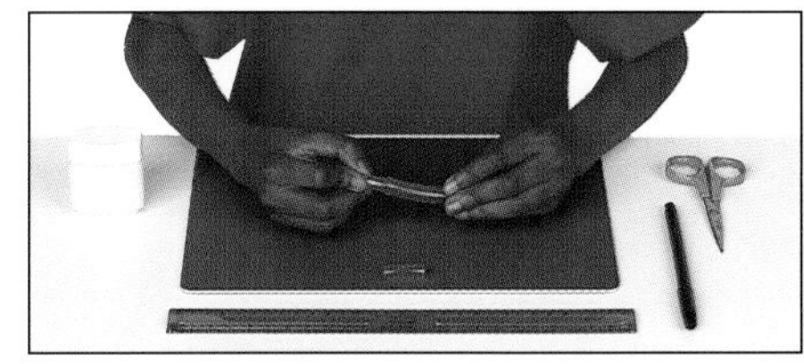

1 CUT THE narrow tube into two 2-in (5-cm) lengths. Grease one end of one of these lengths and insert it about 1 in (2.5 cm) into the wide tube.

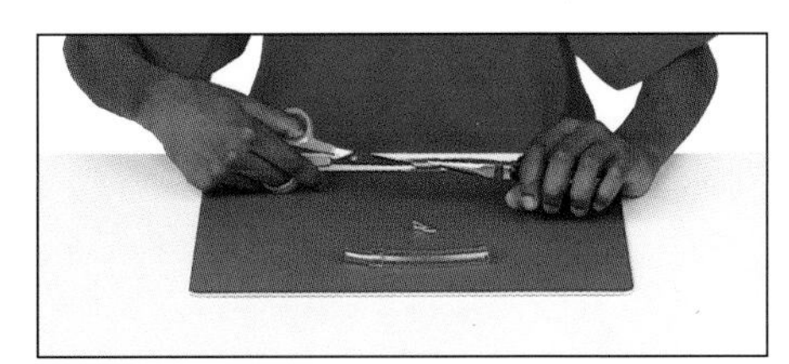

2 MAKE A DIAGONAL cut across one end of the other narrow tube at about 45°. Then place the ball bearing inside the wide tube.

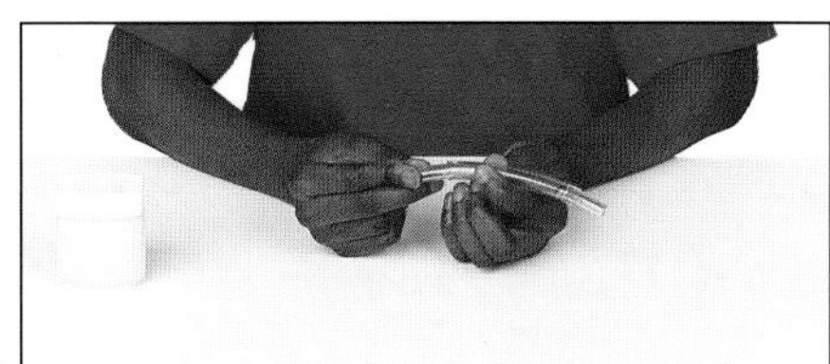

3 GREASE THE diagonally-cut end of the narrow tube. Insert it about 1 in (2.5 cm) into the open end of the wide tube, with the ball bearing inside.

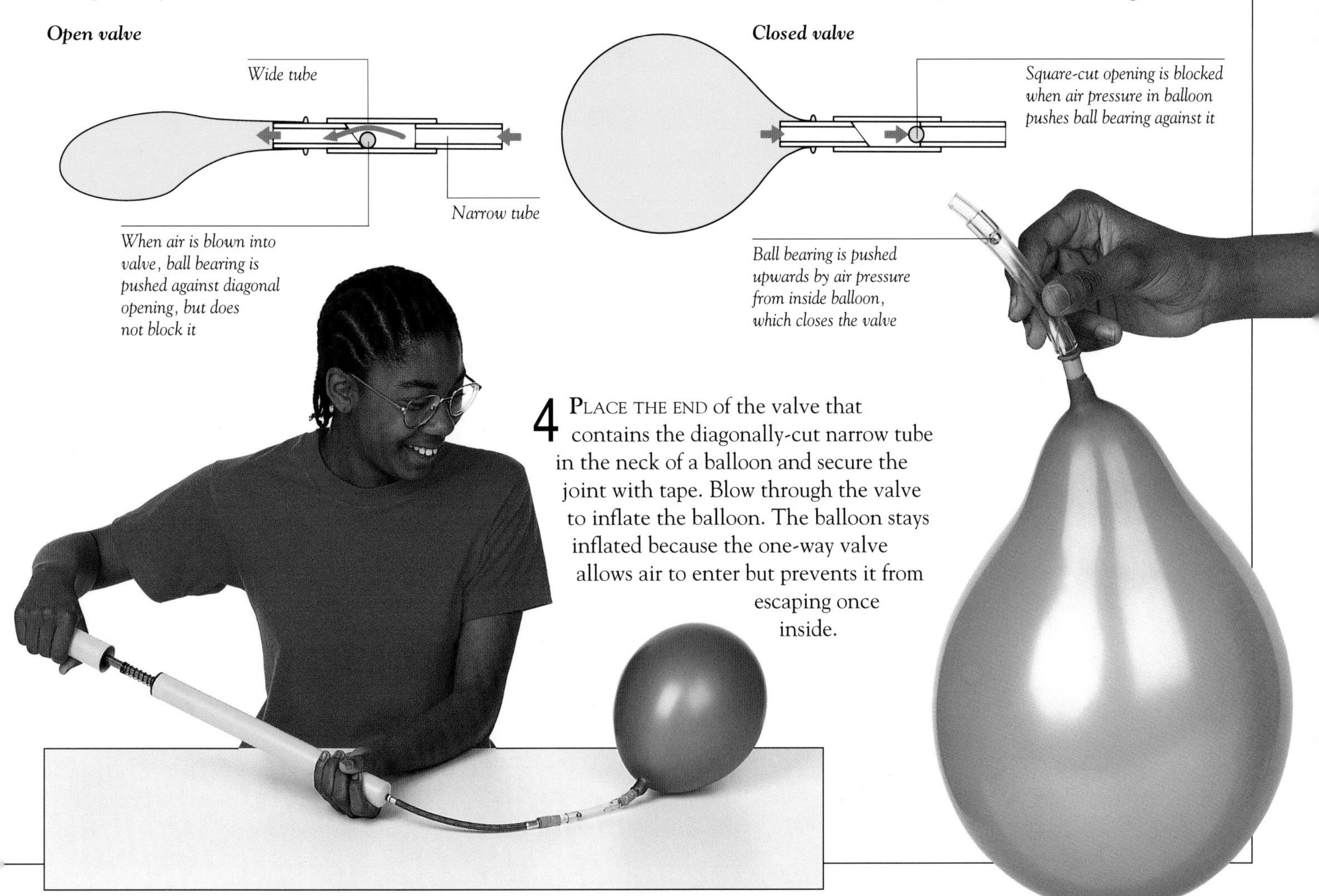

4 PLACE THE END of the valve that contains the diagonally-cut narrow tube in the neck of a balloon and secure the joint with tape. Blow through the valve to inflate the balloon. The balloon stays inflated because the one-way valve allows air to enter but prevents it from escaping once inside.

Pumps

WITH THE AID OF PUMPS, fluids—gases such as air, and liquids such as water and oil—are able to transmit great power and drive machines. A backhoe, for example, contains a pump which raises the pressure of oil fed to the digging bucket. The pressurized oil exerts a strong force on the bucket and drives it through the soil. As well as raising pressure, pumps are also used to deliver gases or liquids through pipes, for example to inflate tires and to fill fuel tanks. Cars contain pumps to move fuel to the engine and to send oil and cooling water around it. Some liquid pumps work by first lowering pressure in the pump. The greater air pressure outside forces liquid into the pump, which then raises the pressure to drive out the liquid.

How a gasoline pump works

Many pumps contain rotating parts that compress and move gases or liquids. A gasoline pump at a service station has a rotary vane pump that draws gasoline from an underground tank and delivers it to a car. The rotary pump consists of a round chamber with a rotor (rotating part) mounted off-center. The rotor has slots containing sliding vanes. As an electric motor turns the rotor the vanes are thrown outward against the chamber wall, dividing the chamber into a series of compartments that rotate and change size. At the pump inlet, the compartments expand in size to suck in the gasoline, and as the rotor turns, they carry the gasoline around the chamber. They then contract and push the gasoline into the outlet of the pump, then up the hose to the nozzle.

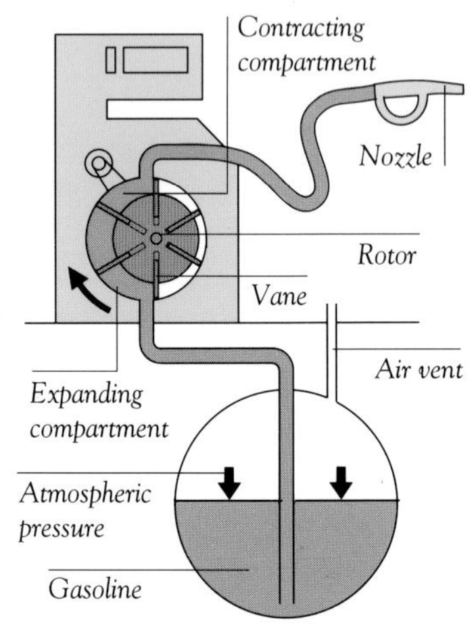

EXPERIMENT

Hand-powered water pump

Adult help is advised for this experiment

Our supply of water depends on pumps to bring water from reservoirs or rivers to our homes. Water supplies may have powerful pumps that raise water to high tanks, from which it can flow down to homes all around. Build a pump that raises water from one container to another. It contains inlet and outlet valves to control the flow of water so that it only moves in one direction.

YOU WILL NEED

• round file • 3 plastic beakers about 5 in (12 cm) in diameter • cutting surface • scissors • craft knife • 12 in (30 cm) length of flexible plastic tube • 3 ft (90 cm) length of narrow flexible plastic tube that fits tightly inside the wider tube • petroleum jelly • large balloon • adhesive tape • ruler • 2 ball bearings that move freely in the wider tube but cannot fit in the narrow tube

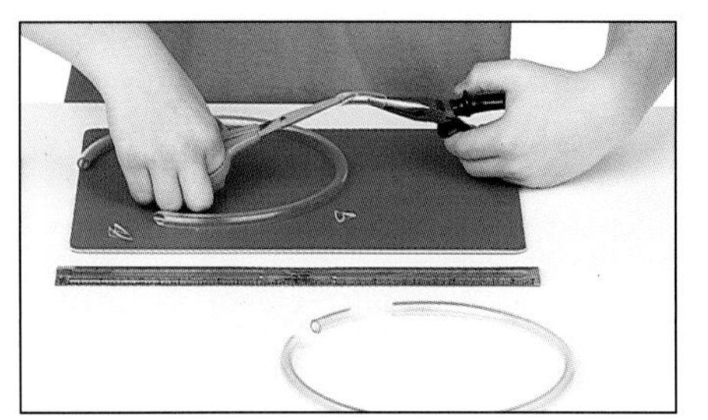

1 CUT TWO NARROW tubes 2 in (5 cm) long and two more 16 in (40 cm) long. Cut a diagonal section from the tip of one short tube and one long tube. Grease both tips of each tube.

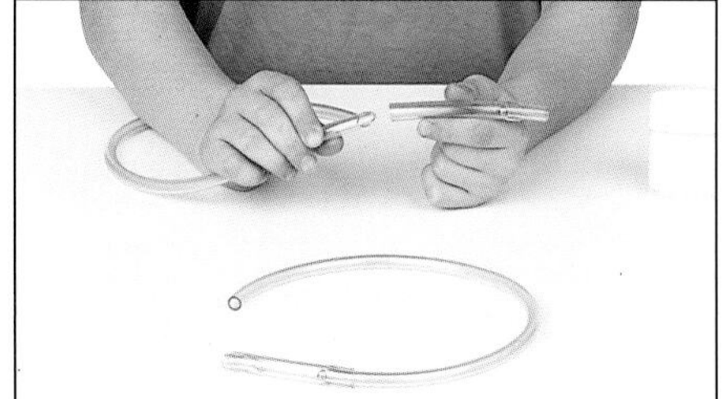

2 CUT TWO WIDE tubes 6 in (15 cm) long. Make two valves as shown in the illustration opposite. Each valve has a short and a long narrow tube, one of which has a diagonal section removed.

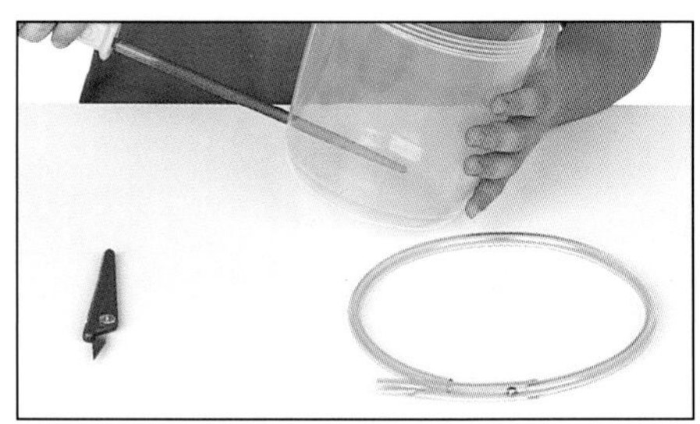

3 USING A CRAFT KNIFE, cut two small crosses on opposite sides of a plastic beaker. Enlarge the crosses with a file to make two holes just wide enough for a narrow tube to squeeze inside.

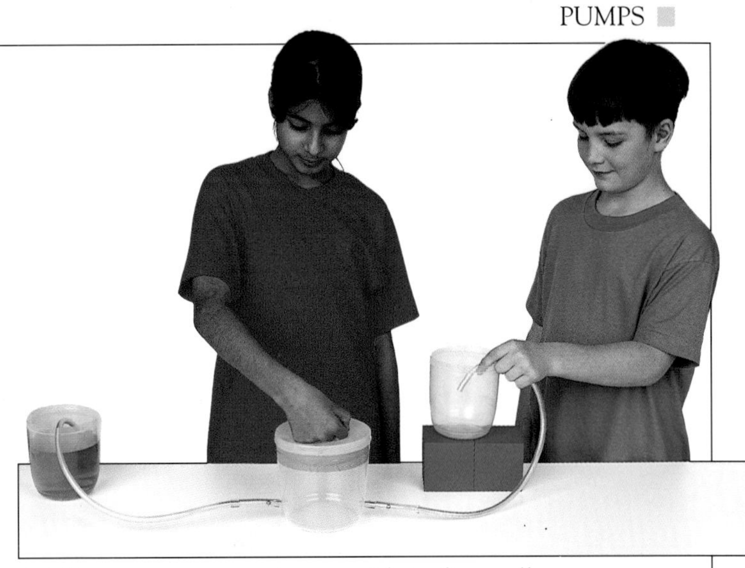

4 GREASE THE HOLES in the beaker with petroleum jelly to produce an airtight seal. Squeeze a narrow tube into each hole in the beaker. Cut the neck from the balloon and discard. Stretch the remainder of the body of the balloon tightly over the beaker. Secure with adhesive tape.

5 THE INLET VALVE is the one with its diagonally cut narrow tube inserted in the beaker. Place the end of the inlet tube connected to this valve in a beaker full of water. Place the end of the other tube in an empty beaker raised above the table, to make more work for the pump.

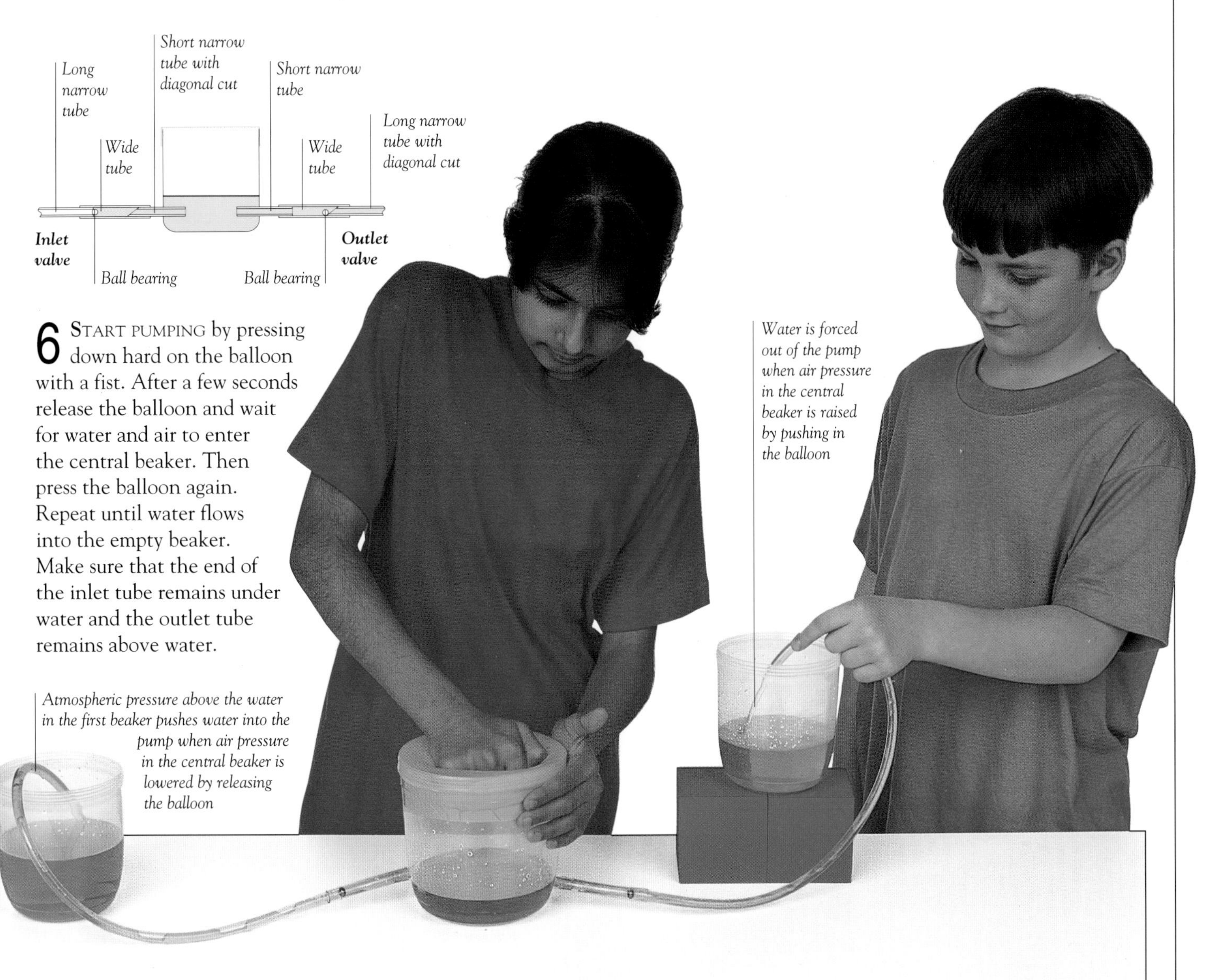

6 START PUMPING by pressing down hard on the balloon with a fist. After a few seconds release the balloon and wait for water and air to enter the central beaker. Then press the balloon again. Repeat until water flows into the empty beaker. Make sure that the end of the inlet tube remains under water and the outlet tube remains above water.

Water is forced out of the pump when air pressure in the central beaker is raised by pushing in the balloon

Atmospheric pressure above the water in the first beaker pushes water into the pump when air pressure in the central beaker is lowered by releasing the balloon

Hydraulics and pneumatics

HYDRAULIC MACHINES such as powerful excavators use pressurized liquid to transmit and increase force. A hydraulic system is simple and robust. It is basically a liquid-filled pipe of any length or shape, with a piston at each end. One piston is pushed in by muscles or a motor, and the liquid transmits this force to the other piston, which moves out and pushes or lifts something. In pneumatic machines such as road drills (p.186), force is transmitted by pressurized air.

Hydraulic car brakes

To slow and stop a car, the driver presses on the brake pedal. A hydraulic system transmits this force equally to the wheels on both sides of the car so that the car slows without swerving. The brake pedal pushes a master piston into a set of pipes that contain brake fluid and go to the brakes. The piston increases the pressure of fluid equally throughout the pipes. In each brake the brake fluid pushes out a pair of pistons that force the brake against a disk or drum fixed to the wheel to slow the wheel down. Each rear brake piston is usually the same size as the master piston and exerts the same force as that applied to the master piston by the driver. The front brakes require greater power and have wider pistons than the rear brakes. A wider piston exerts a greater force because more fluid pushes on it, but the piston moves out a shorter distance than the master piston moves in.

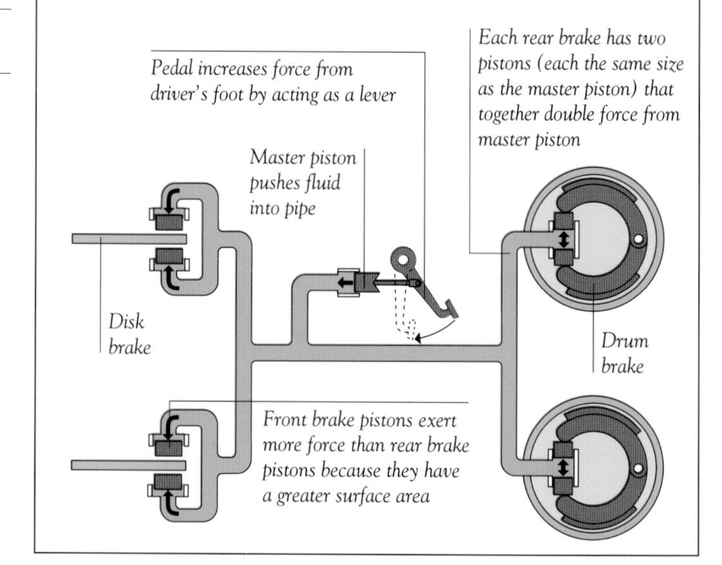

EXPERIMENT

Water-powered weight lifter

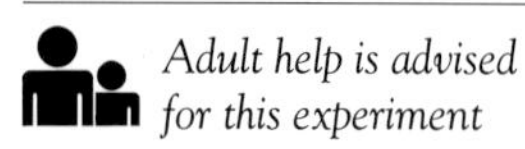

Adult help is advised for this experiment

Make a hydraulic lift that can raise a heavy weight with just a little effort. The lift uses hydraulic pressure to increase the amount of force you apply to it.

YOU WILL NEED

• balloon • rubber band • pitcher • cut-off plastic bottle 3 in (8 cm) taller than can • empty can slightly narrower than bottle • round file • ruler • cutting surface • petroleum jelly • scissors • cork slightly wider than tube • books • cardboard • craft knife • pen • tape • knitting needle • funnel • 18-in (45-cm) length of flexible plastic tube

1 STICK A SMALL piece of tape 2 in (5 cm) above the base of the cut-off bottle. Using a craft knife, cut a small cross through the tape and bottle. Enlarge the hole with a round file until the plastic tube fits tightly inside.

2 SECURE THE balloon to one end of the plastic tube using a rubber band. Blow into the tube to check that the balloon is sealed tightly around the tube. Push the other end of the tube through the hole in the bottle from the inside.

3 PUSH A knitting needle into the cork to make a piston. Use a craft knife to trim the cork until it can be pushed up and down the tube when greased but still keeps a tight seal with the tube sides. Remove the cork from the tube.

4 USING A FUNNEL, fill the balloon and tube with water. Make sure all the air in the balloon escapes. Pull the tube through the hole until the neck of the balloon touches the hole. Grease the cork again with petroleum jelly, and push the piston just inside the tube.

Hovercraft

Floating just above the water, a hovercraft can skim over the sea—or any other flat surface—at high speed. The vessel is raised off the surface on a cushion of air pumped beneath the hull by fans and held in by a flexible skirt around the sides. The air is pumped to a pressure slightly higher than atmospheric pressure; this provides sufficient force to lift the hovercraft off the land or water. The largest type of hovercraft, shown below, can carry over 400 passengers and 60 cars and travel at a speed of 75 mph (120 kph). It is driven and steered by propellers and rudders similar to those on aircraft.

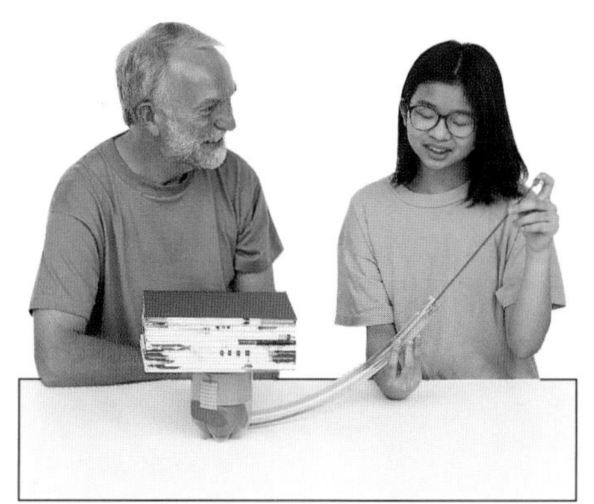

5 PLACE THE CLOSED END of the can on the balloon. Trim the bottle until the can sticks out of the top by 1 in (2.5 cm). Place some books on the can. Make a cardboard scale with lines ¼ in (5 mm) apart. Stick it to the bottle so that the bottom line is by the bottom of the can.

The balloon swells as you push water into it, and pushes on the can, which lifts the books. Because the can is wider than the cork, the water in the balloon pushes it up with more force than you use to push in the cork. Only a little effort is needed to raise the heavy books.

6 PUSH THE SMALL piston (the cork) as far down the tube as it will go. The books are lifted by the big piston (the can) as the small piston pushes water into the balloon. The small piston moves a much greater distance than the big piston, but it is easy to push the small piston down the tube to lift the heavy books.

Automatic machines

AN AUTOMATIC MACHINE operates by itself, without any human attention. Machines that work unaided range from simple devices, such as automatic doors, to complex machines, such as robots. Automatic control systems relieve humans of much of the labor involved in operating machines. As well as saving us work, an automatic system may control a machine much more efficiently than any human can. Automatic machines work in two basic ways. Some, such as washing machines, follow a set routine of tasks when started. Others, such as traffic lights, respond to changes in their environment or check their own performance and control their operations accordingly.

EXPERIMENT

Automatic light

Adult help is advised for this experiment

Modern streetlights go on and off automatically. Build a circuit that turns on an LED as darkness falls and turns it off as light returns, like a streetlight. An LDR, which detects the light level, decreases its resistance to the flow of current as the light dims and increases its resistance as the light brightens. The changing current goes to a transistor and NAND chip, which turn the LED on and off at a preset level. *Read pages 10–11 before starting this experiment.*

The nonstop mill

Automatic control dates back to 1745, when a British inventor, Edmund Lee, built a device called a fantail to turn the top of a windmill so that the sails always pointed into the wind. Before this, millers had to turn the tops of their windmills manually whenever the wind changed direction. The fantail is an example of an automatic control system that senses a machine's environment, in this case the direction of the wind.

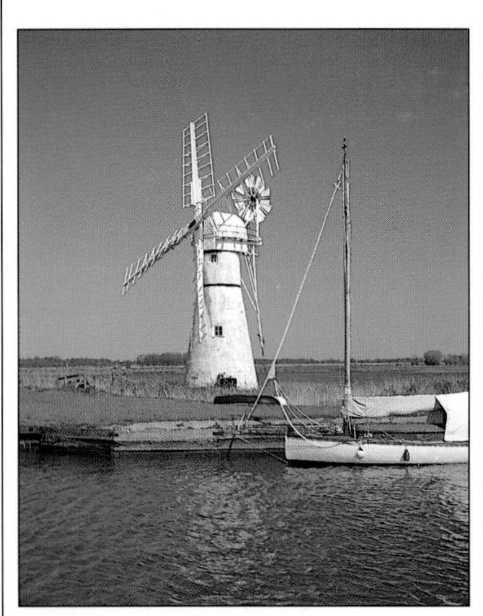

Fantail
The fantail sits behind the sails of a windmill. Its blades start turning when the wind changes direction. The blades are linked to gears that rotate the top of the mill. When the sails point into the wind, the fantail stops turning.

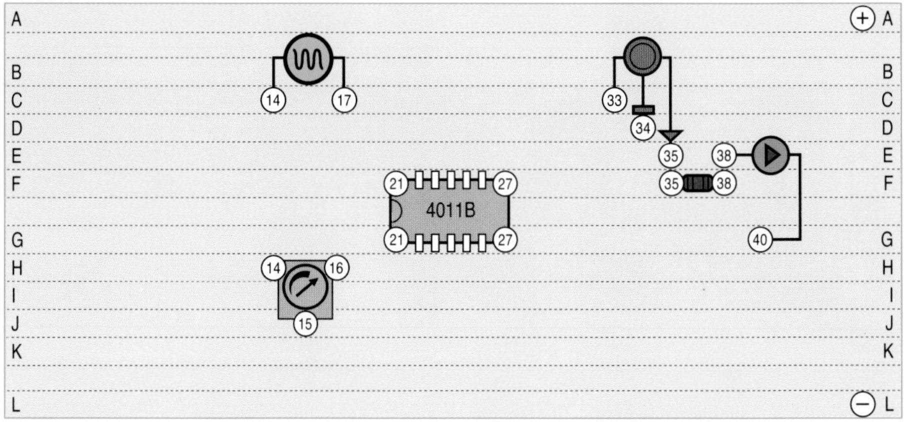

220R resistor
F35–F38

Wires
A17–B17 A21–B21 A33–B33 C22–C23 E14–E22 E24–E34 F14–G14 K15–L15 K27–L27 K40–L40

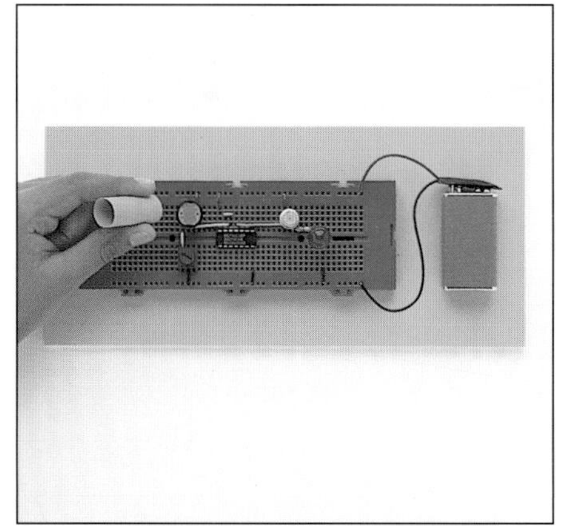

1 CONSTRUCT THE electrical circuit as shown in the diagram above. Make a small hood for the light-dependent resistor (LDR) by rolling up a strip of cardboard and gluing it to form a cylinder. Slide the hood over the LDR.

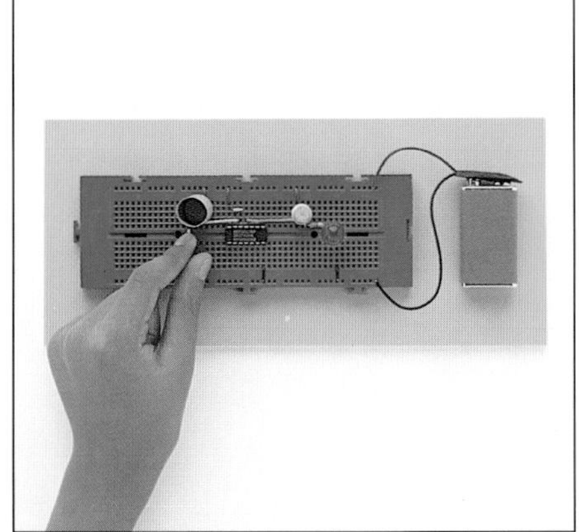

2 PLACE THE CIRCUIT in an area brightly lit by either natural or artificial light. Ensure that the top of the LDR hood is fully exposed to the light. Adjust the dial on the variable resistor until the light-emitting diode (LED) just blinks off.

You Will Need

• 220R resistor • light-emitting diode (LED) • light-dependent resistor (LDR) • 4011B quad NAND chip • NPN transistor, BC441 or equivalent • 5K variable resistor • cardboard • breadboard and base • 9V battery and connector • pliers • glue • wire strippers • breadboard wire

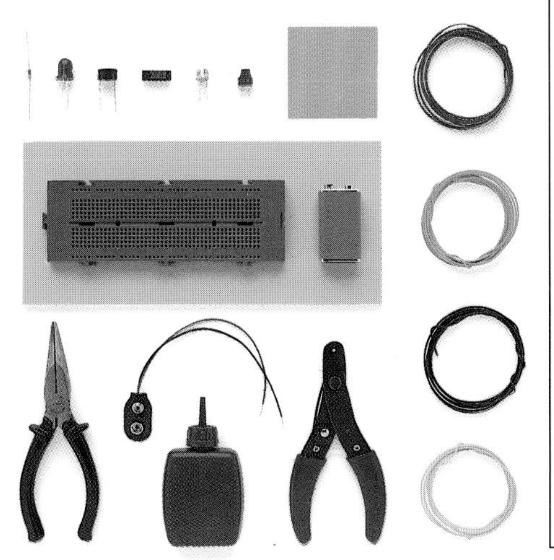

Smart wall

The Arab Institute in Paris, France, has a "smart" wall that automatically controls the amount of daylight entering the building, so that the light level is always right for viewing the institute's exhibits. Behind the windows are iris diaphragms with apertures that narrow in bright light and widen when the sky is overcast.

Automatic aperture
Every window has a large iris diaphragm in the center and a set of small diaphragms around the edge. Under the control of a light sensor, a ring on each diaphragm turns to open or close the blades.

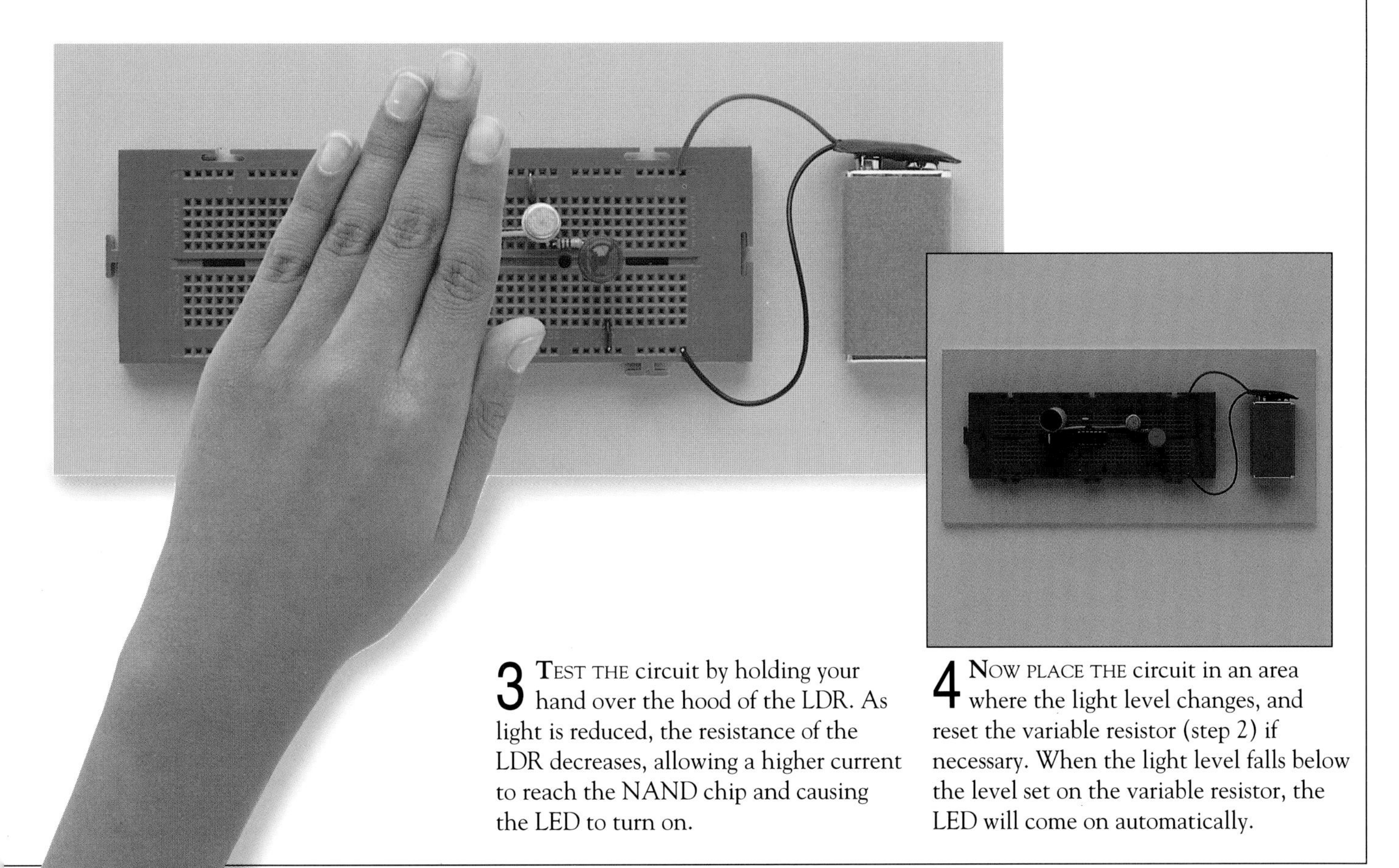

3 Test the circuit by holding your hand over the hood of the LDR. As light is reduced, the resistance of the LDR decreases, allowing a higher current to reach the NAND chip and causing the LED to turn on.

4 Now place the circuit in an area where the light level changes, and reset the variable resistor (step 2) if necessary. When the light level falls below the level set on the variable resistor, the LED will come on automatically.

CONSTRUCTION and BUILDINGS

Strength and style
Structures can be built to last for centuries, as proved by the continued existence of many old buildings. A model (above) of the seventeenth-century dome of the Church of the Sorbonne in Paris, France, shows how the curved beams link to form a very strong structure. Architects also aim to enhance places by designing attractive buildings in a style characteristic of their time, as in the Museum of Contemporary Art, Los Angeles, California (left).

STRUCTURES WORK FOR US, just as machines do. Bridges, dams, and oil rigs play important parts in our everyday lives. Buildings shelter us and their internal systems, such as heating and escalators, make life indoors comfortable. Architects design structures and buildings, while engineers manage their construction. Both use scientific methods and principles to ensure that their structures and buildings will be long-lasting as well as useful.

SHELTER AND SERVICE

WE ALL NEED SHELTER FROM THE WEATHER in order to survive. To make life indoors comfortable, we need buildings equipped with services such as water, electricity, heating, and drains. Architects, engineers, and builders design and construct buildings with these services built in. Modern living also depends on other structures, such as dams; like our shelters, all these structures must efficiently perform their functions. Many buildings and structures are also designed to look attractive in their environment.

These homes in Nigeria have mud walls and roofs made *of reeds, which are adequate building materials for a dry climate and are easy to find when repairs are needed.*

In some parts of the world, there exist buildings that date back hundreds or even thousands of years. Many, such as the temples of ancient Greece, are now in ruins. Earthquakes, fire, bad weather, war, and neglect have taken their toll. But the Great Pyramid of Egypt still stands some 4,500 years after it was built, a massive structure so strong that not even earthquakes can topple it. The pyramid strikes awe into all who see it, as do the centuries-old cathedrals of Europe, still soaring elegantly skyward while almost all other buildings of similar age have long since disappeared. These structures were designed both to impress and to last. They were built of stone, a strong and durable building material, and were designed to resist destructive forces.

Materials

Unlike many large and important buildings, such as stone castles and churches, few ordinary buildings have survived long periods of time. This is because stone is a very expensive building material and many people could build only with materials that were easy to find, transport, and work with—usually mud, reeds, and wood. Mud could be built up or molded around a wooden frame to form walls, while reeds could be tied together to make a thatched roof. Wood was cut into beams, rafters, and planks that could be quickly assembled to build a house. However, mud or reed dwellings quickly disintegrate except in relatively dry climates, where these materials are durable. Wooden homes tend to last longer, especially when protected by paints and preservatives, but they are easily destroyed by fire. Stronger, more durable materials, such as brick and concrete, have also been available for many centuries, and such materials have replaced mud, reeds, and wood in many parts of the world. Bricks, like stone blocks, can be cemented together to build strong and lasting walls. However, brick and stone are too heavy and weak to build most very tall structures such as skyscrapers and towers. These, and other large structures like bridges, need materials that combine strength with lightness, and so are made mainly of concrete and steel. Concrete can be molded in a variety of shapes and reinforced with steel, making it a very strong and versatile construction material. Both reinforced concrete columns and steel beams are used to make strong but light skeletons for buildings. Very tall skyscrapers may be built with steel or concrete frames to which wall panels of glass and other light materials are attached; the frame also supports floors and the roof. Like tall skyscrapers, long bridges need to be both light and strong; such bridges may be made entirely of steel or entirely of concrete, or may use both of these materials.

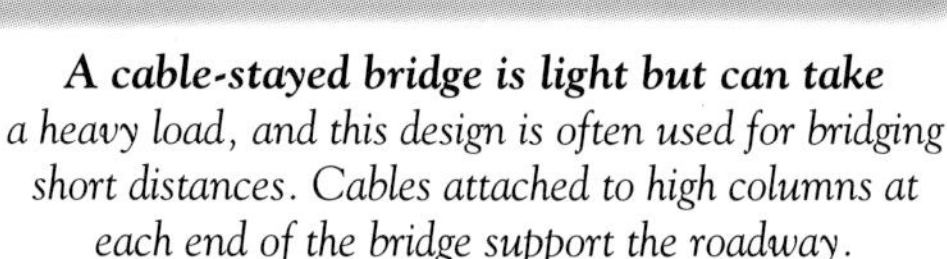

A cable-stayed bridge is light but can take *a heavy load, and this design is often used for bridging short distances. Cables attached to high columns at each end of the bridge support the roadway.*

Elisha Otis demonstrated the *first safe elevator in 1854. Elevators were later used in skyscrapers, so that people did not need to climb lots of stairs.*

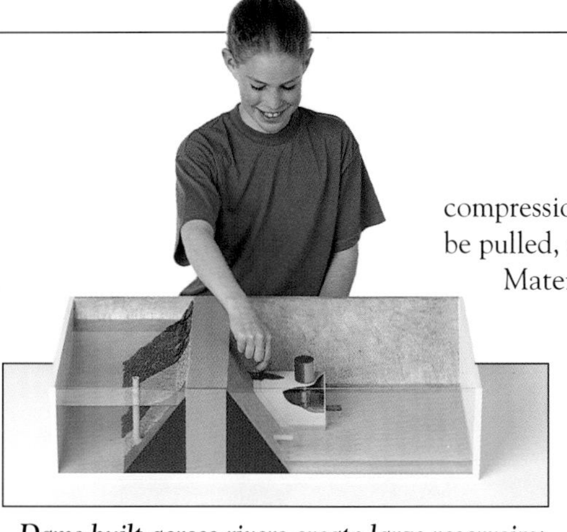

***Dams built across rivers create large reservoirs;** they regulate water supply and may be used to generate electricity. They must withstand the great weight of impounded water as well as possible earthquakes.*

A balance of forces

When designing any building or structure, an architect must finely balance the various forces, or loads, to which the structure will be subjected. The immense weight of the building materials is called a dead load. Live loads are such things as the weight of the building's occupants, furniture, and equipment, which may move around and change. Dynamic loads such as wind and earthquake forces may change suddenly, subjecting a building to potentially destructive forces. Finally, the expansion or contraction of a structure, which occurs when the temperature changes, also constitutes a load. Loads create two different kinds of force: a part of the structure may be squeezed by other parts around it, causing it to undergo compression; or the part may be pulled, undergoing tension.

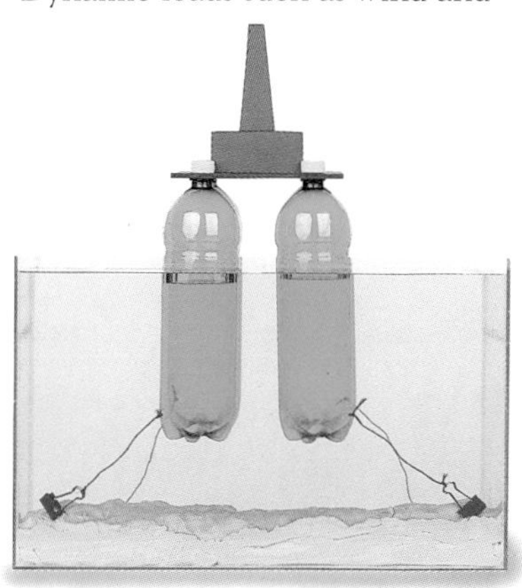

***This model shows how an oil rig** can float on hollow legs at the surface of the sea, secured by cables anchored to the seabed.*

Materials such as concrete are usually strong under compression but may be weak under tension. In contrast, steel has high strength under tension. The architect must ensure that each part of the building is made of materials that can withstand the forces likely to be applied, and that the forces over the whole building always balance each other so that it does not break apart, collapse, or fall over.

Shapes

As well as making these vital calculations, architects must also consider the external and internal appearance of the building. They may seek to build new kinds of shapes that will make a structure look attractive and give it an individual identity. But there is more to the shape of a structure than just its beauty or originality. Certain shapes confer strength upon a structure and help to hold it up. Curved arches and domes do this by placing the construction materials under compression. Many large concrete structures, especially bridges and dams, are composed of curves, making them very strong. The triangle is another very strong shape and resists twisting and deforming when under pressure. The Eiffel Tower in Paris, France, was once the world's tallest structure; it is made of steel girders connected together to form triangles.

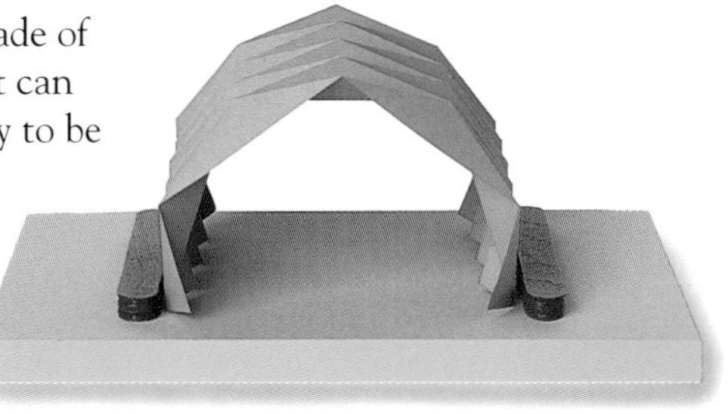

***Architects design buildings in all** kinds of shapes, but must first ensure that the shape is strong and will stand up. This elegant roof can resist forces 400 times its own weight.*

Services

No private or public building would be useful without any services for its occupants. As it is completed, the building must be connected to external supplies of electricity, gas, and water, to the drainage system, and to telephone and data networks. These services, together with internal heating, lighting, and air conditioning systems, must be routed to all the rooms where they are needed. The completed building or structure takes its place in the local infrastructure, which comprises many different kinds of buildings and structures, including those that supply services to buildings and people. For example, dams regulate water supply and may produce electricity; road and rail networks, which often include bridges and tunnels, allow the transport of products and people; oil and gas production rigs supply these fuels from both land and sea; and electricity generated at power stations reaches buildings through a network of power lines carried above and below ground.

***The CN Tower dominates** the skyline of Toronto, Canada. Built of concrete, it rises 1,822 feet (555 meters).*

Foundations

THE FIRST STAGE IN constructing a building involves not construction, but excavation. A deep hole is dug in the ground to make room for the building's foundation. When it is complete, the building can begin to rise above it. The foundation, which often consists of a concrete mass, anchors the building firmly to the ground so that it will neither sink nor tilt during its lifetime. It must support the weight of the building, and may also help to counter the force of strong winds that might blow over a tall building. In some places, buildings rest on solid rock, which easily supports high and heavy buildings. But large buildings now also rise in cities where the ground is composed of weak soil or sand. These buildings require special foundations that will not sink into the soft ground.

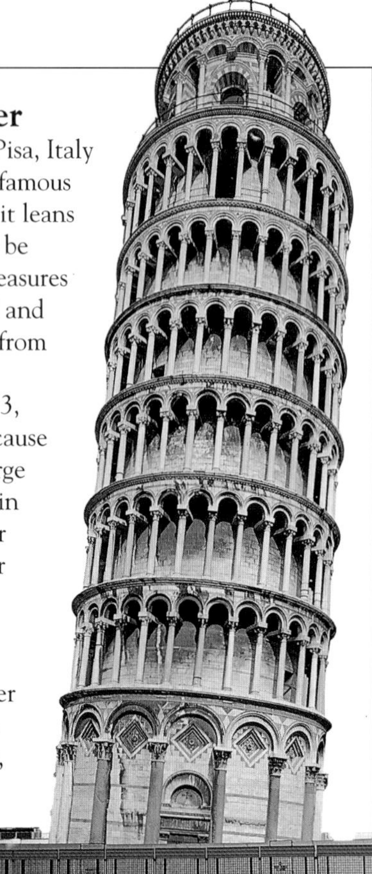

The tilting tower

The marble bell tower at Pisa, Italy is one of the world's most famous buildings, mainly because it leans so much that it appears to be about to topple over. It measures 179 feet (54.5 meters) tall and leans 14 feet (4.2 meters) from perpendicular. Soon after construction began in 1173, the tower began to tilt because the foundation was not large enough to hold it upright in the soft ground. The tower has continued to shift ever since. It was in danger of falling down until work was done to provide the leaning tower with a proper foundation. The firm base has stopped further tilting, and has straightened the tower slightly.

EXPERIMENT

Firm foundations

Adult help is advised for this experiment

Using a bottle for a building, paste for soil, and modeling clay for rock, see how a structure needs a foundation to stay up in soft ground. The building can sit on a solid mass of concrete that is constructed in a pit excavated in the ground, and made as wide as possible. This foundation spreads the building's weight evenly, often over an area wider than the building itself, so that the soft ground does not give way beneath the building. If the soil is still too weak, long shafts called piles can be sunk beneath the concrete block to anchor it and the building to the bedrock, which is hard rock beneath the soil.

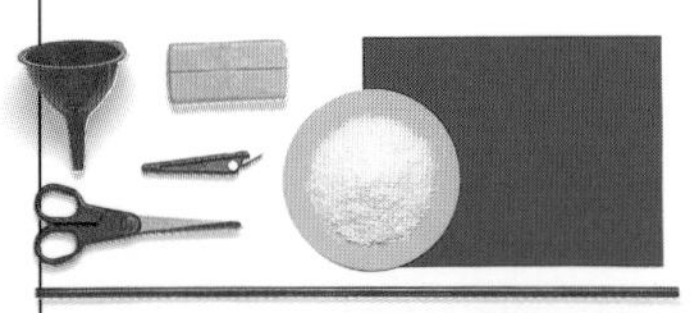

YOU WILL NEED

- *funnel* • *scissors* • *several packs of modeling clay* • *craft knife* • *8 lengths of dowel, each about half the height of the tank* • *wallpaper paste* • *3/16 in (5 mm) foamcore* • *large tank* • *plastic bottle about 5 qt (5 l)* • *ruler* • *steel rule* • *pitcher*

1 LINE THE BASE OF THE TANK with modeling clay, then half fill it with paste. Stand the bottle on the paste. Pour water into the bottle until it starts to tilt. The water represents materials added during construction. The weight of the building increases as it grows, but with no foundation, it soon tilts as the ground gives way beneath it.

Contrasting skylines

Like bones beneath skin, underlying geology once determined the look of a city. During this century, for example, tall pinnacles of skyscrapers have crowded the island of Manhattan at the center of New York, while central London has remained mainly a low-rise city. Manhattan is a rocky island, and the underlying bedrock can support skyscrapers. London lies in a broad valley of soft clay, which does not have such strength.

However, modern methods of building foundations, such as pile-driving and strengthening weak soil, enable high buildings to rise on soft ground. Skyscrapers, though not as high as those in New York, are beginning to dot London's skyline. And modern foundations have benefited New York too: the World Trade Center, the city's tallest building at 1,352 feet (412 meters), is in fact built on weak soil.

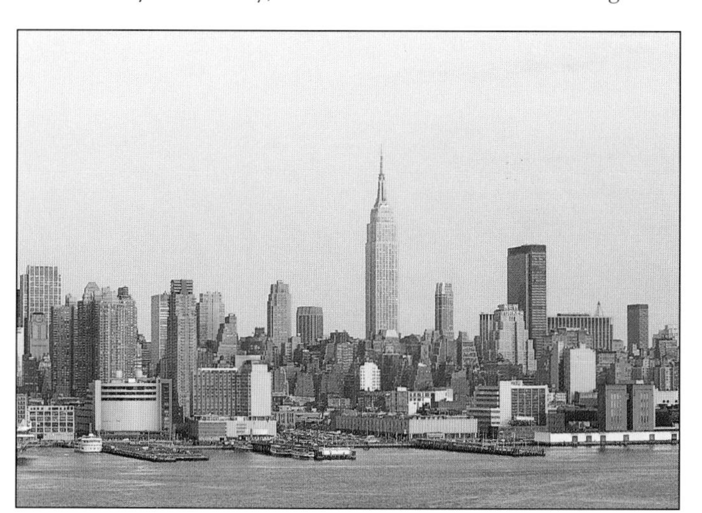

New York
This view of high-rise Manhattan is dominated by the Empire State Building. It was once the world's tallest building at 1,250 feet (381 meters).

London
Few skyscrapers line the River Thames as it flows through London. In the distance is Canary Wharf, Britain's tallest building at 800 feet (244 meters).

2 CUT A FOAMCORE square 2 in (5 cm) larger than the base of the bottle and place it on the paste with the bottle on top. The square represents the foundation, which spreads the building's weight evenly over a larger area. It enables the bottle to hold more water just as foundations enable heavy buildings to rise on relatively weak ground.

3 INSERT EIGHT lengths of dowel into the modeling clay and place the square on top. Replace the bottle and see how much water it can now hold. The dowels represent piles that reach down to the bedrock below ground. Piles can support a very heavy building, and the bottle can be filled with water to the top without sinking or tilting.

Walls and floors

HOUSES AND LOW-RISE BUILDINGS need strong walls that can hold up the building and help keep the air inside as warm (or cool) as required. Stone and brick are two of the traditional materials used for building walls. Some buildings have two walls: an inner wall of wood or concrete blocks, and an outer wall of stones or bricks. The blocks, stones, or bricks are cemented together in patterns that form a strong structure. A central layer of insulating material may be used to prevent the movement of heat. Floors, often made of wood, rest on beams anchored in the walls. Many buildings now have walls and floors made of concrete, which is strong and has the ability to be molded into any shape required.

EXPERIMENT

Stagger for strength

Walls of stone blocks or bricks are built by first laying a course (line) of blocks or bricks, cementing each one to the next, and then adding more courses until the wall reaches the required height. Builders stagger courses so that the ends of the bricks or blocks do not line up with those above or below. Build walls in various patterns and see how some of these patterns can strengthen the wall.

YOU WILL NEED

- *long plastic or wooden blocks*
- *large sheet of paper*

1 BUILD A WALL of blocks on a sheet of paper. Arrange the blocks so that they line up vertically, with all the short edges exactly above each other.

2 GIVE THE PAPER a gentle tug. The wall begins to collapse. This building pattern produces separate columns of blocks, which easily fall over.

3 REBUILD THE WALL, staggering the courses so the end of each block is above the center of the one below. Turn alternate end blocks for even corners.

4 PULL THE PAPER again to make the wall collapse. This time it does not give way so easily. The blocks overlap and help to keep the wall together.

Strengthening concrete

Concrete is not always very strong. A concrete beam bends and may crack if too large a load is placed upon it. The beam's upper edge is under compression (squeezed), making it strong; the lower edge is under tension (stretched), which makes it weak. However, if a beam is stressed (squeezed along its length), the whole beam becomes much stronger. To stress a beam, stretched steel bars are set into it and anchored at each end. The bars then contract, pulling in the ends of the beam to compress the concrete. The compression overcomes the tension and reinforces the beam.

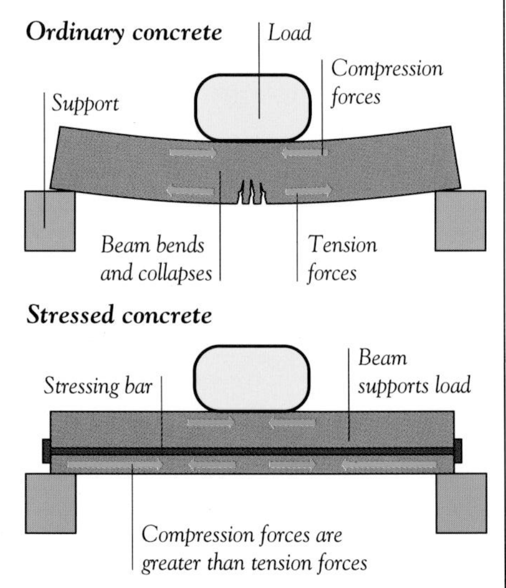

EXPERIMENT

Reinforced concrete

Adult help is advised for this experiment

Concrete is made by mixing cement, sand, and stones with water, and then pouring the liquid mixture into a mould, where it sets hard. Placing steel rods in the mould, so that the concrete sets around the rods, reinforces the concrete and strengthens it. Using plaster instead of concrete, and wooden strips for rods, show how reinforcing increases strength. Make beams of plaster, of wood, and of plaster reinforced with wood, and test the strength of each.

Beam mold

Card is 6 x $2\frac{1}{4}$ in (15.5 x 5.5 cm)

$\frac{1}{4}$ in (7 mm)

Fold along dotted lines

Cut a slit at each corner to make tabs

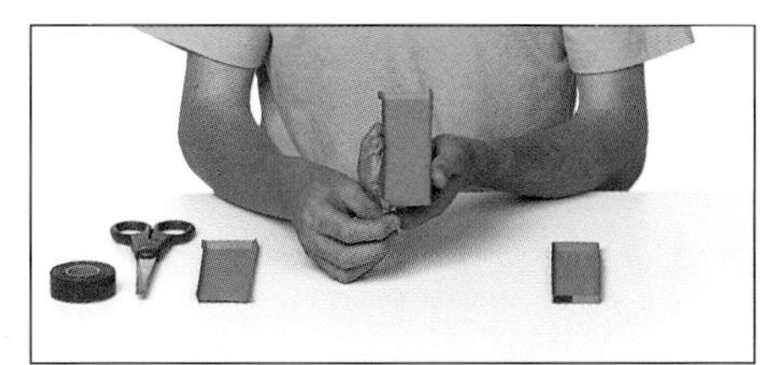

1 MAKE SIX molds from cardboard rectangles (below left). Mark a $\frac{1}{4}$ in (7 mm) square in each corner. Cut a slit on one side of each square. Fold the edges up and tape the corners closed.

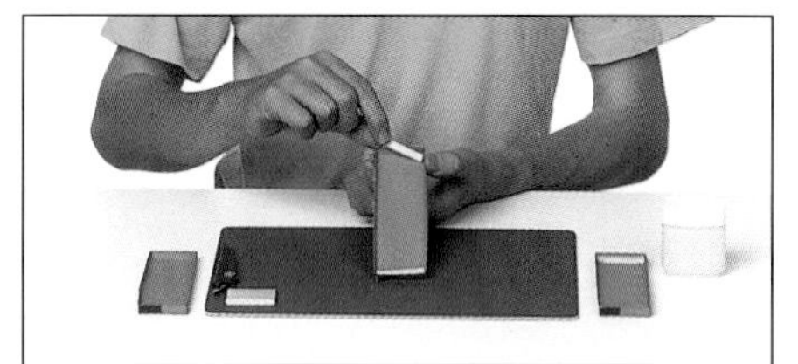

2 MAKE 18 bars of modeling clay $1\frac{3}{4}$ in (4 cm) long, $\frac{1}{4}$ in (7 mm) wide, and $\frac{1}{8}$ in (3 mm) deep. Place a bar in each end of each beam mold. Grease the inside of each mold.

3 CUT THE tongue depressors into 30 strips $5\frac{1}{2}$ x $\frac{1}{8}$ in (14 cm x 3 mm). Place five strips each in three beam moulds, securing the strips in the clay bars. Mix eight beakers of plaster of Paris with six of water. Immediately fill all six molds with plaster, and leave to set for 45 minutes. Then peel away the molds and leave to harden for 24 hours. Make three wooden beams by securing five wood strips in each of the remaining three pairs of clay bars.

YOU WILL NEED

• plaster of Paris • tape • petroleum jelly • craft knife • about 2 ft (60 cm) electrical wire • beaker • jug • steel ruler • cutting mat • plastic bag • about 10 wooden tongue depressors or ice-lolly sticks • modeling clay • scissors • stiff cardboard • sand • 2 supports

4 LAY A wooden beam on the supports. Loop the wire through the bag's handles, then suspend the bag from the beam. Pour beakers of sand into the bag until the beam collapses. Note how much sand the beam supports before it breaks. Test the other wooden beams.

5 TEST THE plaster beams, and then the reinforced plaster beams. Work out the average amount of sand that each type of beam can support, by adding the three amounts and dividing the total by three. You should find that reinforced beams are stronger.

Roofs

A BUILDING WEARS A ROOF for the same reasons that a person wears a hat: both provide shelter from rain, wind, cold, or heat, and both contribute to overall appearance. Most homes today have either a flat roof or one shaped like an inverted V (a gable roof). Public buildings, by contrast, may be topped with domes, barrels split in two, zigzags, dishes, saddle-shaped or wing-like structures, or even updated versions of an ancient shelter—the tent. Whatever a roof looks like, its shape gives it the strength to withstand storms and high winds.

Structures for shelter

The slates or tiles of a steep-sided gable roof rest on a stiff framework of beams firmly fixed to the tops of the walls. Flat roofs rest on parallel beams or a grid of beams, though concrete slabs may be strong enough to require no such support. Wide spaces are often covered by curved roofs, their strength based on the same principles as the arches or suspending cables of bridges (pp.66–67). A dome, for example, is like a whole set of thin arches linked to form a hemisphere; it is very stiff and very strong.

Cable roof
This elegant roof is supported by cables hanging between the tops of two rows of leaning columns.

Folded barrel roof
This roof is semicircular, like half of a barrel; its zigzag folds lend it strength.

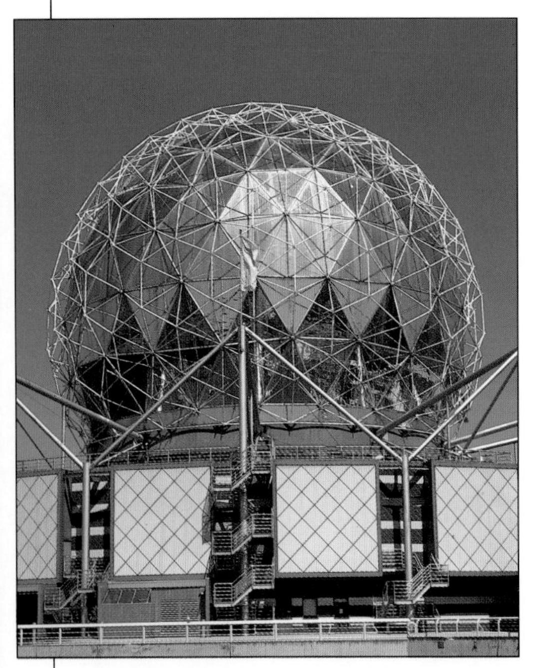

Geodesic dome
Light but strong, a geodesic dome is a spherical frame of straight bars connected in five-sided and six-sided shapes. A roof can be attached to the inside of a geodesic dome.

Membrane roof
Fabrics made of reinforced plastics can be hung from cables and stretched taut to produce a light but stiff structure. This membrane roof resembles a row of tents suspended in mid-air.

EXPERIMENT

Cantilever roof

Flat roofs are not very strong and need support to cover wide areas. Roofs that curve upward at their ends are stronger than flat roofs because their curves resist bending. Show this effect using a sheet of paper. When flat, the paper bends easily, but when curved it stiffens so that it does not buckle. This kind of roof, which is supported at one edge just as the paper juts out from your hand, is called a cantilever roof. Strong cantilever roofs are made of thin, curved-up layers of concrete.

YOU WILL NEED

- *sheet of stiff paper*
- *large pen for a weight*

1 HOLD THE EDGE of the paper flat between your fingers and thumb. It sags, unable to support even its own weight.

2 NOW CURVE THE sheet upwards slightly. In addition to supporting its own weight, it can hold up the weight of a large pen.

EXPERIMENT

Folded barrel roof

Using just a sheet of paper, build a special roof that can support up to 400 times its own weight. The design combines a barrel roof with a folded plate roof. A barrel roof is shaped like a half-cylinder, which is strong because it acts like an arch to support a load. A folded plate roof is constructed of zigzags, which make the roof stiff so that it does not buckle. Two buttresses prevent the sides of the roof from moving outwards.

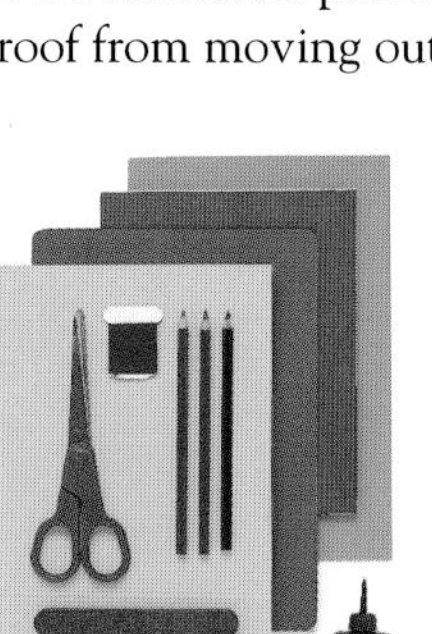

YOU WILL NEED
- *scissors*
- *thread*
- *pencils*
- *wooden tongue depressors*
- *wooden base 12 x 8 in (30 x 20 cm)*
- *cutting mat*
- *books for weights*
- *stiff paper 12 x 8 in (30 x 20 cm)*
- *glue*
- *ruler*

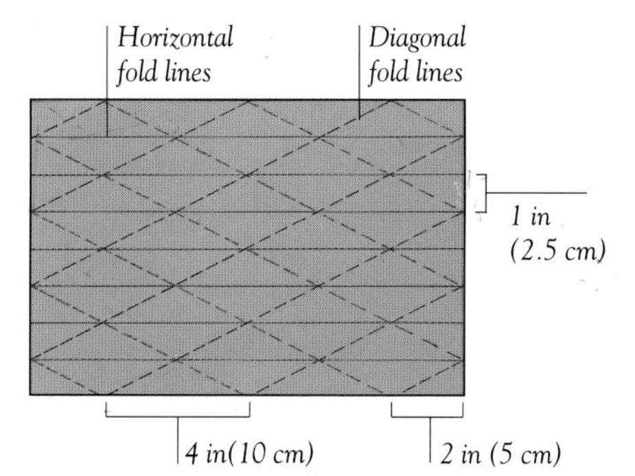

1 DRAW THE PATTERN shown below, enlarged to fill a stiff paper sheet. Using a ruler, fold upward along the seven horizontal lines. Rub the creases with a pencil to make them sharp.

2 TURN THE SHEET over. Now fold upward along the diagonal lines so that the second set of folds is in the opposite direction to the first set. Sharpen these creases with the pencil.

3 PINCH TOGETHER all the folds along one short edge of the paper. Work along the sheet toward the other short edge, pressing the folds together to make an accordion shape.

4 GLUE TWO STACKS of tongue depressors 6 in (15 cm) apart on the base. Let the folded sheet expand into a barrel shape. Place it between the tongue-depressor buttresses you have made.

5 TIE THREAD between the ends of two pencils so that they are 2 in (5 cm) apart and parallel. Rest the pencils on the barrel. The tops of the pencils should be above the apex of the roof.

6 CAREFULLY PLACE books one by one on top of the frame. How many books will the roof support? Compare the load supported by this roof, made from a single sheet of paper, to that supported by the curved paper sheet in the experiment opposite.

Electricity supply

FLIP A SWITCH in your home, and immediately a light comes on or an electrical machine springs into action. The electricity is generated in a power station and is carried to your home by power lines supported by pylons or poles, or laid underground. Once in your home, the electricity flows through wires in the walls, floors, and ceilings to lights and outlets. But it first enters a safety device (a fuse box or a circuit breaker box) that cuts off the power if a machine or wire is faulty and unsafe.

EXPERIMENT

Circuit breaker

Usually, the electrical machines in your home work well and safely. However, defects may occasionally appear. In most instances, the machine simply will not work. But sometimes the defect may cause it to consume more power; the machine, or the wire carrying electricity to it, may then heat up and could catch fire. A circuit breaker prevents accidents, responding to the sudden extra flow of electricity by immediately cutting off the supply. The circuit breaker can be reset later, when the faulty machine has been removed or made safe. Build a model circuit breaker that uses a solenoid like a real circuit breaker. Do not connect the model to an electrical outlet.

YOU WILL NEED

• 12V solenoid with plunger • battery connector • pliers • foamcore base about 8 x 12 in (20 x 30 cm) • 220R resistor • paper clip • LED • 4 alligator clips • tape • aluminum foil • wire • wire strippers • scissors • 9V battery • six 1 in (2.5 cm) foamcore squares • double-sided tape

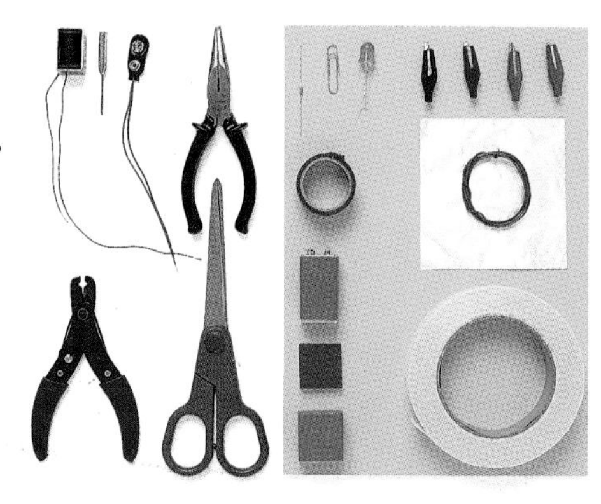

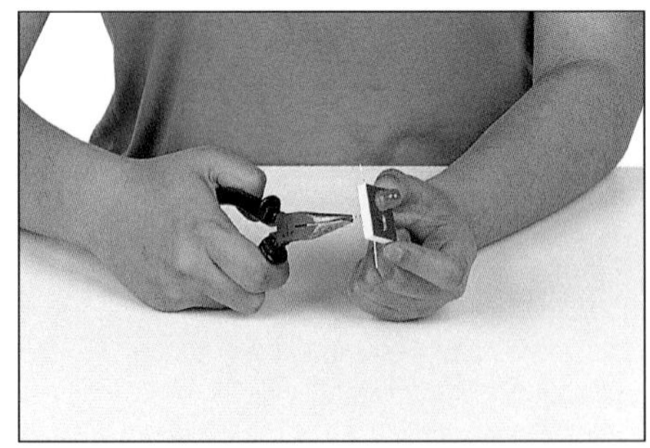

1 PUSH THE legs of the LED and the resistor through a foamcore square so that the longer leg (anode) of the LED is next to a resistor leg. Using pliers, twist the two legs together. Bend the free legs so that they project beyond the square.

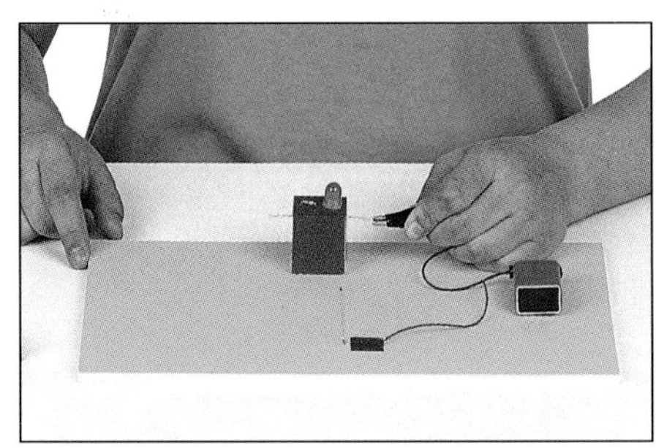

2 BEND THE paper clip at a right angle to make a contact. Tape the horizontal part to the base, then connect it to the positive battery terminal. Mount the LED square atop two more squares fastened vertically to the base. Connect the free LED leg to the negative terminal.

Overhead power
Power lines are suspended by insulators beneath the arms of tall pylons.

Distributing power

The electricity supplied to homes usually has a strength of about 120 volts. However, electricity flowing along power lines from the power station may be 1,000 or more times stronger. This is because electricity traveling at a very high voltage loses less power in the form of heat than electricity traveling at a low voltage. Between the power station and homes, the voltage is changed in substations. These contain transformers, in which electricity changes voltage when passing from one coil of wire to another.

Generators at the power station produce electricity at about 25,000 volts

A substation steps up the voltage to between 132,000 and 400,000 volts

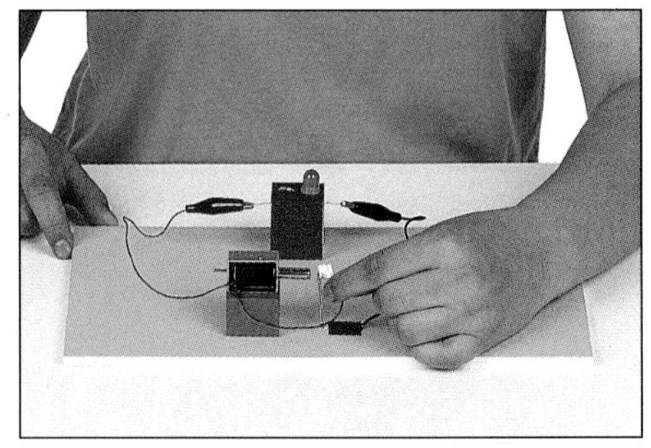

3 MOUNT THE solenoid on a frame of squares so that the plunger strikes the paper-clip contact when it is halfway out of the coil. Connect the free resistor leg to one wire from the solenoid, and a small square of aluminum foil to the other. Tape this foil to the end of the plunger.

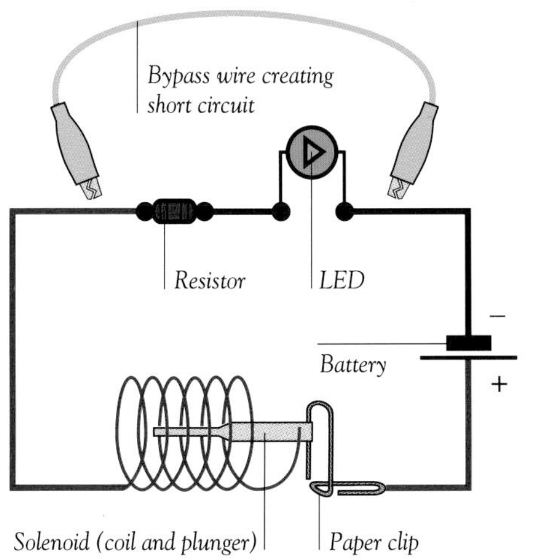

4 PULL THE solenoid plunger out until it touches the paper-clip contact and makes the LED light up. This represents a situation in which an electrical appliance (the LED) is working normally.

5 SIMULATE A fault by short-circuiting the LED. Connect the exposed LED and resistor legs using wire. The electricity that was used by the LED now bypasses it, so a higher current goes to the solenoid. The coil's magnetic field is increased, which pulls the plunger in, breaking the circuit.

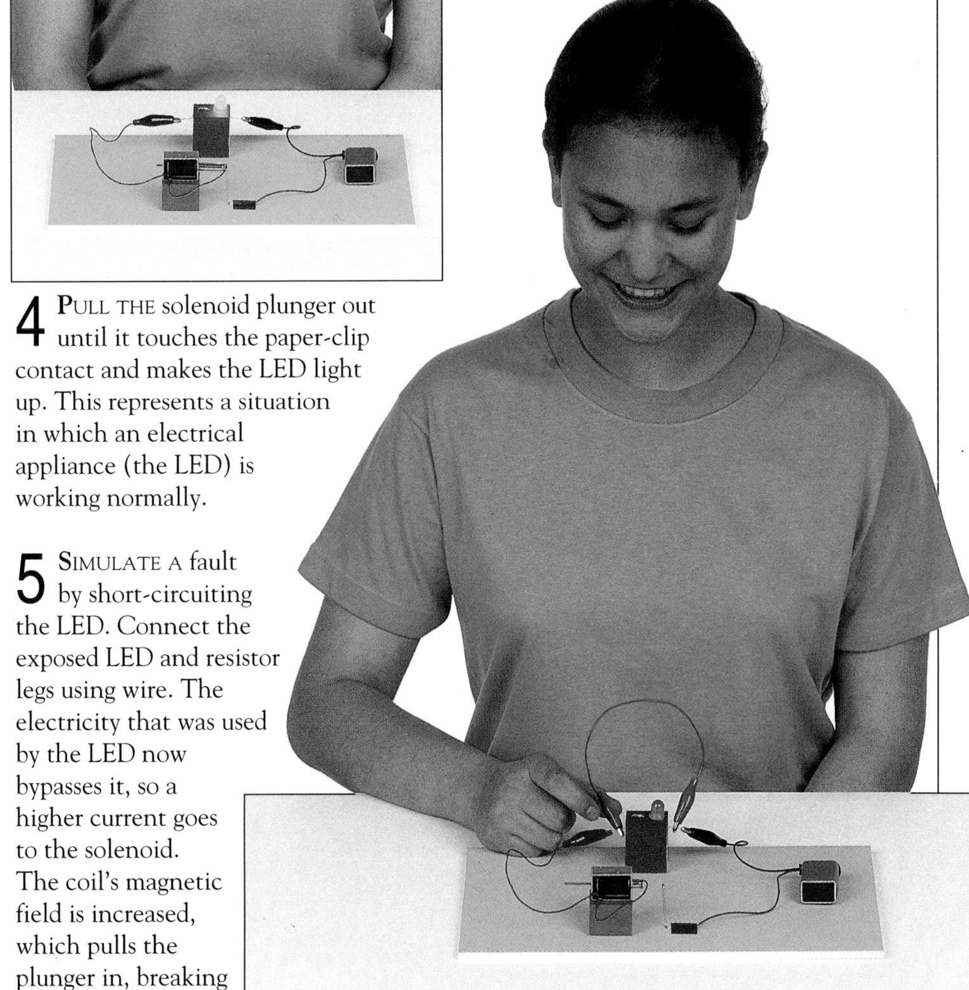

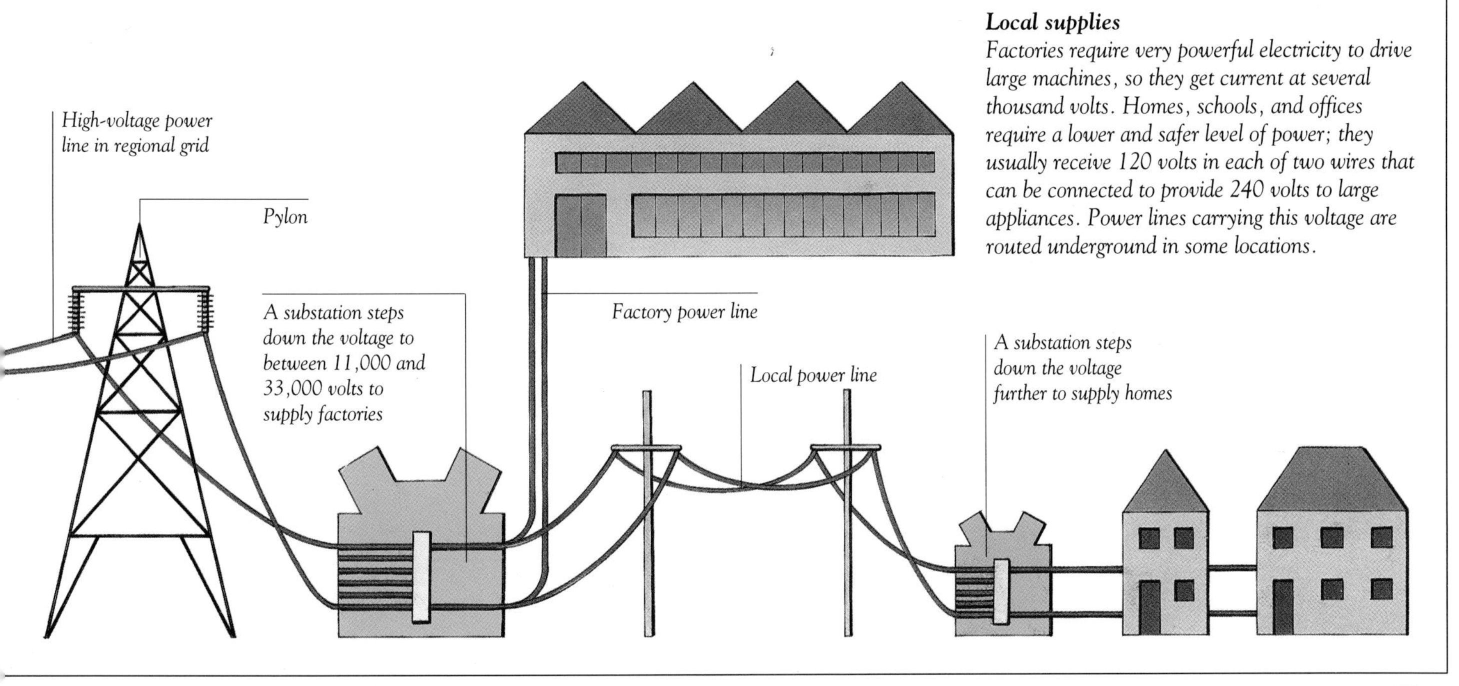

Local supplies
Factories require very powerful electricity to drive large machines, so they get current at several thousand volts. Homes, schools, and offices require a lower and safer level of power; they usually receive 120 volts in each of two wires that can be connected to provide 240 volts to large appliances. Power lines carrying this voltage are routed underground in some locations.

Water system

TURN ON A FAUCET in your home, and water flows immediately. The water may be hot or cold and it is usually pure enough to drink. The water comes from a reservoir, lake, river, spring, or well, and it is usually purified before it reaches your home. It arrives through an underground pipe. The water may then go directly to pipes that feed cold-water faucets in your home, or it may first go to tanks that store the water. A water heater fueled by electricity, gas, coal, or solar energy (p.58) heats the water that goes to hot-water faucets in your home.

EXPERIMENT

From tower to tap

Adult help is advised for this experiment

Water usually comes from a nearby water tower or service reservoir that is higher than your home. Gravity causes the water to flow down to your home; in some buildings it flows up a pipe to a cistern, or storage tank, beneath the roof. Every time you open a tap, water flows from the tower or the tank through pipes to the faucet. A storage tank has a ball-cock valve that allows it to refill with water to a maximum level without overflowing. Build a working model of this system.

YOU WILL NEED

• cutting mat • 12 in (30 cm) flexible plastic tube • craft knife • glue • faucet to fit plastic tube • long balloon • scissors • skewer • petroleum jelly • tape • vise • round file • drill with 1/8-in (3-mm) and 5/16-in (8-mm) bits • 14-in (35-cm) length of 5/16-in (8-mm) dowel • four 5 qt (5 l) square-section plastic bottles • 3 large blocks for bottles, of heights 7 1/2 in (19 cm), 12 in (30 cm), and 15 in (38 cm) • steel ruler • saw • plywood piece 6 1/4 x 2 1/4 x 3/16 in (16 x 6 x 0.5 cm) • wood piece 2 1/2 x 1 x 3/16 in (6.5 x 2.5 x 0.5 cm) with a 5/16-in (8-mm) hole drilled 1/2 in (1.5 cm) from a short edge • rubber ball, 3 in (8 cm) diameter

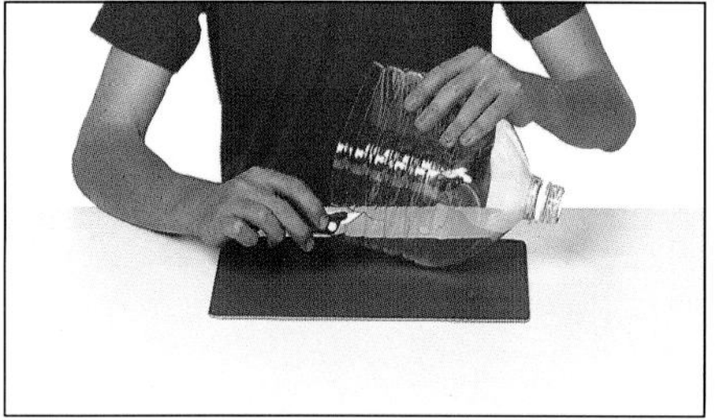

1 CUT THE top from a bottle. The top is a valve unit, the bottom a sink. Using the diagram opposite, cut opposing central holes in the valve unit for tubes. Cut a slot below one hole. Cut skewer holes and plywood slots in the other faces.

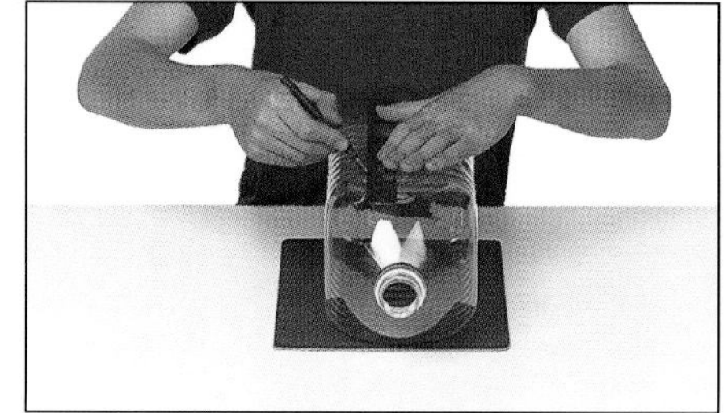

2 CUT AND discard the neck from a bottle to make a supply tank. Cut a tube-sized hole near the bottom of the tank. Cut a side panel from a third bottle to make a cistern. On the other side of the cistern, cut a tube-sized hole near the top.

3 CUT THE NECK from a long balloon to make a rubber tube. Slide a 4-in (10-cm) plastic tube into each end of the rubber tube and secure with tape. Leave a 1 1/2-in (4-cm) length of rubber tube in the middle, to make a valve.

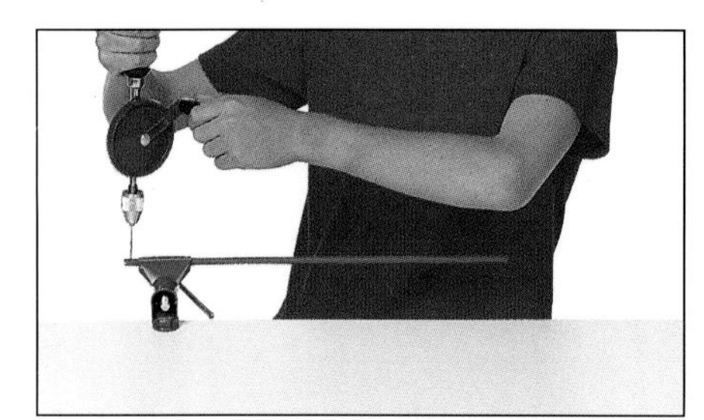

4 USE THE 1/8-in (3-mm) drill bit to make a hole through the dowel, 3/16 in (5 mm) from one end. This end of the dowel will form a hinge. Drill a 5/16-in (8-mm) hole through one side of the ball, then glue the ball to the other end of the dowel.

5 FILE TWO 45° bevels on the wood piece, along the short edge farthest from the hole. Slide the dowel through the hole. Glue the wood to the dowel, 2 in (5 cm) from the dowel's hinge end, with the beveled edge parallel to the hole in the dowel.

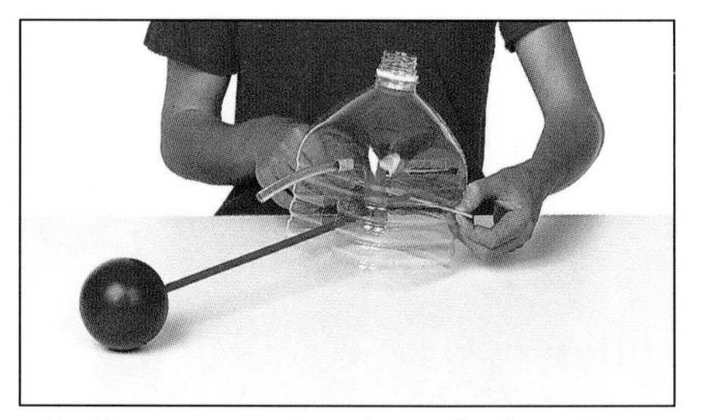

6 SLIDE THE valve through the tube-sized holes in the valve unit so that the rubber tube is held inside the bottle. Hinge the dowel on a skewer that passes through the skewer holes in the bottle, so that the beveled wood piece points up toward the valve.

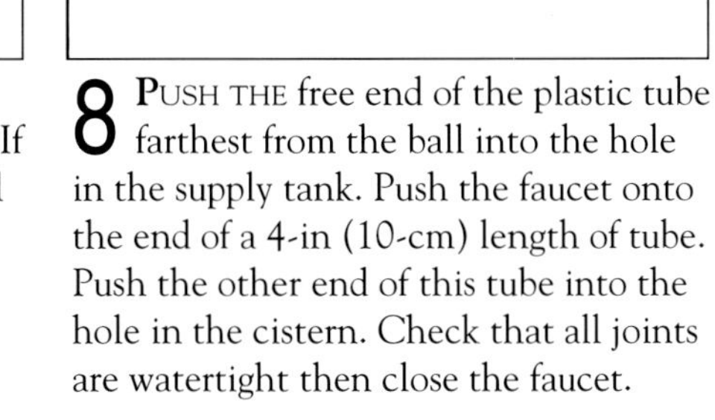

7 SLIDE THE PLYWOOD rectangle through the slots in the valve unit. If you lift the ball on the end of the dowel arm, the beveled piece of wood should squeeze the valve's rubber tube against the plywood rectangle before the dowel reaches a horizontal position.

8 PUSH THE free end of the plastic tube farthest from the ball into the hole in the supply tank. Push the faucet onto the end of a 4-in (10-cm) length of tube. Push the other end of this tube into the hole in the cistern. Check that all joints are watertight then close the faucet.

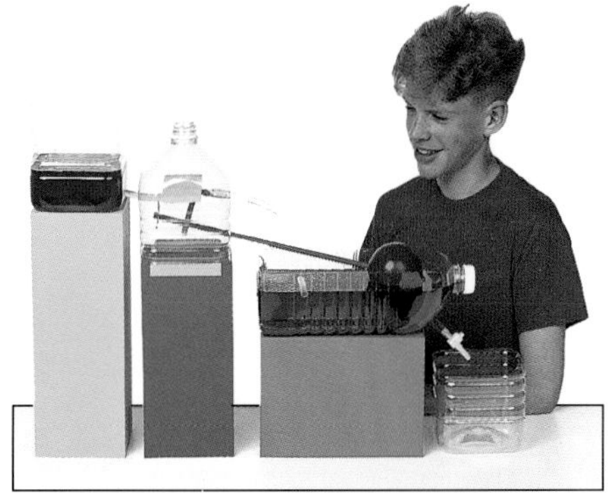

9 PLACE THE supply tank on the 15-in (38-cm) block, the valve unit on the 12-in (30-cm) block, and the cistern on the small block, with the ball in the tank and the sink under the faucet. Fill the supply tank with water.

10 WATER FLOWS from the supply tank to the cistern, causing the ball to float higher. The beveled wood piece is lifted toward the rubber tube, squeezing it shut when the cistern is full.

Sewage disposal

Waste water from sinks, showers, baths, and toilets leaves most homes through a drainage pipe before entering a sewer pipe. It then goes to a sewage plant, which removes all the impure matter from the water, and then discharges clean water.

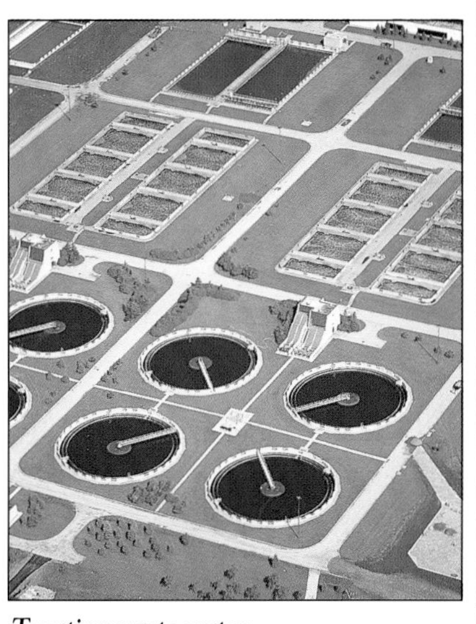

Treating waste water
A sewage plant first filters out solid objects, then destroys impurities by treating the dirty water with bacteria.

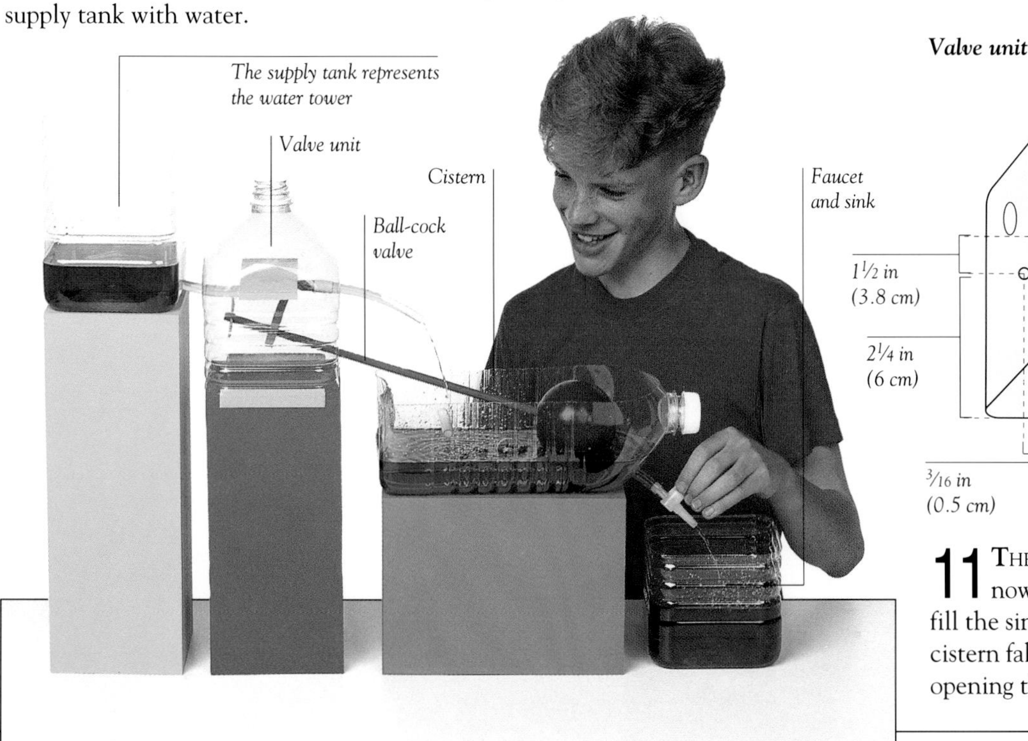

Valve unit

1 in (2.5 cm)
$1\frac{1}{2}$ in (3.8 cm)
$2\frac{1}{4}$ in (6 cm)
$\frac{3}{16}$ in (0.5 cm)
$2\frac{1}{4}$ in (6 cm)
$\frac{3}{4}$ in (2 cm)
$2\frac{1}{2}$ in (6.4 cm)

11 THE WATER supply to the cistern is now cut off. Turn on the faucet to fill the sink. As the water level in the cistern falls, the ball floats lower, opening the valve to refill the cistern.

Heating system

BUILDINGS WHERE people live or work need to be heated during cold weather. Fires or small heaters can warm individual rooms but central heating can heat an entire building. A central heating system may have one boiler that uses electricity, gas, oil, or a solid fuel such as coal. The boiler heats water that circulates through pipes to radiators in rooms and may also heat water to be used in baths and sinks. Some systems use the sun's rays to heat water inside solar panels. Other systems heat air that flows through the building, or use electric elements concealed in the floor or ceiling. Heaters and heating systems are controlled by timers and thermostats (p.77), which turn the heat on or off to maintain a set temperature.

All-season system

Many people live in places that have cold winters and hot summers, and their buildings need systems that can both heat and cool the rooms. An all-season air-conditioning system can deliver either warm or cool air to the rooms. It contains a fan that circulates air through ducts to all the rooms in the building. The blower fan also pulls in fresh air from outside to mix with the air inside the building, and a moisture unit controls the air's humidity. In winter, an element heated by electricity, steam, or hot water heats the air as it comes from the rooms. The warm air flows back to the rooms, where the temperature is controlled by thermostats. In summer, the air flows through a cooling unit that works like a refrigerator (p.78). This unit transfers heat from the rooms to the outside of the building. The heat in warm air inside the building is absorbed by a refrigerant in an evaporator. The refrigerant is then pumped into a condenser where it releases the heat into the atmosphere outside.

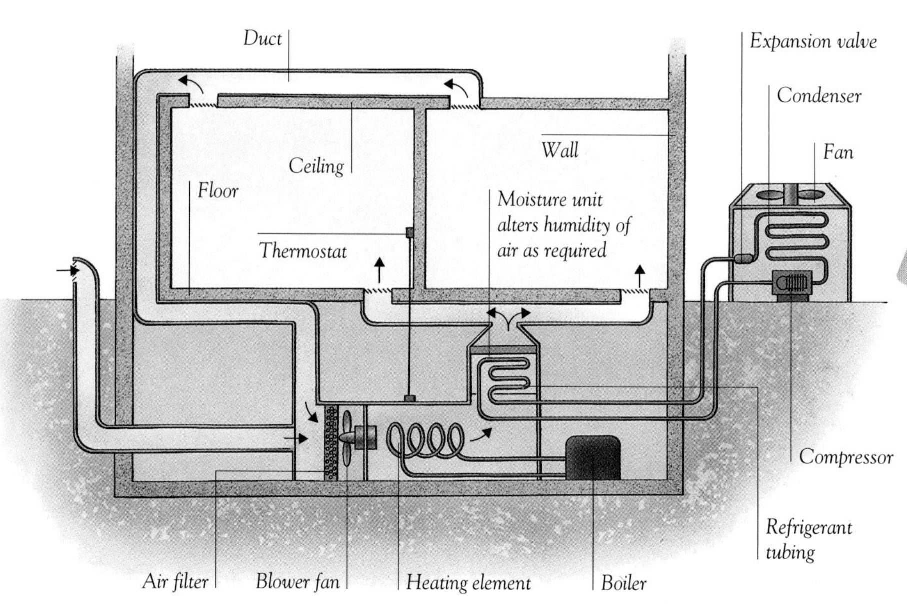

EXPERIMENT

Solar heater

Adult help is advised for this experiment

Using a lamp for the sun, make a "solar" water heater. Solar panels on roofs use the sun's energy in the same way.

YOU WILL NEED

• plastic bottle • ruler • round file • 2-ft (60-cm) length of 3/8-in (1-cm) diameter flexible copper pipe • sandpaper • pitcher of room-temperature water • cutting surface • shallow box about 10 x 8 in (25 x 20 cm) • foamcore, 9½ x 3 in (24 x 8 cm) • plastic wrap to cover box • scissors • desk lamp • masking tape • two 20-in (50-cm) lengths of 3/8 in (1 cm) diameter flexible plastic tube • thermometer • craft knife • wooden block • funnel • cork to fit bottle • tape • string • glue

1 STICK A PIECE of masking tape above the base of the bottle. Place another piece 2 in (5 cm) above the first. Cut a $\frac{3}{8}$-in (1-cm) cross through each into the bottle. Using a file, enlarge the crosses into holes that fit the plastic tube tightly.

2 USE A CRAFT KNIFE and sandpaper to pare down one end of each length of plastic tube until each fits tightly inside each end of the copper pipe. Secure with tape. Bend the pipe into a gentle W shape that fits inside the box.

3 PLACE THE COPPER pipe inside the box. Use scissors to cut two holes in one side of the box for the plastic tubes to pass through. Cover the box with plastic wrap with the pipe inside to make a "solar panel."

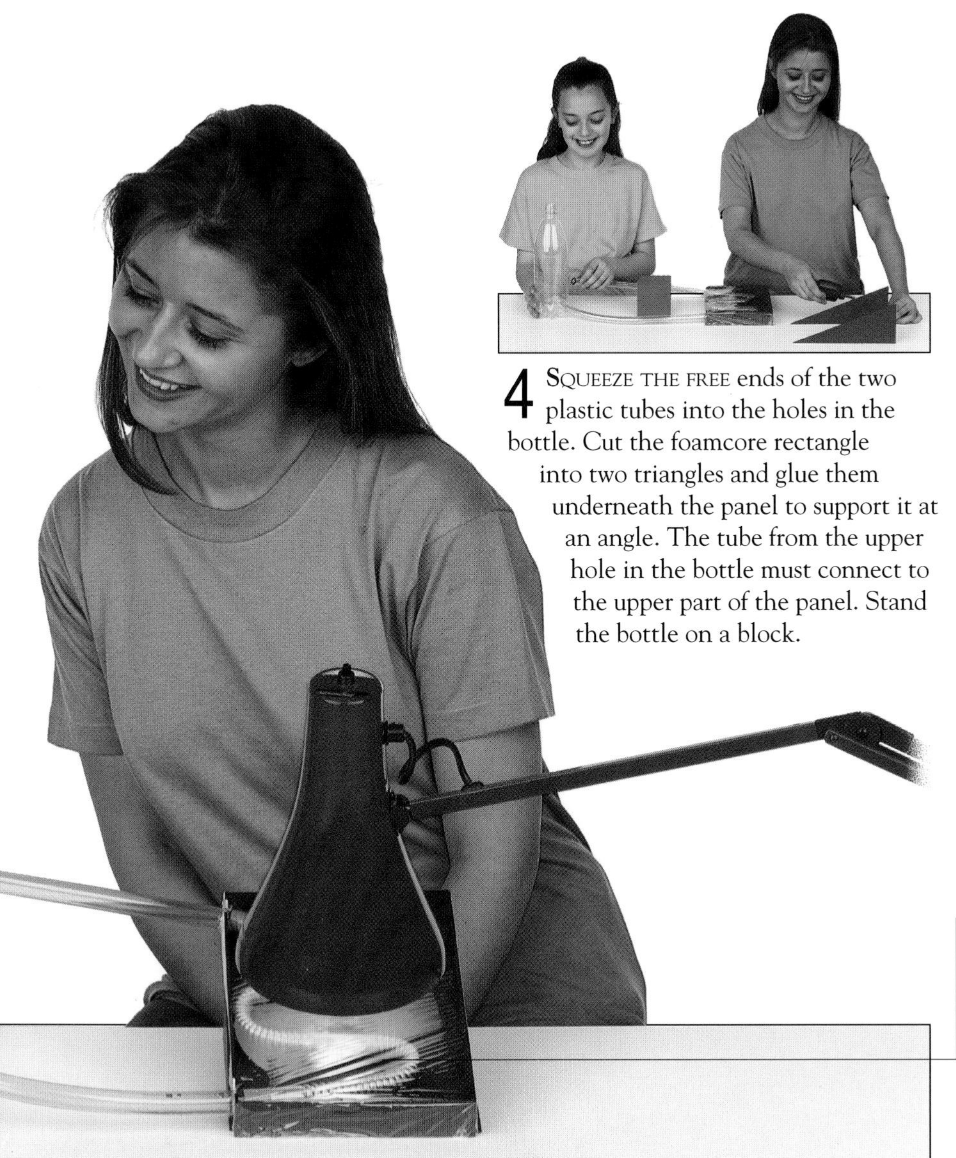

4 SQUEEZE THE FREE ends of the two plastic tubes into the holes in the bottle. Cut the foamcore rectangle into two triangles and glue them underneath the panel to support it at an angle. The tube from the upper hole in the bottle must connect to the upper part of the panel. Stand the bottle on a block.

5 POSITION THE PANEL by the bottle and check the tube connections. Slowly pour water into the bottle so t hat it fills the tubes as well. The water level must be above the upper hole. Suspend the thermometer on string so that its bulb is just beneath the water's surface. Put the cork in the bottle, trapping the string.

6 PLACE THE LAMP over the panel and turn it on. Watch the thermometer to see how much your solar panel can heat the water.

A solar panel contains a pipe through which water circulates from a tank; water in the pipe is warmed by the sun's rays and rises, flowing up to the top of the tank; cooler water sinks, going out of the bottom of the tank and flowing back down to the heater where it is warmed before returning to the tank

Escalators

ESCALATORS TAKE the effort out of climbing long flights of stairs in public buildings. The moving stairs form a continuous loop of circulating steps with the lower, returning steps hidden beneath the upper, visible steps. The weight of the hidden steps counterbalances the weight of the visible steps, so the escalator's motor has to move only the weight of the people on the escalator, which may go up or down. A skyscraper or any building with many floors would be impractical without safe elevators to carry people up and down. An elevator (p.183) has a counterweight that balances the weight of the car and half the passengers. The motor moves the cable holding the car and has to raise little, if any, weight.

EXPERIMENT

Model hand-powered escalator

Adult help is advised for this experiment

Build a model escalator to see how the steps are level at the top and bottom and move up or down to become stairs in between. The steps are joined together in a chain, and each step has a pair of wheels at the front and a pair at the back. One pair is higher than the other, and they run on two different pairs of rails beneath the steps. An outer pair of rails carries the high wheels, and an inner pair of rails bears the low wheels. At the top and bottom of the escalator, the inner rails drop below the outer rails so that the steps level out. In between, both pairs of rails are aligned at 45° to the ground so that the steps rise or drop to form a moving flight of stairs.

YOU WILL NEED

- *3/16-in (5-mm) foamcore*
- *double-sided tape*
- *12 beads 3/16 in (5 mm) wide and 24 smaller beads, all able to rotate on a skewer*
- *string*
- *cutting surface*
- *poster board*
- *pen*
- *scissors*
- *craft knife*
- *wood glue*
- *compass*
- *three 4 1/2-in (12-cm) skewers and three 3-in (7.5-cm) skewers*
- *steel ruler*
- *ruler*

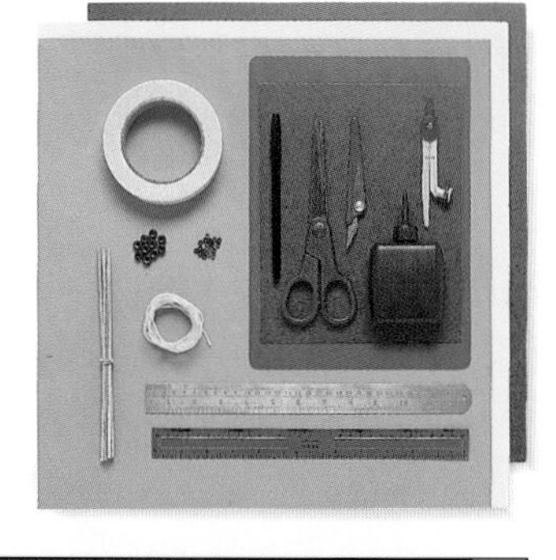

1 CUT THE foamcore into the shapes shown opposite. Tape the inner rails together, with spacers between, using double-sided tape. Fasten the outer rails to the inner rails with spacers between.

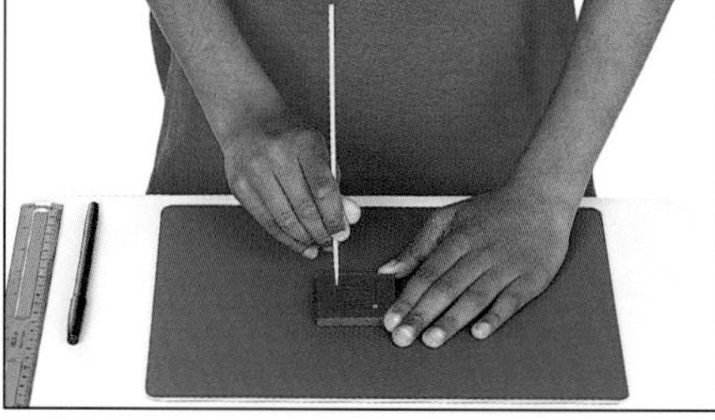

2 DRAW A 1-in (2.5-cm) square in the center of each of the two side panels of each step. Using a skewer, make a hole through two diagonally opposed corners of each square.

3 GLUE A small panel between a pair of side panels as shown in the diagram opposite. Glue a top panel to the top of the side panels. Make two more steps in this way.

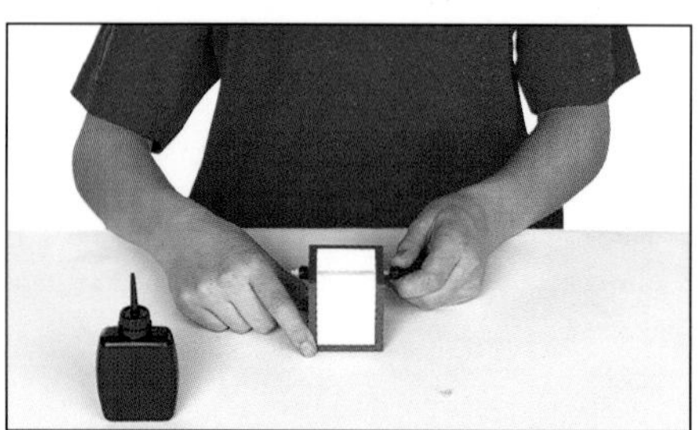

4 PLACE A short skewer through one pair of holes on each step to make the lower axles. Add large beads, making sure they can rotate on the inner rails. Glue a small stopper bead on each end.

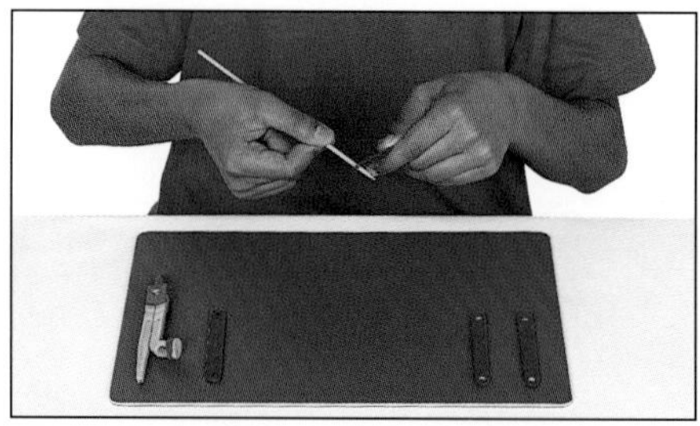

5 CUT FOUR 3 3/8 x 3/8 in (8.5 x 1 cm) strips of poster board. Using a compass, make two holes 3 in (7.5 cm) apart in each strip, then enlarge the holes with skewers.

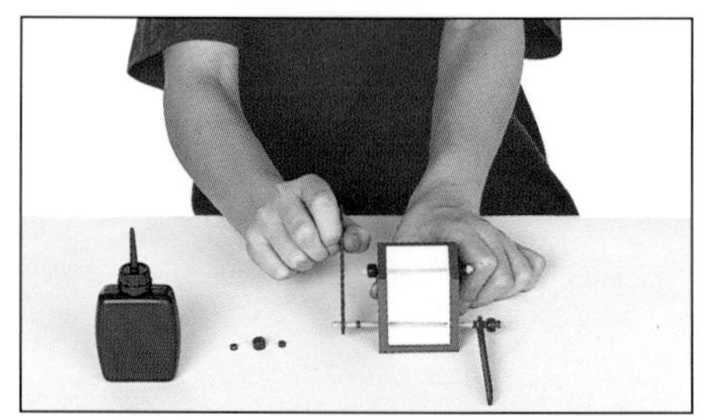

6 ADD UPPER axles (long skewers) to the steps. Connect them with poster-board strips. Add large beads, making sure they can rotate on the outer rails. Glue the small spacer and stopper beads.

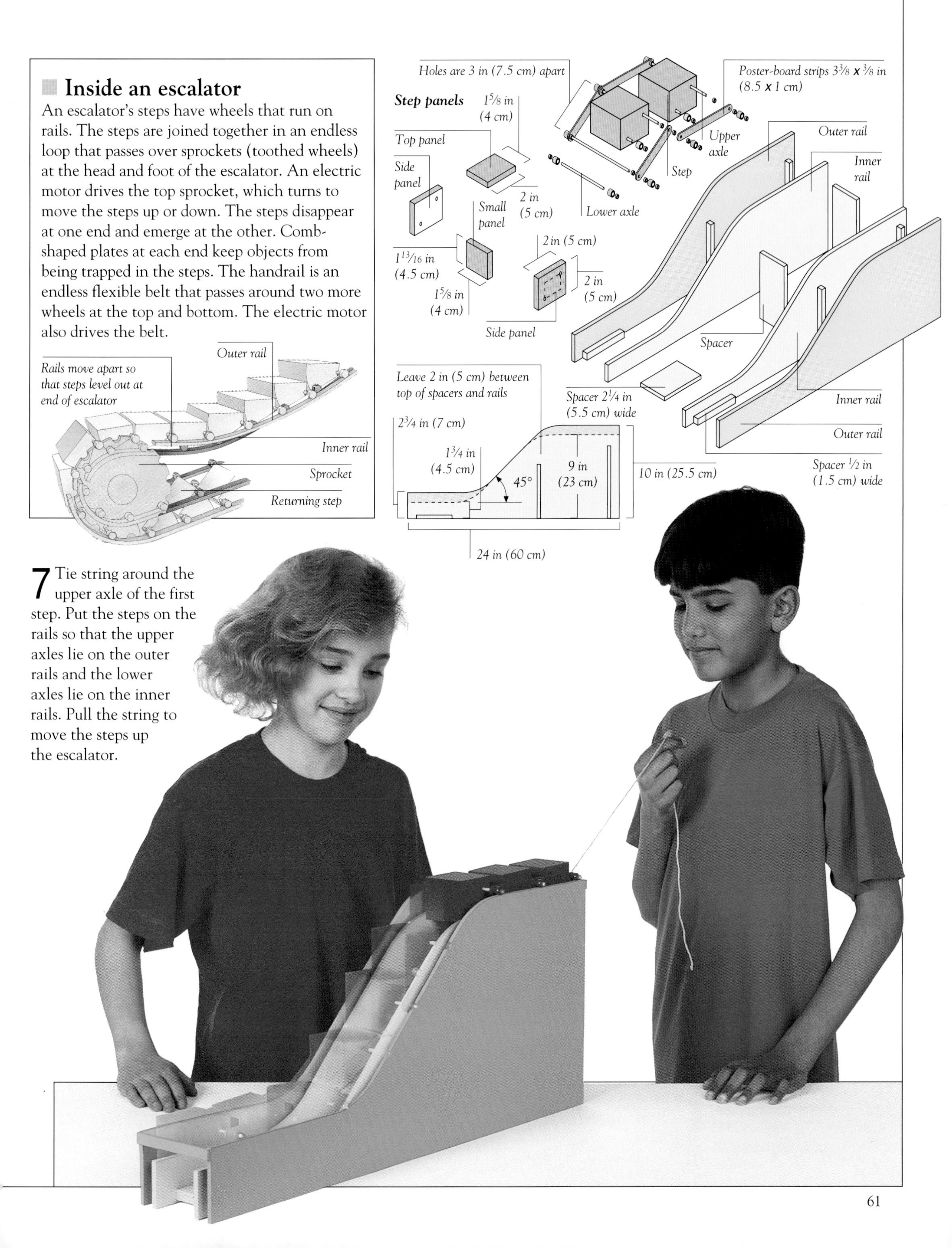

Inside an escalator

An escalator's steps have wheels that run on rails. The steps are joined together in an endless loop that passes over sprockets (toothed wheels) at the head and foot of the escalator. An electric motor drives the top sprocket, which turns to move the steps up or down. The steps disappear at one end and emerge at the other. Comb-shaped plates at each end keep objects from being trapped in the steps. The handrail is an endless flexible belt that passes around two more wheels at the top and bottom. The electric motor also drives the belt.

7 Tie string around the upper axle of the first step. Put the steps on the rails so that the upper axles lie on the outer rails and the lower axles lie on the inner rails. Pull the string to move the steps up the escalator.

Skyscrapers

THE WEIGHT OF a skyscraper is so enormous that it cannot be built in the same way as a low building, in which the walls hold up the structure. To support the immense load of the building above, the lower walls would have to be impossibly thick. A skyscraper therefore has an internal frame of steel or concrete beams, to which lightweight wall panels and floors are fixed. However, such a frame is flexible; the taller it is, the more it sways in the wind, and the more vulnerable it is to damage by earthquakes. To reduce the swaying, skyscrapers are stiffened either by a rigid central core, or by inflexible triangular or X-shaped braces around the frame.

EXPERIMENT

Miniature skyscraper

Build a model skyscraper by first constructing an internal frame and then stiffening the frame to make it rigid. Straws represent the beams that are connected to form a building's frame. The floors are made of cardboard.

YOU WILL NEED

- *cardboard tube, about 2 in (5 cm) diameter and 26 in (65 cm) tall*
- *cutting mat*
- *scissors*
- *foamcore 10 x 8 x 3/16 in (25 x 20 x 0.5 cm)*
- *thin drinking straws cut to 6 in (15 cm), into which matches fit tightly*
- *cardboard*
- *compass*
- *craft knife*
- *matches with heads cut off*
- *pencil*
- *steel ruler*
- *double-sided tape*

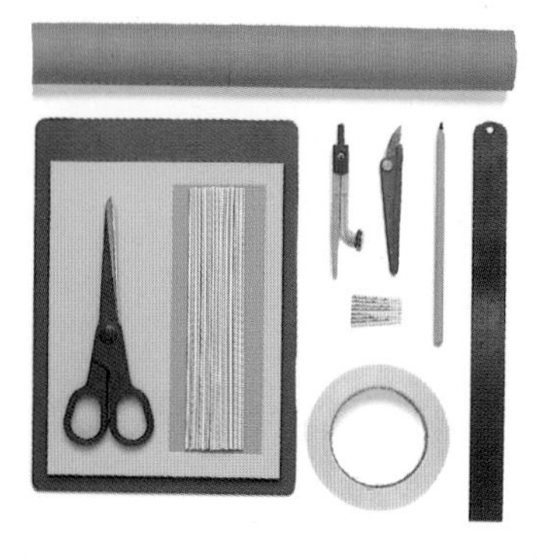

1 CUT A central hole in the foamcore base, into which the cardboard tube fits snugly. Make four tight holes to fit matches at the corners of a 6-in (15-cm) central square.

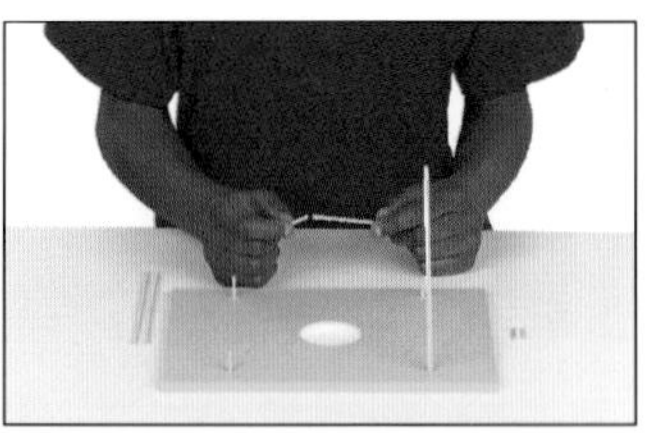

2 PUSH A match into each hole in the base. Push a straw onto each one. Insert a match halfway into the open end of each straw to make a vertical support.

3 COPY THE floor template opposite onto cardboard. Cut slots in the ends of four straws that allow the straws to slide onto the corner tabs around the square. Make four floors in this way.

4 CUT 16 cardboard strips 8¾ x ⅜ in (22 x 1 cm). Bend a ⅜-in (1-cm) tab at each end of each strip to make a "Z" shape. Stick double-sided tape to each tab, on opposite sides of the strip.

5 PUSH THE holes in the first floor over the matches in the vertical supports. Add a straw and a match to each support, then add the second floor. Build up the structure in this way.

6 HOLD THE base with one hand while you apply a gentle sideways force to the top of the skyscraper. Note how easily it bends, due to the lack of stiffening elements.

7 CAREFULLY SLIDE the cardboard tube through the central holes in the floors, down to the base. In real skyscrapers, a rigid shaft like this usually houses the elevators.

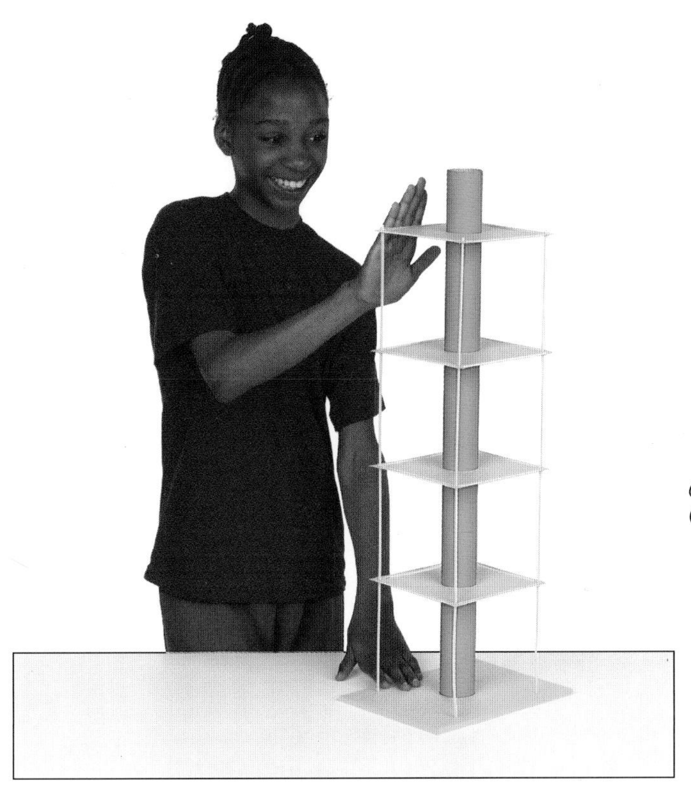

8 HOLD THE base with one hand and gently try to bend the skyscraper again. Notice how the rigid central core stiffens the skyscraper's frame.

Floor template

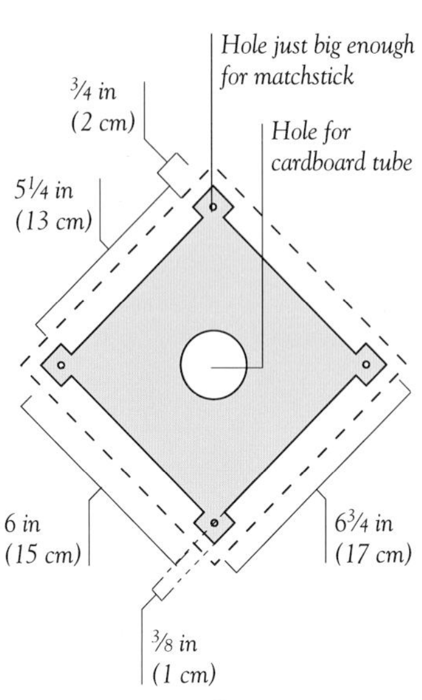

9 REMOVE THE central core from the skyscraper. Stick a cardboard strip diagonally across each face between adjacent floors. Arrange the strips in a zigzag pattern both horizontally and vertically.

10 HOLD THE base and try to bend the skyscraper again. The diagonal strips form rigid triangles with the straws, and these prevent the skyscraper bending.

High-rise island

Hong Kong is a small island with limited building space, and the city is filled with skyscrapers. Gales and typhoons lash the buildings during the monsoon season. The skyscrapers must be built to withstand wind damage.

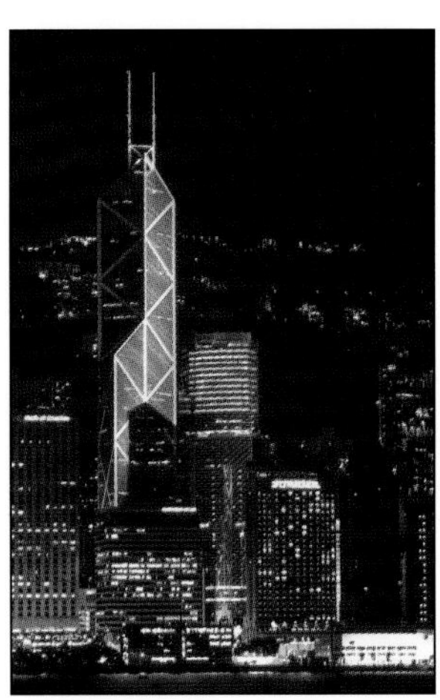

X for strength
This skyscraper has an X-shaped external framework to stiffen the structure and resist storms.

Dams

WATER IS PERHAPS the most precious of all our resources. We need it to drink, to grow crops, and to make paper and many other products. We also use water to produce power. In order to obtain a constant supply of water, dams have been built across river valleys throughout the world. A huge barrier of soil, rock, stone, or concrete cuts off the river so that a large artificial lake or reservoir of water builds up behind the dam. The water may then be piped to homes, farms, and factories. The dam may also supply water to a hydroelectric power station to generate electricity. Dams can also help prevent floods by controlling the amount of water flowing down a river. These structures are designed to be very strong so they can resist the great weight of water behind them. The actual design depends on the shape and size of the valley, and on the kind of soil and rock found there.

Model dam

This is a model of an embankment dam. It is built inside a plastic tank, which represents a valley. The dam has a central core of modeling clay that keeps water from seeping through. The dam's sides are made by piling up earth and covering it with gravel. Plastic pipes carry water through the dam when necessary. At the foot of the dam is a structure representing a power station.

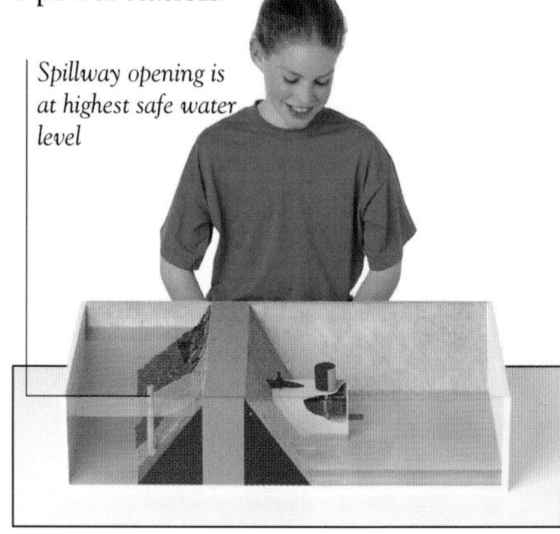

Preventing overflow
Heavy rainfall may cause the water in the reservoir to rise to the top of the dam. An overflow could be dangerous, so a pipe called a spillway releases the extra water before it can flow over the top of the dam.

Filling the reservoir
When the dam is complete, water is poured into the tank behind the dam to represent the flow of the river down the valley. The pipes are closed so that no water can pass, and the reservoir fills. Water is then piped from the reservoir to supply the surrounding area. The level rises and falls as water enters and leaves.

Inflow pipe to turbine

Turbine blades

Generating power
The pipe leading from the reservoir down through the dam to the power station is opened. Water flows to turbines in the power station, and the turbines drive electricity generators (p.16). The water then leaves the power station, and continues down the river. The electricity produced by the generators is distributed by power lines to towns and cities.

Embankment dam

A valley that is shallow and wide requires an embankment dam, which consists of a huge pile of soil or rock built across the valley. A central core of clay extending down into the valley floor prevents water from seeping through the dam. The immense weight of the dam keeps it in position.

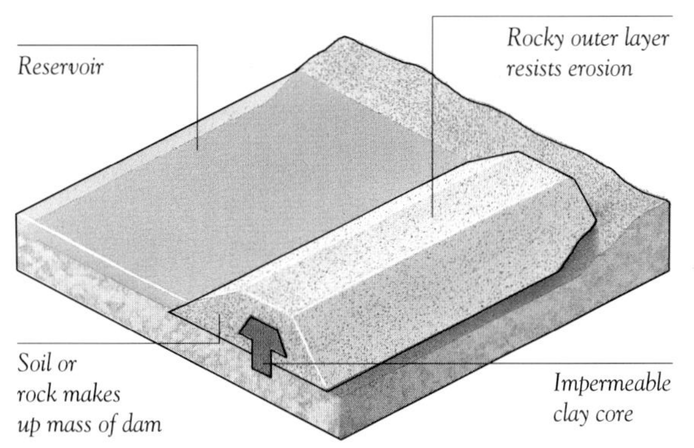

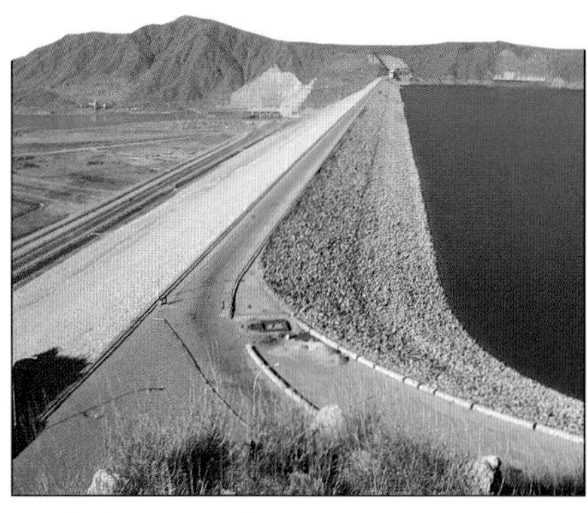

Tarbela Dam, Pakistan
The world's largest soil embankment dam.

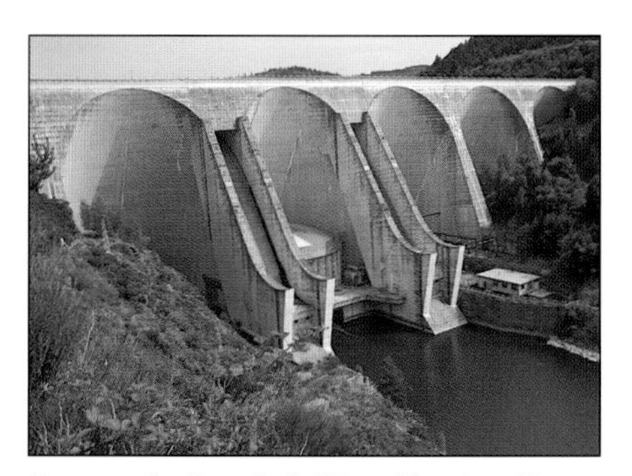

Barrage de Grandval, River Truyère, France
Spillways are built into two of the buttresses.

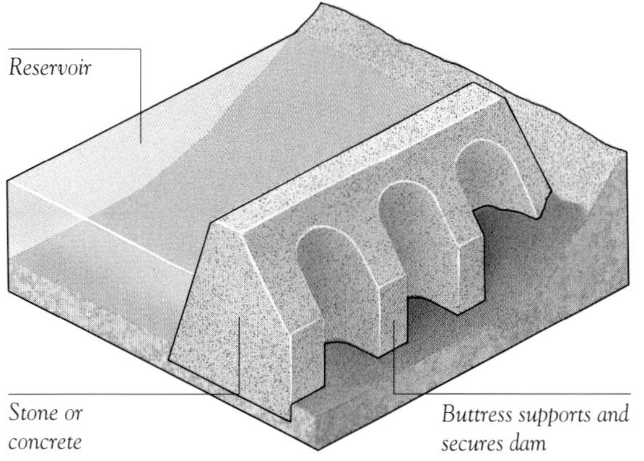

Buttress dam

Narrow valleys that are not very high may be dammed by building a thick barrier of stone or concrete. The amount of material in the barrier can be reduced by using buttresses to support the dam and hold back the weight of water. Buttresses also hold the dam securely to the valley floor.

Gravity dam

A gravity dam is made entirely of concrete or stone, and is built across a narrow valley. Like an embankment dam, a gravity dam holds back the water because of its great weight. It does not need to be as firmly secured to the valley sides and floor as buttress dams and arch dams, which are lighter than gravity dams.

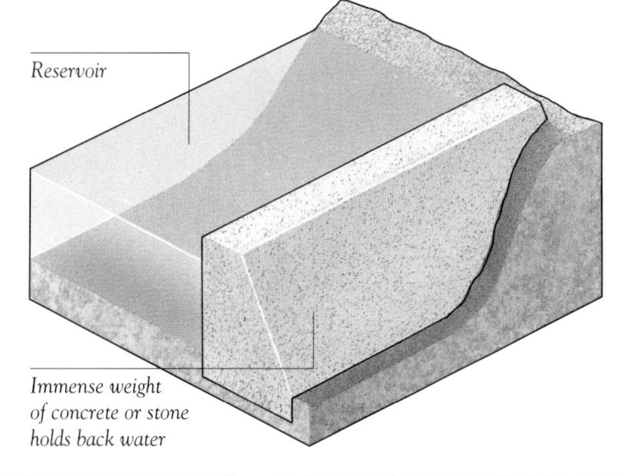

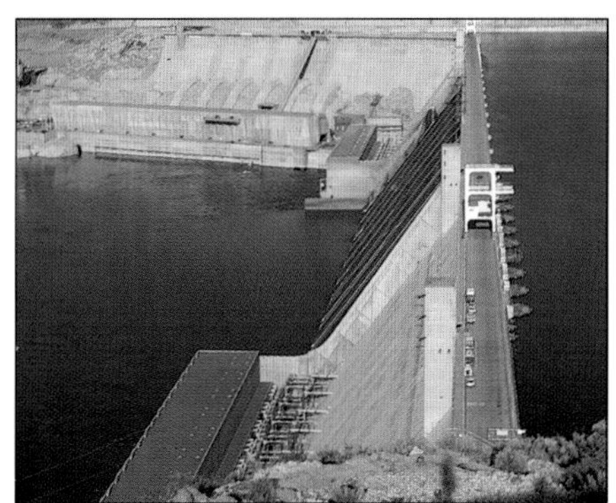

Grand Coulee Dam, Washington
This gravity dam holds a reservoir 150 miles long.

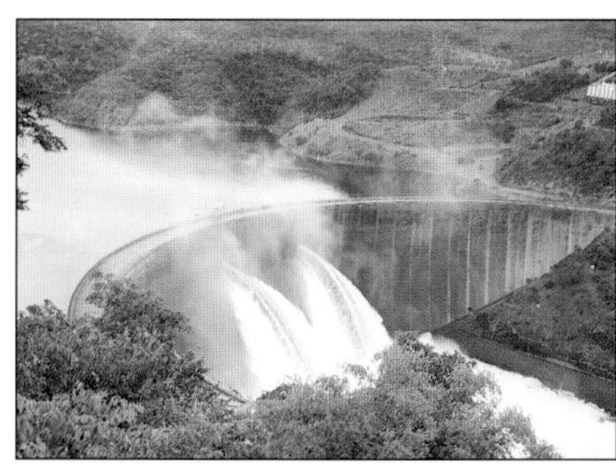

Kariba Dam, Zimbabwe
This concrete arch dam is on the River Zambezi.

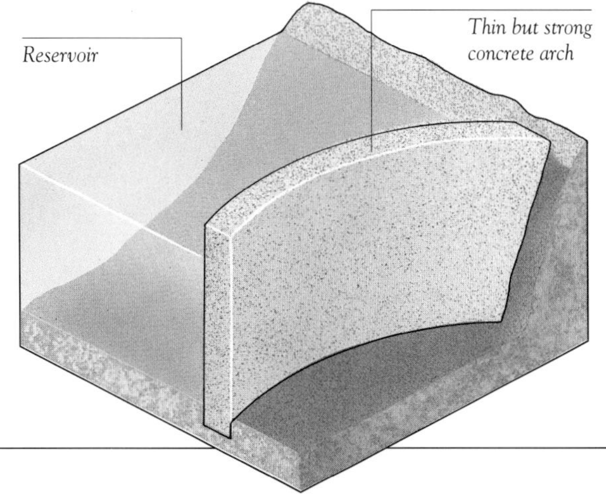

Arch dam

Narrow and high valleys require arch dams, which are thin curved barriers made of concrete and firmly fixed to the valley sides and floor. The water pushing against the dam compresses the concrete, greatly increasing its strength. As a result, the dam can hold back the weight of water even though it is relatively thin.

Bridges

THE DESIGN OF A BRIDGE depends on its height, its length, and the load it has to carry. Beam bridges carry a roadway (or railway) on a straight beam supported at each end, sometimes with columns in the middle. Arch bridges support the roadway on or under an arch. Suspension and cable-stayed bridges hang the roadway from steel cables or bars. Bridges are subject to two different forces: compression (a squeezing force) and tension (a pulling force). Arches and columns are compressed by the weight of the roadway and traffic, but compression makes them stronger so that they can support this load. Steel cables and bars undergo tension, but they are strong enough to carry the load and do not snap.

Four main types of bridge

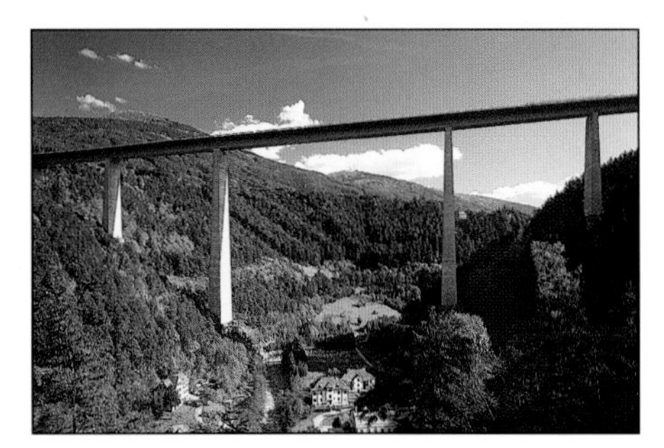

Europa Bridge, Brenner Autobahn, Austria
This beam bridge consists of several short spans supported by high columns.

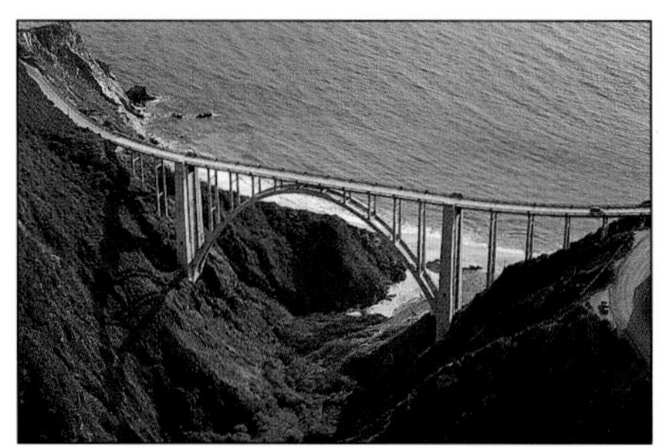

Bixby Bridge, Highway 1, California
This arch bridge leaps over a deep, wide valley in a single span supported by a tall arch.

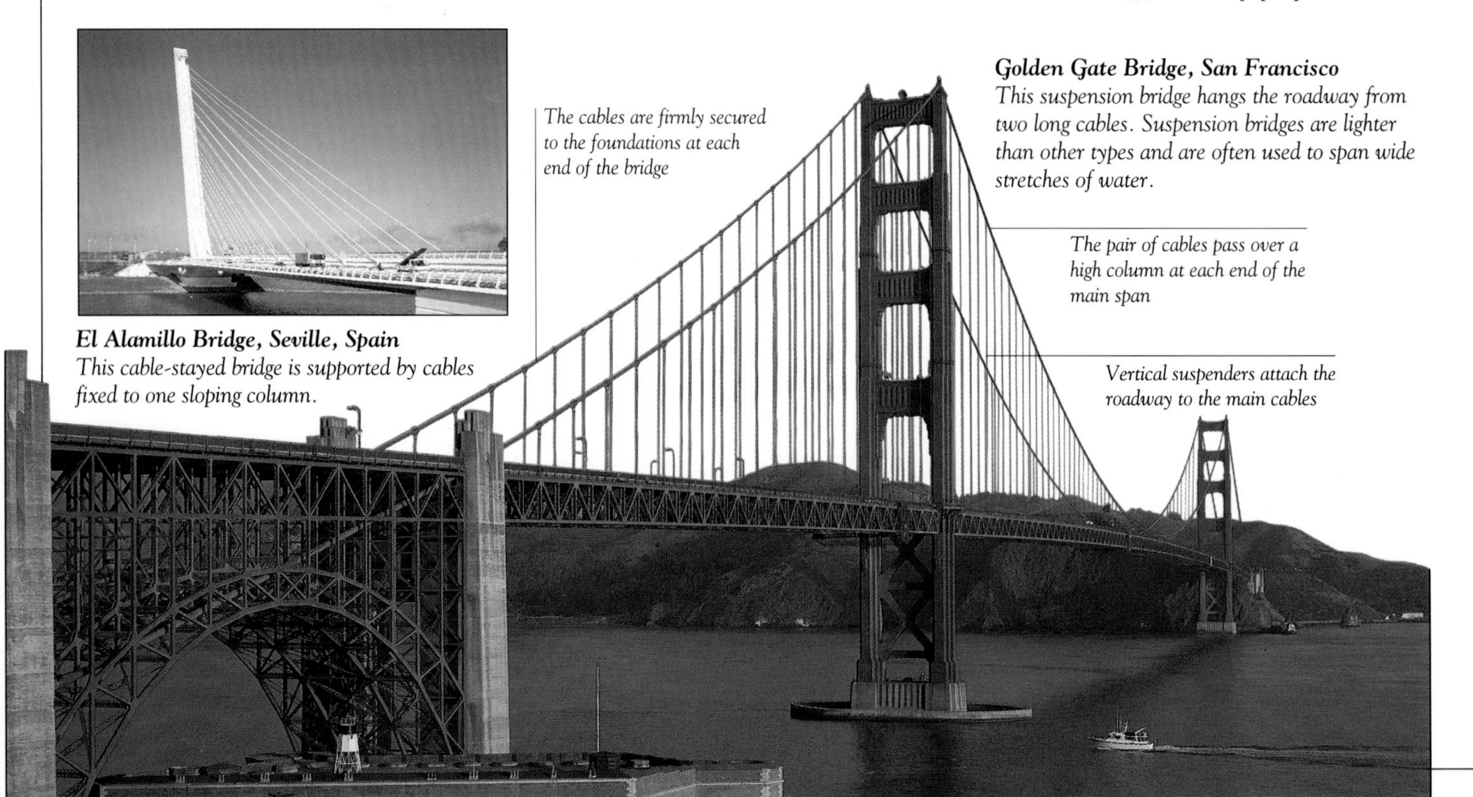

El Alamillo Bridge, Seville, Spain
This cable-stayed bridge is supported by cables fixed to one sloping column.

The cables are firmly secured to the foundations at each end of the bridge

Golden Gate Bridge, San Francisco
This suspension bridge hangs the roadway from two long cables. Suspension bridges are lighter than other types and are often used to span wide stretches of water.

The pair of cables pass over a high column at each end of the main span

Vertical suspenders attach the roadway to the main cables

EXPERIMENT

Model bridges

Adult help is advised for this experiment

Build models of a beam bridge, an arch bridge, and a cable-stayed bridge, and test their strengths. Beam bridges are supported at each end and are not very strong. Long beam bridges need supporting columns in the middle, so they cannot be built very high. Arch bridges are stronger and need no central supports, so they can be long and high. Cable-stayed bridges have no arches or central columns and are light as well as strong. The supporting cables stiffen the bridge.

YOU WILL NEED

- *drill and 3/16-in (5-mm) bit*
- *thin file*
- *C-clamp*
- *ten 2 oz (60 g) pieces of modeling clay*
- *wooden base 20 x 6 in (50 x 15 cm)*
- *four 12-in (30-cm) lengths of 3/16-in (5-mm) dowel*
- *pen*
- *two 20 x 4 in (50 x 10 cm) pieces of poster board*
- *two 4-in (10-cm) wooden cubes*
- *scissors*
- *string*
- *glue*
- *drilling base*
- *paper punch*

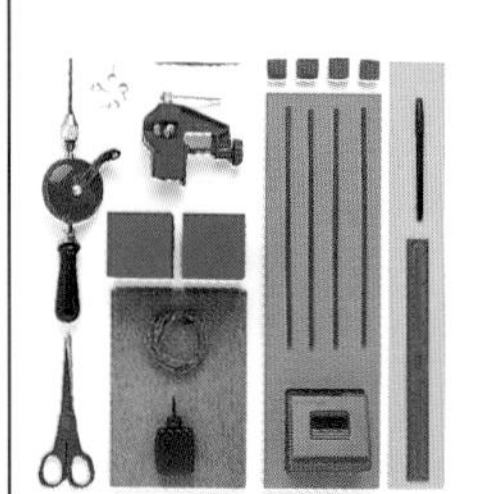

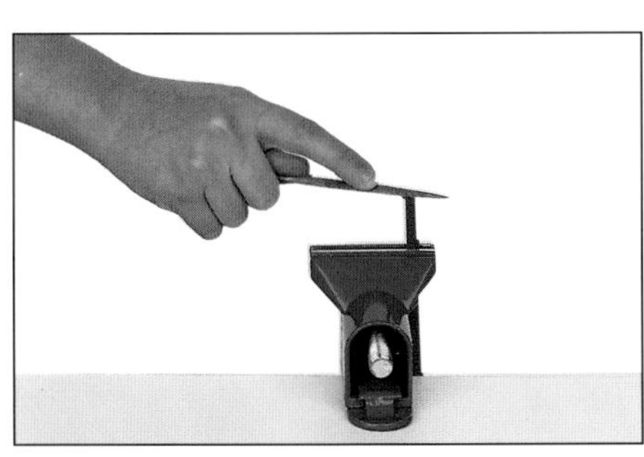

1 USING A THIN FILE, make a groove across the top of each length of dowel. The grooves should be wide enough for the string to fit inside them.

2 DRILL A PAIR of dowel-sized holes 5 in (12 cm) from and parallel with each end of the base. The holes in each pair should be 4 in (10 cm) apart.

3 GLUE A WOODEN CUBE behind each pair of holes in the base. Insert a length of dowel into each hole and secure it with glue.

4 BUILD A BEAM BRIDGE by laying two poster-board strips over the blocks. The upper strip represents the roadway and the lower one the beam. Load the bridge with modeling clay. Even a small load bends the beam because it only rests on the blocks, with no other support.

5 NOW MAKE AN ARCH BRIDGE by curving the lower strip between the blocks with the roadway (upper strip) on top. This bridge holds much more weight than the beam bridge before it begins to bend. The arch becomes strong when compressed and can transfer the load to the wooden blocks.

6 MAKE A cable-stayed bridge. Punch two holes at the center of each side of the roadway. Tie strings to the tops of the dowels at one end of the strip. Pass the strings over the grooves in the dowels, through the holes in the strip, and attach to the dowels at the other end.

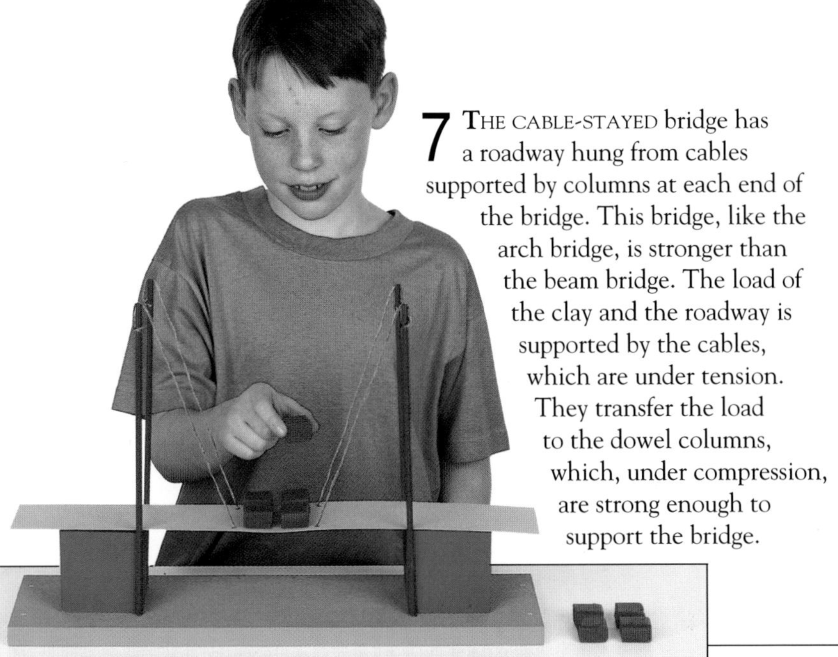

7 THE CABLE-STAYED bridge has a roadway hung from cables supported by columns at each end of the bridge. This bridge, like the arch bridge, is stronger than the beam bridge. The load of the clay and the roadway is supported by the cables, which are under tension. They transfer the load to the dowel columns, which, under compression, are strong enough to support the bridge.

Oil rigs 1

CRUDE OIL (PETROLEUM) and natural gas are vital sources of energy often found in rock deep under land or sea. Crude oil is the source of the fuels that run most forms of transportation, and both oil and gas are used in homes for cooking and heating. Crude oil is also used to produce oil for lubricating engines, and chemicals for making plastics, medicines, and many other products. To find oil and gas, exploration rigs are used to drill deep wells in places geologists believe have oil and gas deposits. For exploration on land, the rig is erected at the site to be tested. For offshore exploration, the complete rig is first built in a shipyard and then towed to the site. Rigs at sea stand on tall legs resting on the seabed, or they float on the surface. The drill goes down through the water and into the seabed. On land and sea, once the search is over, the rig is usually moved to a new exploration site. A production well or production platform is built to tap any oil that is discovered.

Platforms and flares

This production platform is pumping oil from beneath the seabed into a tanker moored alongside it. Natural gas and water mixed with the oil are removed on the platform. The gas is used to generate electricity for the rig. Unused gas is burned in a flare at the side of the rig. The large, heavy platform can withstand high waves and violent storms.

Drilling for oil

An oil rig is like a huge power drill that bores down into rock. It has a tall tower called a derrick, which supports a hollow drill pipe with a sharp rotary bit at the end. The pipe rotates and is lowered into the ground while the bit bores a shaft. As it gets deeper, more drill pipes are added. A special liquid called mud is pumped down to the rotary bit to cool it. The mud then flows back up the shaft carrying pieces of rock with it. When oil is struck, it may rush up the shaft and blow out of the rig like a black fountain. This is dangerous, so most rigs have a series of valves called a blowout preventer to block any sudden uprush of oil.

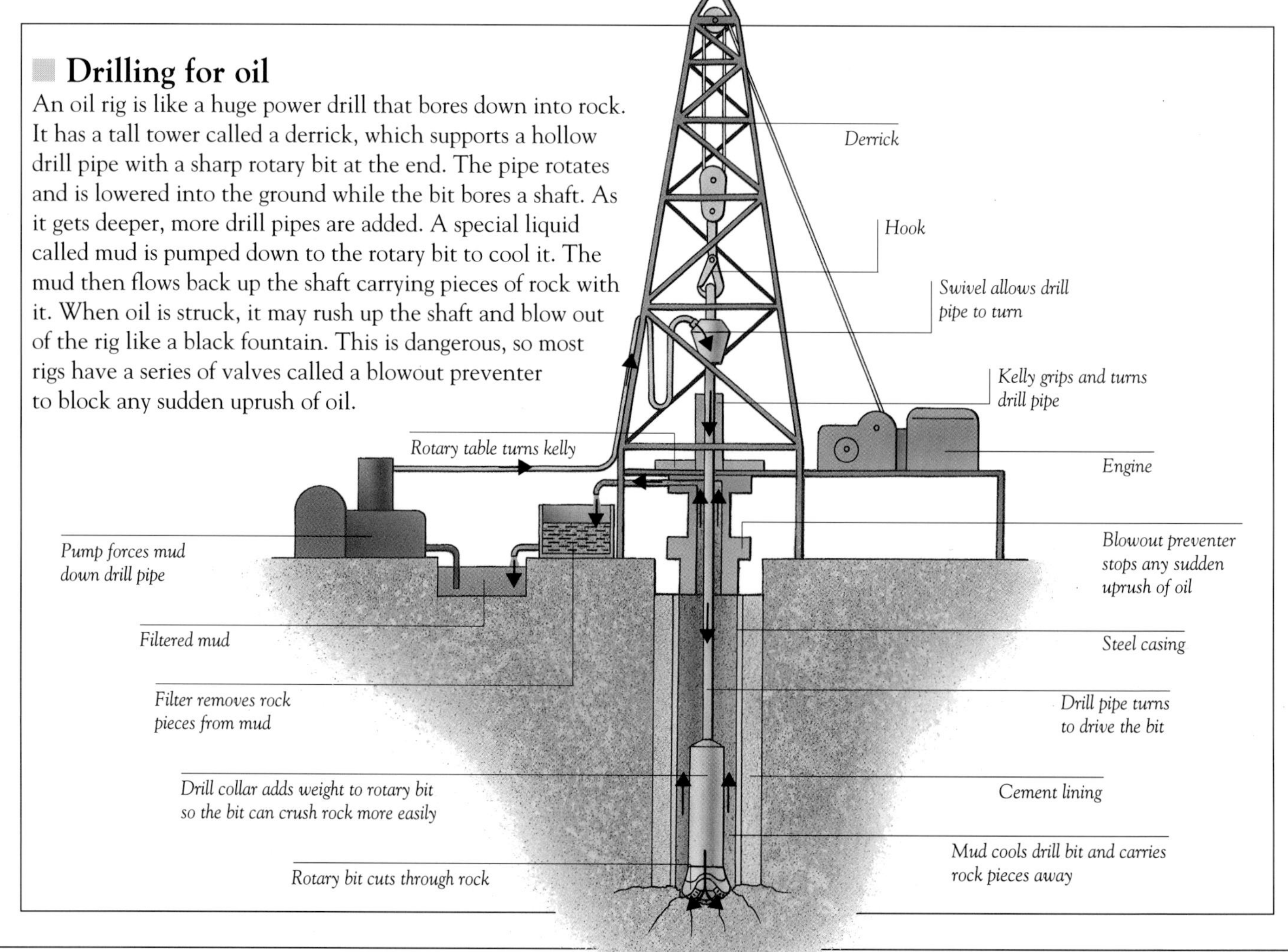

EXPERIMENT

Floating and sinking

Make models of two different kinds of oil rigs used at sea. Both kinds are built in shipyards and towed out to the site by barges. The first kind of rig, called a submersible rig, has massive concrete legs that are hollow inside. The legs are initially full of air and enable the rig to be floated out to the site. There, valves are opened to allow water into the legs. The rig slowly sinks until the legs rest on the seabed, with the platform held above the waves. In deeper sea a semi-submersible rig is used. Hollow tanks under the platform enable it to float above the site. Cables are used to anchor the legs to the seabed so that the rig remains steady.

You Will Need

- *round file*
- *masking tape*
- *string*
- *craft knife*
- *4 binder clips*
- *glue*
- *4 large plastic bottles with tops*
- *foamcore*
- *scissors*
- *enough modeling clay to make a 2 in (5 cm) layer on bottom of tank*
- *large glass tank about 18 in (45 cm) deep*

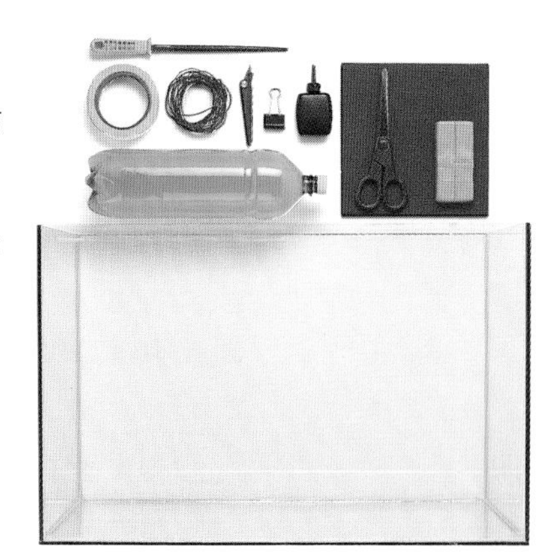

1 Place two pieces of masking tape about 1 in (2.5 cm) apart, just above the base of each bottle. Cut two crosses into each bottle through the tape. File the crosses into ½ in (1.5 cm) holes.

2 Unscrew the bottle tops. Cut a neck-sized hole in each corner of a foamcore square that is big enough to fit over the four bottles. Glue a bottle neck into each hole, then screw on the tops.

3 Put modeling clay and water in the tank to represent the seabed and sea. Add an optional foamcore derrick to the model of a submersible rig. Float the rig in the tank.

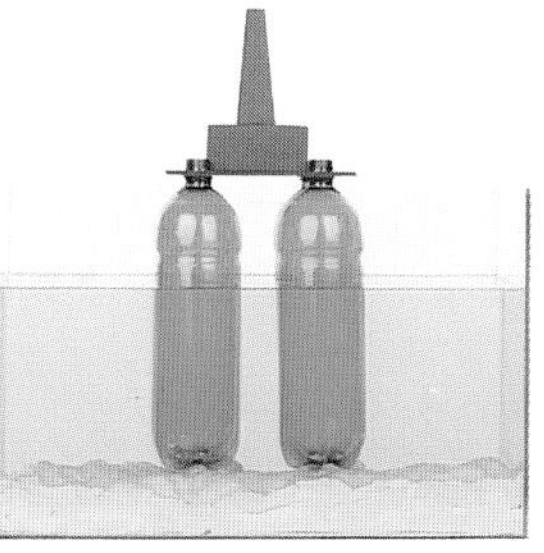

4 Hold the rig steady, then remove all the bottle tops and let go. Watch as the bottles, or legs, fill with water. The rig sinks to the "seabed," supporting the platform above the surface. This represents a submersible rig in place at its site.

5 Now make a model of a semi-submersible rig. Remove the rig from the tank and add water to the tank to represent deeper sea. Tie a binder clip to each bottle using lengths of string passed through the holes in the bottles. Float the rig on the sea. Remove the bottle tops to allow the rig partially to sink, then replace them. Tether the rig in place using the binder clips as anchors.

Oil rigs 2

ONCE A DRILLING RIG has struck oil or gas, it is usually taken down and moved to a new site. Several wells may be drilled into a large deposit. On land, the crude oil or natural gas flows up the well and then into a pipeline that takes it to an oil refinery or gas terminal. At sea, a large kind of oil rig called a production platform is moved to the undersea well or built around the drilling rig. Oil or gas flows up into the production platform, which contains equipment to clean and process it and to bore more wells. The platform also has living quarters for over one hundred workers. After cleaning and processing, the oil or gas is either pumped to shore through a pipeline or shipped in huge tankers.

EXPERIMENT

From rig to refinery

Adult help is advised for this experiment

Crude oil flows from an underground deposit up an oil well and then into a pipeline. The oil may have enough pressure to flow unaided. But it often needs help. Pumps may raise the oil, or pressurized water, steam, or gas may be sent down another shaft to force the oil up the well. In this experiment, a jar of cooking oil represents the deposit, and plastic tubes the shaft, well, and pipeline. Forcing air (a gas) or water down to the oil pushes it up and into the pipeline to an oil tanker or oil refinery, represented by a beaker.

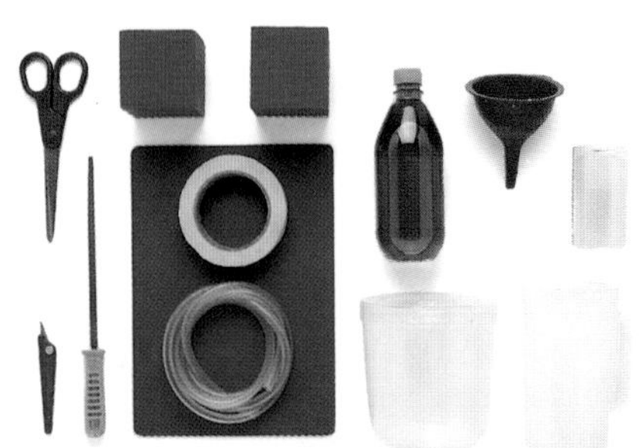

YOU WILL NEED

- *airtight plastic jar with lid* • *beaker* • *two 30-in (75-cm) lengths of flexible plastic tubing* • *cooking oil* • *craft knife* • *wooden blocks* • *funnel* • *cutting mat* • *round file* • *scissors* • *pitcher* • *masking tape*

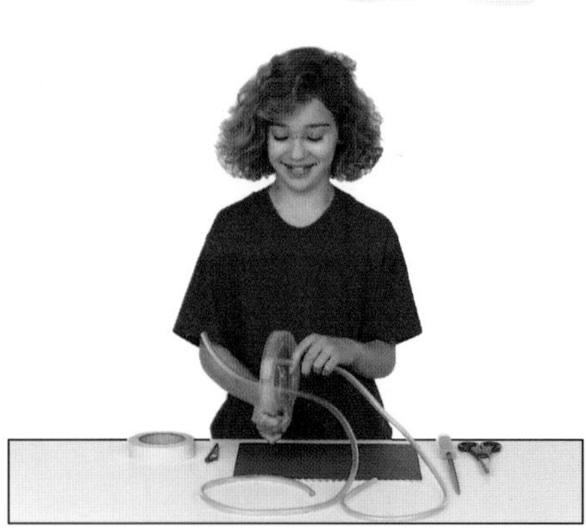

1 STICK MASKING TAPE to the plastic jar lid. Cut two small holes through the tape and lid. Enlarge them with the file until the tube fits snugly. Push a length of tube through each hole.

2 HALF-FILL THE JAR with cooking oil and attach the lid. Rest the beaker on a pile of blocks higher than the lid. Ask a friend to lead one tube into the beaker to form an outlet pipe.

Production

Large offshore production platforms recover oil and gas from deposits beneath the seabed. Many stand on steel towers, some higher than the world's tallest building. Others rest on hollow legs or float on hollow tanks. A platform may inject gas or water into the oil deposit below to raise the oil up the well pipe to the platform.

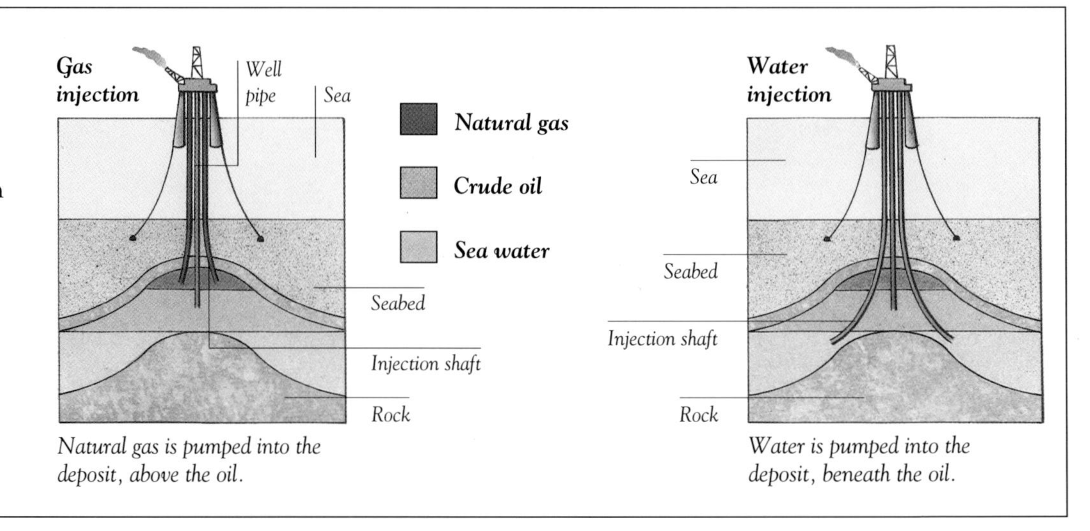

Natural gas is pumped into the deposit, above the oil.

Water is pumped into the deposit, beneath the oil.

3 PUSH THE END of the outlet pipe into the oil in the jar. The other tube (the injection pipe) may be in or out of the oil. While your friend holds the outlet pipe, blow into the injection pipe to force oil into the beaker.

■ DISCOVERY ■

The oil boom

The first oil rig was built by Edwin Drake in Pennsylvania, in 1859. It was an immediate success because crude oil was a good source of kerosene, used to fuel lamps. But the oil industry really took off with the invention of the automobile in 1885. After a boom as prospectors scrambled for oil, the industry was regulated and came to be dominated by large companies and governments.

Oil town
In the early 1900's, prospectors built haphazard clusters of oil rigs over deposits so that each could get as much oil as possible.

4 NOW REARRANGE the apparatus. Pull the outlet pipe up until its end is just below the surface of the lid. Push the end of the injection pipe below the surface of the oil. Attach a funnel to the end of the injection pipe.

5 ARRANGE THE BEAKER as before, raised on wooden blocks above the level of the jar lid. Hold the end of the outlet pipe over the beaker. Ask your friend to hold the funnel above the level of the beaker and slowly pour water into it. Watch as the water sinks beneath the oil and then forces the oil up through the outlet pipe and into the beaker.

HOUSEHOLD MACHINES

Time servers
Timing is important in household machines, many of which have to complete a task in a set amount of time, as does a toaster (above). Many household machines whose activities were once timed by mechanical clocks containing gear wheels (left) now have electronic timing systems.

A HOUSEHOLD MAY POSSESS a whole army of machines, each ready to serve its owners at any moment. These many mechanical and electronic hands not only make light work, but may also keep a home safe and secure. The development of sophisticated control systems allows many of our machines to work on their own under minimal human supervision. In the future, with centralized computer control, household machines will cooperate with each other, creating the "smart" home that looks after itself.

HELP IN THE HOME

HOUSEHOLD MACHINES MAKE OUR LIVES EASIER by taking over hard work, such as washing dirty clothes, as well as helping us with other tasks like cooking and cleaning. In reducing our workload they provide us with more time and energy to spend on such activities as leisure and education, so that we can lead more enjoyable and more fulfilling lives. In addition to as making our homes comfortable, machines can be used to protect and safeguard them.

The origins of some basic household machines and equipment are to be found far in the past. Door locks, instruments that measure time, and devices that weigh quantities accurately were among the very first mechanical inventions of the ancient world. However, many household machines are of more recent origin and date back only a century or so.

Power to the people

Until the nineteenth century, almost all work in the home was done by hand. Hard jobs such as sweeping and washing, intricate tasks such as sewing, and time-consuming chores such as cleaning were done by the people themselves. (The wealthy and privileged paid servants to do these jobs.) Inventors began to develop some devices such as the sewing machine and lawn mower that were powered by hand or foot but used mechanical parts to reduce human effort. Although these machines made tasks easier, household work still had one essential ingredient—muscle power. This was because there was no other ready source of energy that could drive machines in the home. Water and steam powered the industrial revolution, but such aids to work did not extend beyond the factory gate.

At the beginning of the last century the introduction of fuel gas (at first made from coal and later supplied from natural gas deposits) allowed an external energy source to be supplied to individual homes for the first time. Gas was a ready source of bright light and heat for cooking, but it could not be used to drive machines. Powered domestic machines only became possible toward the end of the nineteenth century, when electricity began to be supplied to houses and the first practical electric motors were invented.

Here at last was a clean, silent, and comparatively safe energy source that could be brought directly to the home and could supply as much power as a machine required. Electricity revolutionized the home as machines began to perform all kinds of jobs; electricity is still essential in our homes today.

Effortless existence

Many homes have machines that take all or most of the physical work out of domestic chores. Some gadgets such as electric toothbrushes and can openers are hardly essential, but other machines are important. Washing dirty clothes requires the soiled garments to be agitated thoroughly in water in order to force out the dirt. A washing machine does this job much more effectively than human hands. Sweeping floors and carpets is likewise not very efficient when done with a broom, but a vacuum cleaner easily gets out

***The sundial is a forerunner of the** household clock. The time is shown by the position of the shadow of the gnomon (pointer) on the scale of hours. Sundials are not totally accurate; the sun's apparent motion varies, and corrections must be made to tell the exact time.*

***The vacuum cleaner gets its name** because it uses a partial vacuum, created by a fan, to suck air into the cleaner. Dust and dirt are carried by the air from the floor into the bag. Modern vacuum cleaners work in much the same way as this machine of the early 1900's.*

***Locks are a home's main defence against** burglars and other intruders. The cylinder lock with its serrated key is a common design, often used for cars as well as buildings, but it is not highly secure.*

the dirt and dust, sucking it up and storing it for later disposal.

Eating is a central part of our lives and many different household machines are used in the kitchen. For example, food processors help us to prepare food, and although such technology may not otherwise save much labor, it certainly improves the quality of food that we consume. Stoves of various kinds are to be found in almost all homes. They provide a constant source of heat, usually from gas or electricity, to make cooking easier, faster, and more reliable than it would be over a fire. Refrigerators and freezers, most powered by electricity but some by gas, keep food fresh for a long time. They improve our lives by allowing us to store a wider range of foods than we would otherwise enjoy.

Heating systems and stoves are controlled by *thermostats that keep rooms and ovens at the required temperature. Thermostats make use of temperature-sensing devices, which complete or break electrical circuits to switch the heat on or off. This model thermostat senses temperature-related expansion and contraction of the air inside the glass jar.*

Safety and security

In addition to labor-saving machines, many modern homes have devices that help keep their occupants safe. One of the most important safety devices is the smoke detector, which raises an alarm if smoke from a possible source of fire begins to spread through the house. And if a fire does break out, fire extinguishers give the occupants a chance of fighting the fire before it takes hold and causes severe damage.

Security is mainly provided by locks on doors and windows to provide protection against intruders. Burglar alarms, some of which can even detect an intruder's body heat, provide a warning if someone does succeed in gaining entry.

Keeping control

Many household machines can work on their own, without much human supervision. In order to do this, a machine needs an automatic control system. An oven, for example, has a thermostat that controls the production of heat, maintaining the oven's interior at a constant temperature so that the food cooks evenly and thoroughly. A stove burner or oven may also have a timer so that the cook need only put the food in a pan, press a few buttons, and come back later to find the food ready.

Automatic control is also important in equipment that does more complex operations, such as washing machines or sewing machines. In these devices a set of control routines is programmed into the machine, and the machine follows the routine chosen by the user. Such control systems are increasingly provided by electronic rather than mechanical means, with microchips storing the complex sequences of instructions required to carry out different tasks. Electronics can also extend the range of activities of a particular type of machine. For example, electronic scales are not only faster and more accurate than their mechanical forebears, but some can even do their own calculations. A scale might subtract the weight of a container from the total to give the weight of the contents only—and then display the result in any of several kinds of units.

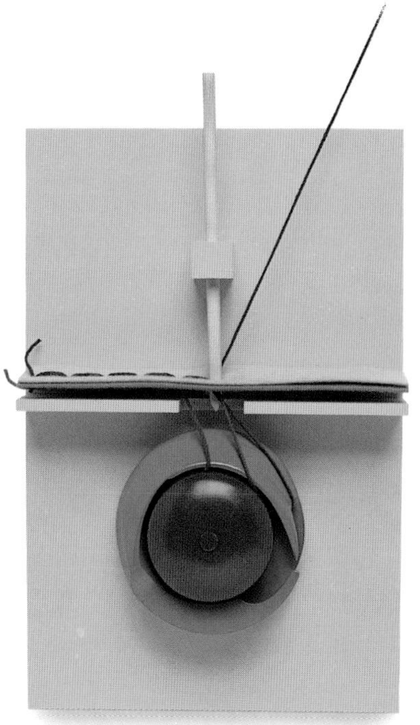

A model of the stitching mechanism *of a sewing machine demonstrates how two lengths of thread are interlocked to form stitches. One length is held by a needle that pierces the fabric, while the other is contained in a spool placed beneath the fabric.*

Machines can lighten labor in the garden as well as *in the home. Lawn mowers, often powered by electric motors or small combustion engines, have rotating blades that cut the grass and bags or containers to hold the clippings. The mechanism can be adjusted to vary the height to which the grass is cut.*

Stoves and toasters

ONE OF THE MOST important machines in any household is the stove. Stoves may use electricity, gas, or solid fuel. Electric and gas burners produce heat at a steady rate, set by the cook. In the oven, which becomes a sealed chamber once the door is shut, heat gradually builds up to the temperature set by the cook. Once this temperature is reached, a thermostat controls the flow of electricity or gas to maintain a steady temperature. Microwave ovens don't need thermostats because they cook the food very quickly. They simply have timer switches and heat settings to control the amount of heat applied to the food. Another common cooking machine is the toaster. Toasters produce heat at a constant rate and have timers to control the amount of heat produced.

Microwave oven

The main advantage of microwave ovens is that they cook food very quickly. In the oven, a device called a magnetron converts electricity into a beam of invisible microwaves, which is reflected by a fan and by the sides of the oven. The microwaves bombard the food from all directions. They penetrate the food and rapidly cook it by heating the moisture inside the food.

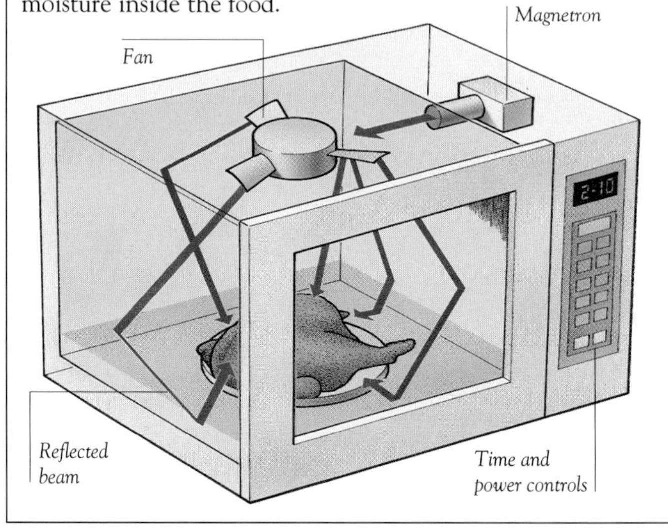

How a toaster works

Pushing down a toaster's lever lowers a spring-loaded rack on which the bread sits. The lever turns on heating elements that toast the bread. Meanwhile, a bimetallic strip bends as it heats up, because one metal in the strip expands more than the other. After a period set on the toaster's timer, the bending strip touches a contact that activates an electromagnet. The magnet attracts a catch that releases the rack, which pops up with the toast and switches off the heating elements.

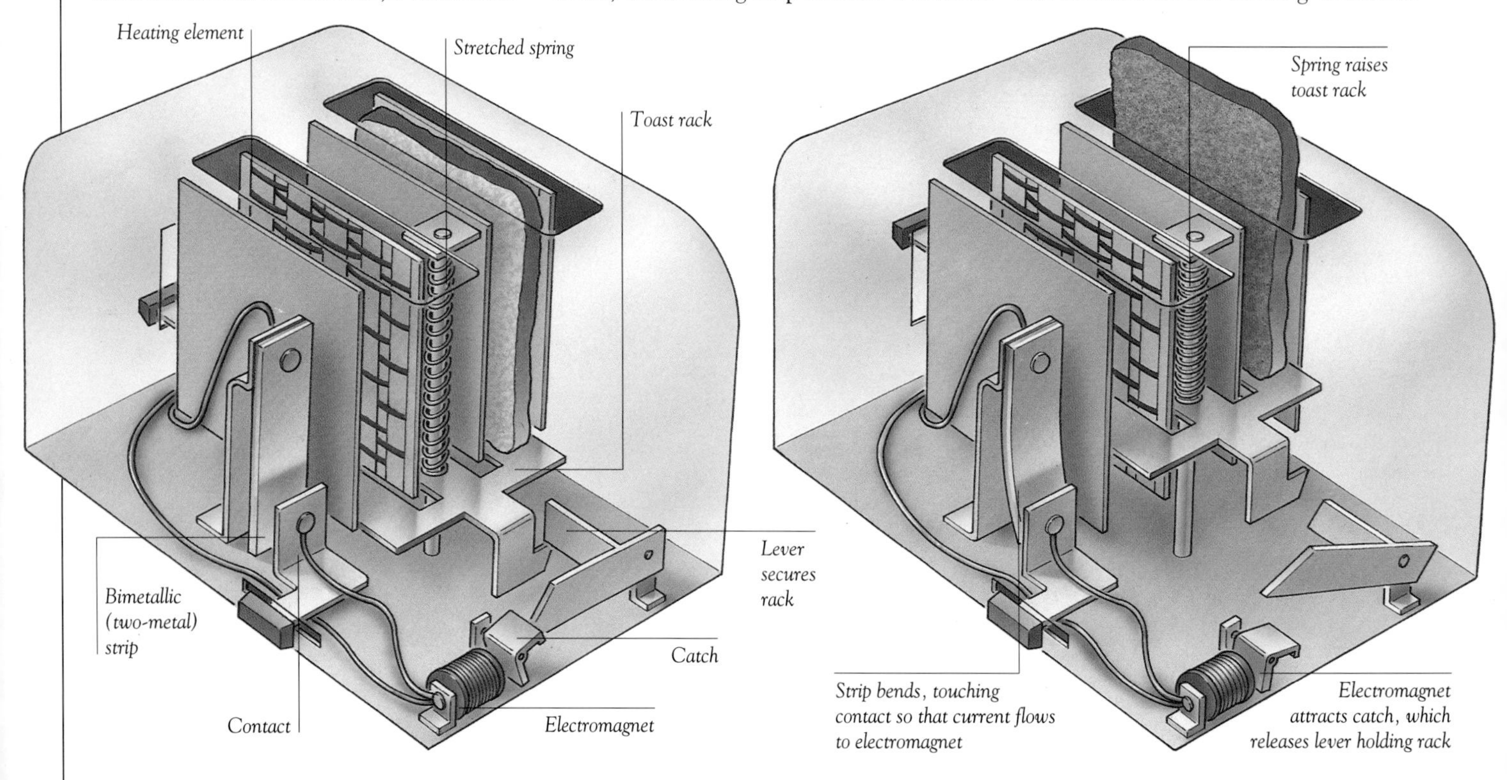

EXPERIMENT

Thermostat

Adult help is advised for this experiment

Stoves have thermostats to control the temperature of the oven. When the oven heats up and exceeds the set temperature, the thermostat turns off the source of heat. The temperature drops slightly, and the thermostat switches the heat on again. By repeatedly switching the heat on and off, the thermostat maintains the set temperature. Build a thermostat to control the temperature in a heated jar.

You Will Need

• desk lamp • drill and 1/8 in (3 mm) bit • two wood strips (arms) 8 in (20 cm) and 10 in (25 cm) long and 1 1/4 in (3 cm) wide • double-sided tape • scissors • screwdriver • insulated wire • heatproof jar • teaspoon • thermometer • long and short screws • 4.5V battery • rubber bands • wire strippers • drilling baseboard • bulb and bulb holder • tape • small piece of modeling clay • vise • balloon

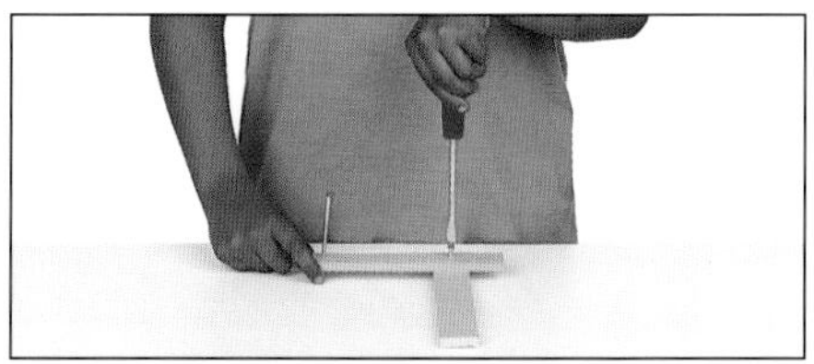

1 Drill screw holes 5 in (13 cm) apart in the short wood arm and a hole 2 in (5 cm) from one end of the long arm. Join the arms at 90° with a short screw. Put a long screw in the remaining hole.

2 Stick the thermometer inside the jar with modeling clay, so that it can be read. Cut the neck and half the body from a balloon. Stretch the remainder tightly over the jar and secure with tape.

3 Fix the long arm to the jar with rubber bands so that the short arm is above the balloon. Tape a teaspoon upside down on the balloon so that the handle reaches the long screw.

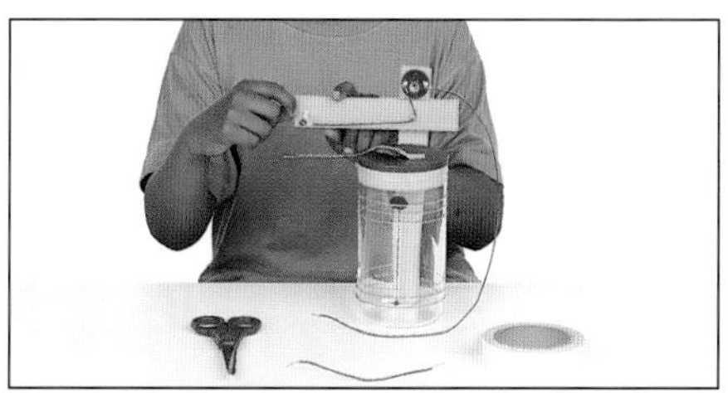

4 Screw the bulb holder to the top end of the long arm. Strip the ends of two 10-in (25-cm) wires. Attach one wire to each bulb terminal. Wind the end of one wire around the long screw.

5 Strip the ends of a third wire, and tape one end to the teaspoon shaft. Connect the other end to a battery terminal. Connect the remaining loose wire to the other battery terminal.

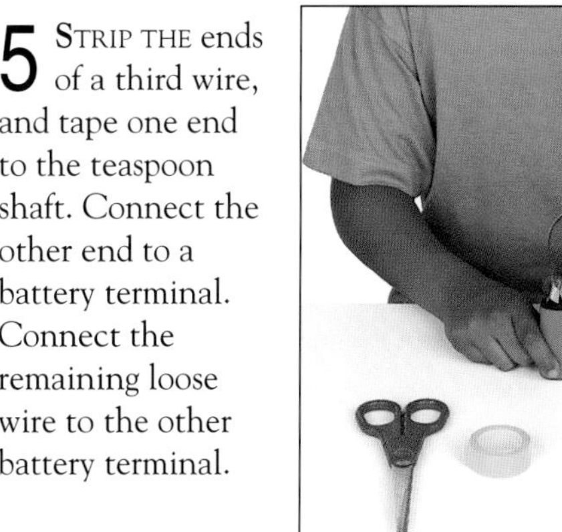

The heat of the lamp expands the air in the jar and pushes out the balloon. This makes the teaspoon handle drop so that the bulb goes out. The lamp is then turned off, causing the air to contract. This sucks in the balloon and raises the handle so that the bulb lights up again, signaling that it is time to turn the lamp on again.

6 Note the temperature in the jar. Shine a desk lamp into the jar. Wait until the temperature in the jar has risen by about 5–10° C (10–20° F). Place the short arm so that the long screw is as close as possible to the teaspoon without touching. Turn the lamp off. Watch the bulb. When it comes on, turn the lamp on. When it goes out, turn the lamp off. Take regular temperature readings. The thermostat should keep the temperature steady to within a few degrees.

Refrigerators and vacuum bottles

AT ROOM TEMPERATURE, hot drinks cool rapidly and ice cubes soon melt. This happens because heat normally flows from a warmer place to a cooler place. Heat flows up through the surface of the hot drink, and through the cup, into the cooler air. The air surrounding the ice cubes is relatively warm, so heat flows from the air into the ice and melts it. This natural flow of heat is reversed by refrigerators and freezers and slowed by vacuum bottles. In a refrigerator or freezer, heat is removed from inside and released at the back. The interior gets cold and remains cold, keeping food fresh and ice cubes solid. A vacuum bottle has an insulating layer that slows down the flow of heat across it, so it can keep drinks hot or cold for a long time.

How a refrigerator works

A refrigerator contains two connected pipes through which a substance known as a refrigerant is pumped. As the refrigerant circulates around the refrigerator, it alternately evaporates (changes from liquid to vapor) and condenses (changes from vapor back to liquid). Liquid refrigerant at high pressure passes from the first pipe (the condenser) through a tiny hole into the second pipe (the evaporator). Here, the pressure of the refrigerant falls, causing it to evaporate and absorb heat from the food compartment. The refrigerant vapor is then pumped at high pressure into the condenser at the back of the refrigerator. The high pressure causes the refrigerant vapor to condense, giving out heat from the back of the refrigerator as it liquefies.

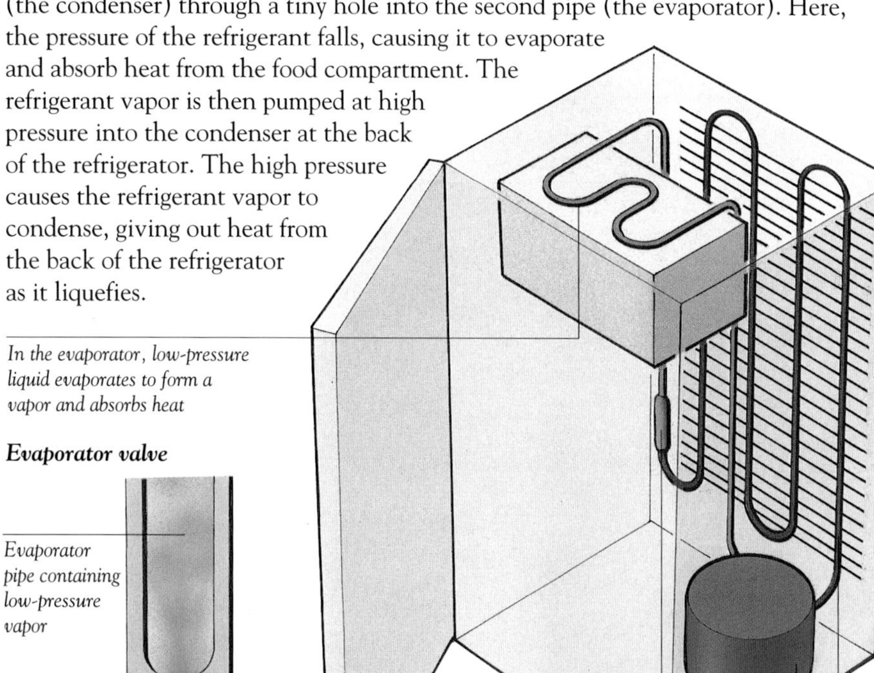

EXPERIMENT

Cooling down

Adult help is advised for this experiment

Liquids must have heat in order to evaporate (change from a liquid to a vapor), and this heat may be taken from the surroundings. A refrigerator uses an evaporating liquid to take heat from the food compartment and cool it down. Compare the cooling effects of two evaporating liquids, rubbing alcohol and water, on your hands. As the liquids evaporate, they take the heat needed to make this change from your hands, which lose heat and get cold.
Never drink or taste rubbing alcohol.

YOU WILL NEED
- *rubbing alcohol*
- *glass of water*
- *2 droppers*

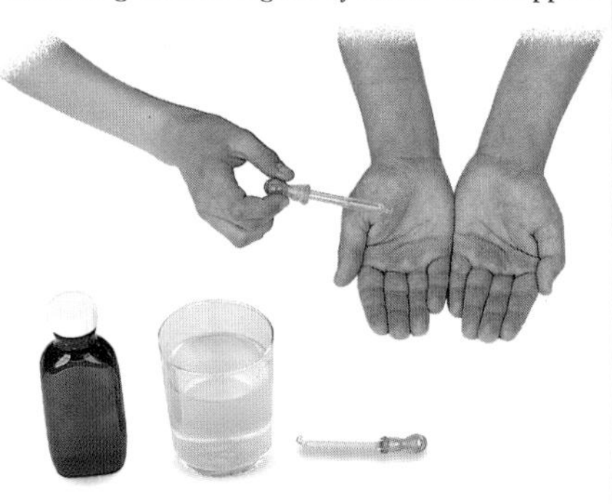

1 PUT A SMALL amount of water in one dropper and a small amount of alcohol in another. Ask a friend to place a drop from each on the center of your open palms.

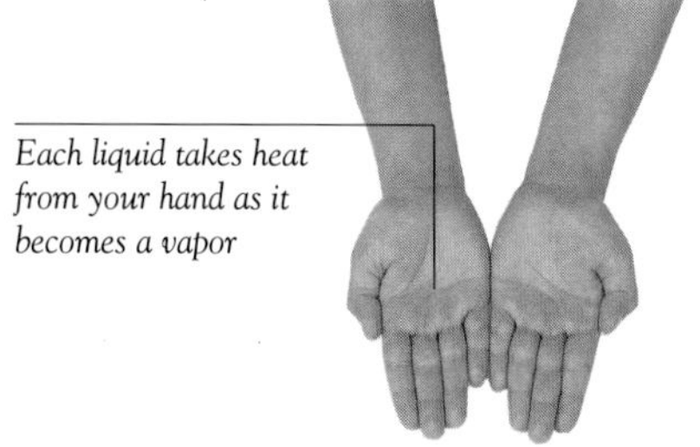

2 BLOW OVER THE liquids on your palms to encourage evaporation. The alcohol will begin to evaporate right away, making your palm feel very cold.

EXPERIMENT

Keeping cool

Make a bottle that keeps things cool. Show how ice cubes take much longer to melt inside the bottle than they do outside. The bottle consists of two containers, one inside the other. The space between them reduces the heat entering the inner container from outside. The containers are covered in shiny foil, and the inner container stands on a cork pad. The foil and cork also help to keep heat out, keeping the bottle cool. A vacuum bottle works in a similar way.

You Will Need

- *1 large and 2 small jars with lids*
- *cork coaster that fits in large jar*
- *aluminum foil*
- *ice cubes*
- *scissors*

1 Wrap foil around the large jar and around one small jar. Fold the ends of the foil together to secure them.

2 Place a cork coaster, or a similar piece of insulating material, in the bottom of the large jar.

3 Place the small foil-wrapped jar on top of the coaster inside the large jar.

4 Put an identical number of ice cubes in each small jar, and put the lids on. Put the lid on the large jar.

5 Wait until the ice in the jar with no foil wrapping has almost completely melted. Then remove the small jar from the large one and compare the amount of melted ice with the amount in the first jar.

Vacuum bottle

When a vacuum bottle is filled, the hot or cold liquid is held in a bottle with double silvered walls. The space between the walls is a vacuum, which means that it is completely empty. Heat passes less easily through a vacuum than through air, and is reflected by the bottle's shiny walls. The bottle has a cork or plastic stopper and is suspended inside an outer container. This also helps to keep heat from entering or leaving the bottle, so the drink stays hot or cold for several hours.

Bottle with double silvered walls with a vacuum in between

Washing machines and dishwashers

CLOTHES and dishes are washed with detergents, which dissolve in the washing water and remove the grease that binds dirt particles to surfaces. The detergent must strike the dirty object with some force to be effective, so washing machines contain a rotating cylinder in which the soiled clothes tumble through the water. The cylinder rotates in alternate directions to prevent the clothes from tangling. Dishwashers clean greasy dishes by spraying them with hot water containing detergent. Both machines may also dry their loads.

■ DISCOVERY ■

Washing without work

People have long tried to lighten the toil of washing clothes. Boiling clothes in a large tub of water was one old method. Hand-powered machines that agitated the wash date back about two centuries. A fully automatic washing machine, however, must perform a complex cycle of operations. This type of machine did not reach the home until the 1930's.

Krauss washing machine
This semi-automatic washing machine was invented in 1923. Clothes were placed in the perforated copper drum, which was spun backward and forward by an electric motor. The water was heated by a fire of coal or wood lit in the base of the machine.

EXPERIMENT

Rotating spray

Adult help is advised for this experiment

Inside a dishwasher are rotating arms that spray hot washing water over the dirty dishes and utensils to reach all parts of the load. The arms are not powered by a motor but by the washing water: as the jets of water spray out, they push the arms around in the opposite direction. Show how this happens by making jets of water shoot from a suspended bottle.

YOU WILL NEED
- *plastic bottle*
- *2 drinking straws*
- *round file with tip narrower than straws*
- *large plastic pan*
- *masking tape*
- *cutting mat*
- *small piece of modeling clay*
- *strong cotton thread*
- *tape*
- *compass*
- *craft knife*
- *pliers*
- *thin wire*
- *scissors*

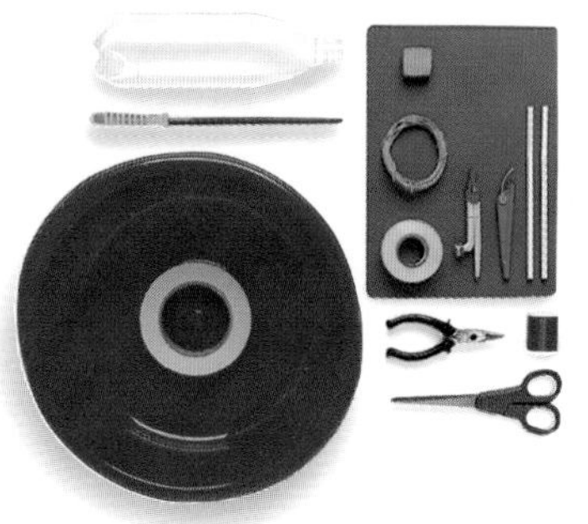

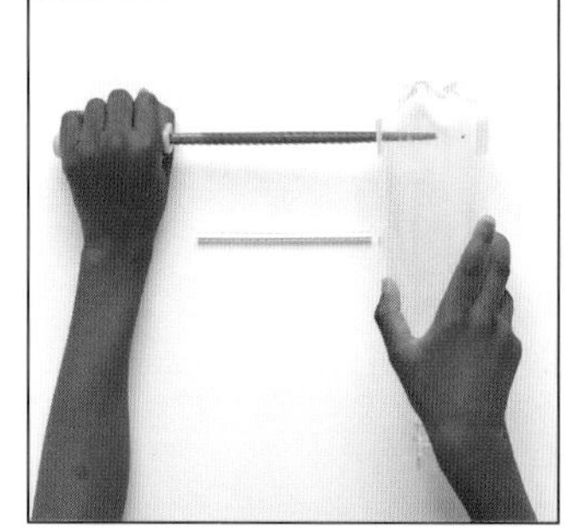

1 STICK TWO patches of masking tape on opposite sides of the bottle, near the base. Cut a 1/8-in (3-mm) cross through each patch into the bottle. Use a file to enlarge holes so the straws fit tightly.

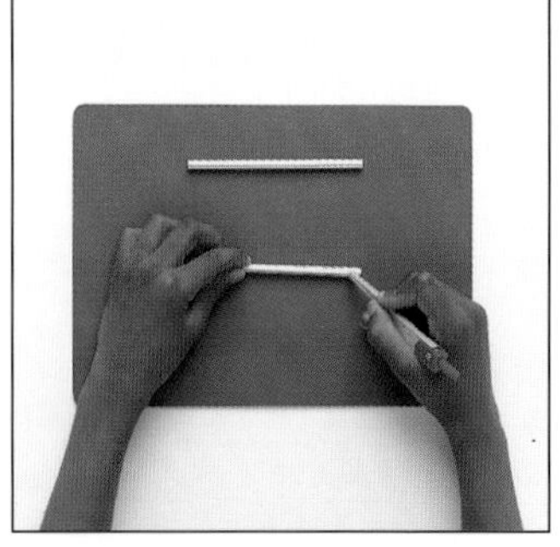

2 CUT TWO straws to 6 in (15 cm) in length. Use the compass to pierce holes at one end of each straw, through one side. Seal this end with modeling clay.

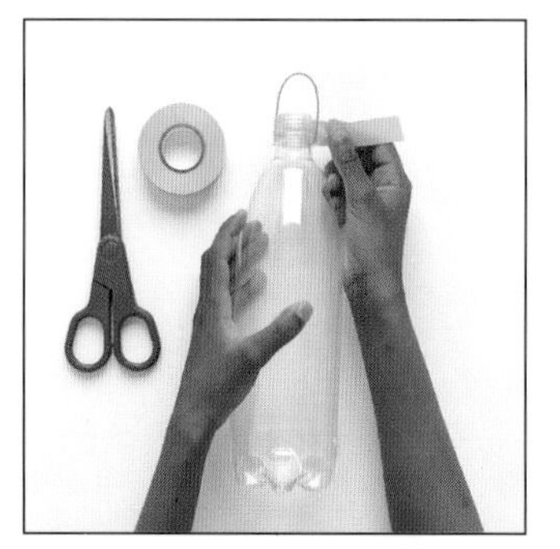

3 CUT A 4-in (10-cm) length of wire. Bend it over and tape it to the neck of the bottle to make a loop. Tie a 6 in (15 cm) length of thread to the wire loop.

4 PUSH THE open end of a straw into each hole in the bottle. The rows of holes in the straws should be pointing horizontally in opposite directions. Immerse the bottle in a pan of water to fill it, then lift it out by the thread.

EXPERIMENT

Spinning the water out

When a washing machine has finished washing and rinsing the clothes, the cylinder starts to spin rapidly. The wet clothes are pressed against the wall of the cylinder, and most of the water in them flies out through holes in the cylinder wall. After spin-drying, the clothes are still slightly wet, but they dry quickly when hung up. See how much water you can get out of a wet cloth by spinning it.

YOU WILL NEED
- *large watertight plastic bag*
- *knitting needle or dowel*
- *net bag*
- *scissors*
- *pitcher of water*
- *12 in (30 cm) square piece of cotton cloth*
- *red and yellow tape*
- *18-in (45-cm) length of string*

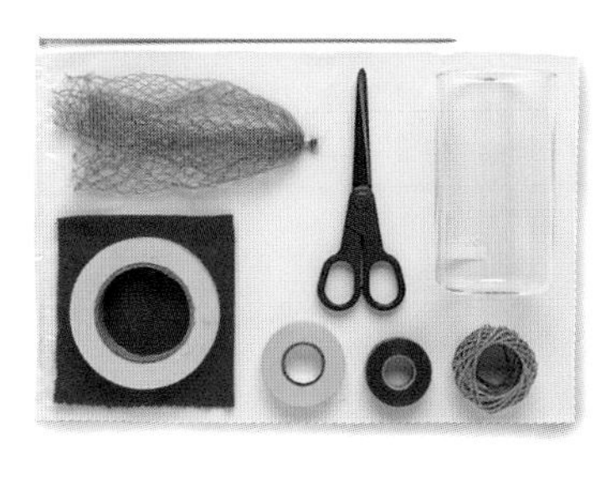

1 HALF-FILL the pitcher with water. Mark the level of the water with a strip of yellow tape.

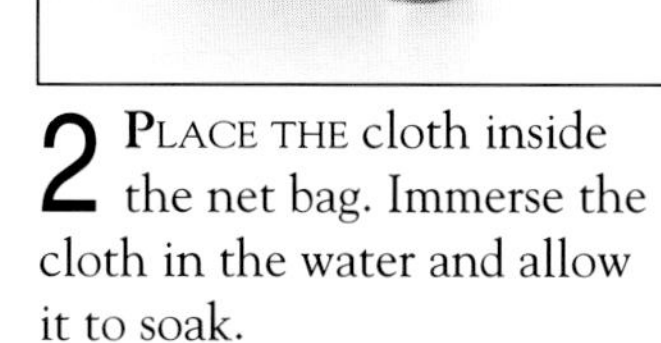

2 PLACE THE cloth inside the net bag. Immerse the cloth in the water and allow it to soak.

3 PASS THE knitting needle through the net bag. Hang the bag over the water in the pitcher. Let the wet cloth drip for 30 minutes. Mark the new water level with red tape.

4 TAKE THE bag from the pitcher. Make two holes at the top of the plastic bag. Put the net bag inside the plastic bag. Tie their tops together with the string.

5 FIND A clear area outdoors, away from any obstructions. Grasp the end of the string and carefully swing the bags in a vertical circle for about two minutes.

6 UNTIE THE bags, keeping the cloth and collected water separate. Pour the water into the pitcher. Note how much water has come out of the spin-dried cloth.

Weighing machines

MEASURING INSTRUMENTS often help us with tasks in the home. Weighing machines and scales are used to weigh quantities in cooking and to check body weight. Basic scales or balances compare the weight of an object, or a quantity of something, with standard weights. More complex spring balances and bathroom scales contain a spring that stretches in proportion to the weight of the object being weighed. The spring moves a pointer on a graduated scale or a dial. An electronic weighing machine contains a strain gauge, which produces an electric signal proportional to the weight. A microchip converts the signal into a number for display.

■ DISCOVERY ■

Ancient but accurate

The balance, invented in Egypt some 5,500 years ago, was the first accurate weighing machine. It consisted of a rod pivoted at its center with a pan hanging from each end, and was used with a system of standard weights. An object was weighed by placing it in one pan, then adding weights to the other pan until the pans balanced. Egyptian balances had to be very precise because they were used to weigh a precious commodity—gold.

Early Egyptian stone weights

Egyptian metal weights

EXPERIMENT

Sensitive microbalance

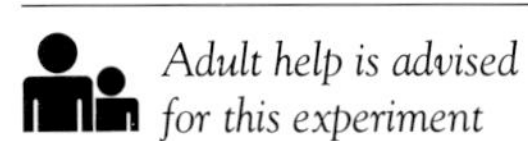

Adult help is advised for this experiment

Make a balance that can weigh objects so light that they seem to have no weight at all, such as a piece of thread. The object balances against a heavier screw, but the screw is near the pivot of the balance. A very light object farther from the pivot can therefore tilt the balance, and its weight is read against a scale.

YOU WILL NEED

• 2 poster board pieces 2 x 1 in (5 x 2.5 cm) • tweezers • scissors • poster board sheet 6 x 2 in (15 x 5 cm) • thin stiff wire • screw to fit inside straw • piece of foamcore 10 x 1 x 3/8 in (20 x 2.5 x 1 cm) • pen • drinking straw • double-sided tape • pliers • 3/8-in (1-cm) cube modeling clay • light objects such as thread, hair, or paper pieces • ruler

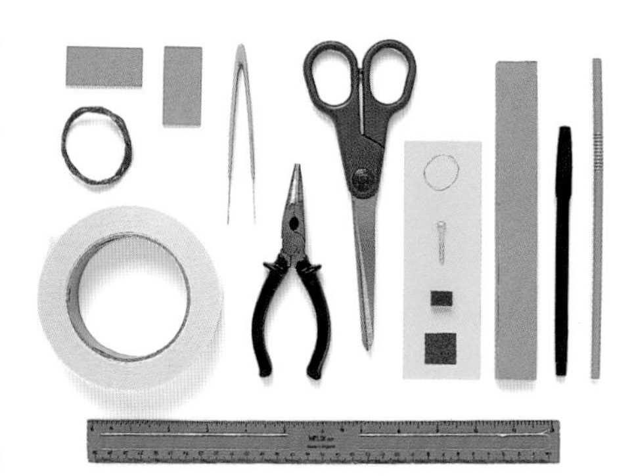

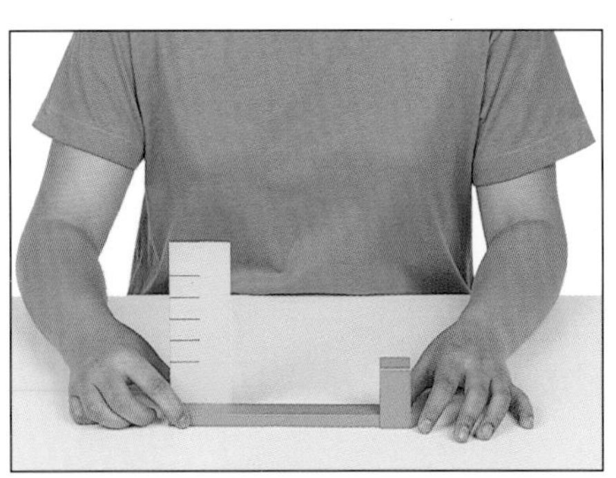

1 DRAW A SCALE of marks spaced 3/4 in (2 cm) apart on the large sheet of poster board. Use double-sided tape to attach the scale to one end of the foamcore base as shown. Glue the small pieces of poster board on either side of the other end of the base to make a support for the pivot.

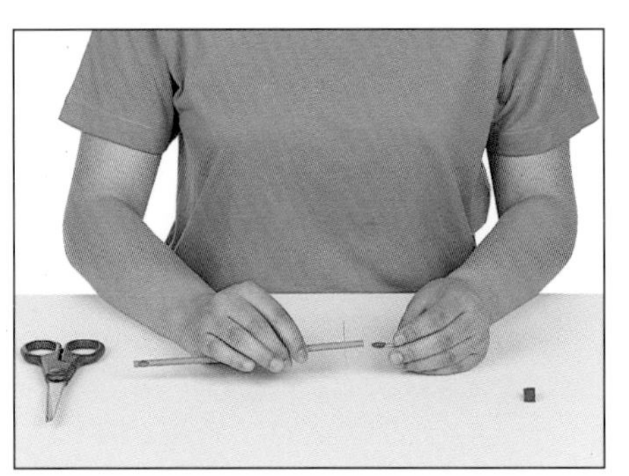

2 CUT A NOTCH IN the straw, 3/4 in (2 cm) from one end. Using the pliers, push a 2-in (5-cm) wire pivot crosswise through the straw, 3/4 in (2 cm) from the other end. Push the screw into this end of the straw, and secure with modeling clay. Pivot the straw on the base by resting the wire on the support.

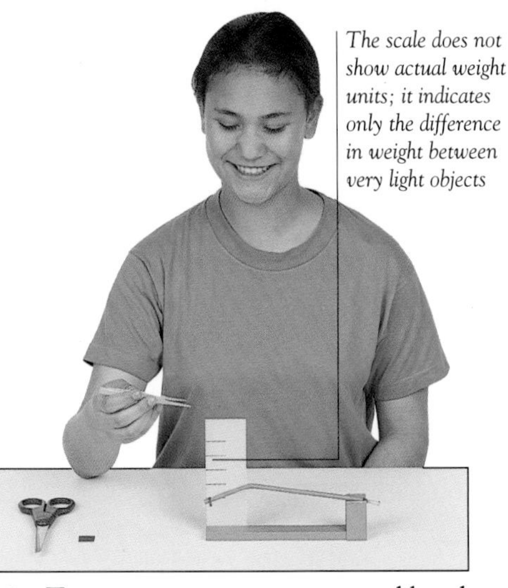

The scale does not show actual weight units; it indicates only the difference in weight between very light objects

3 TURN THE SCREW in or out and bend the straw so that it balances pointing at the top of the scale. Place the object to be weighed in the notch on the straw.

EXPERIMENT

Spring balance

Adult help is advised for this experiment

Make a balance that you can use for weighing ingredients and other amounts. It works because a spring stretches in a regular way when it is pulled: a weight makes the spring extend by a particular amount, twice the weight stretches it twice as much, and so on.

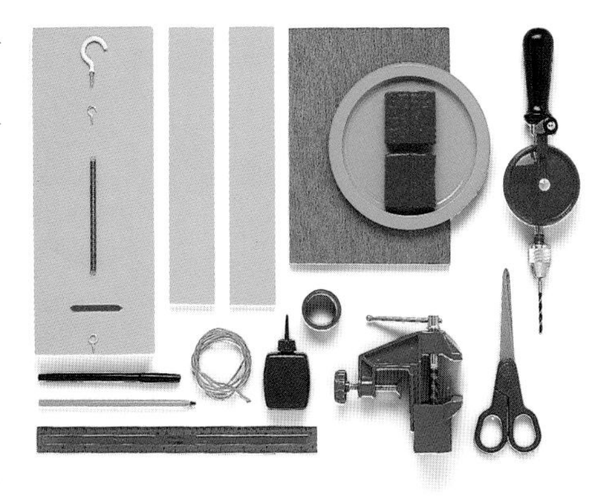

YOU WILL NEED

- *piece of wood 14 x 5 x 3/4 in (35 x 13 x 2 cm)*
- *large screw hook*
- *small screw hook*
- *foamcore strip 12 x 2 x 3/16 in (30 x 5 x 0.5 cm)*
- *drilling board*
- *small plastic saucer*
- *set of known weights, each about 2 oz (50 g)*
- *drill and bits to fit hooks and screw eye*
- *weak extension spring, about 5 in (13 cm) long*
- *cardboard strip*
- *small screw eye*
- *pen*
- *pencil*
- *ruler*
- *string*
- *glue*
- *tape*
- *vise*
- *scissors*

1 DRILL A HOLE 1 in (2.5 cm) from each end of the wood, along the long midline. One hole is for the screw eye, the other for the small screw hook. Screw the large hook into the short edge above the small hook.

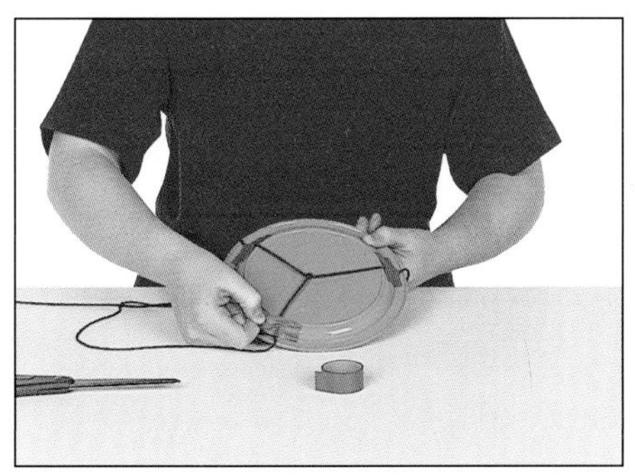

2 TIE THREE 16-in (40-cm) lengths of string together at one end, and tie them to a 6-in (15-cm) length at the other. Cradle the saucer inside the three strings with the short string at the top. Secure the cradle with tape.

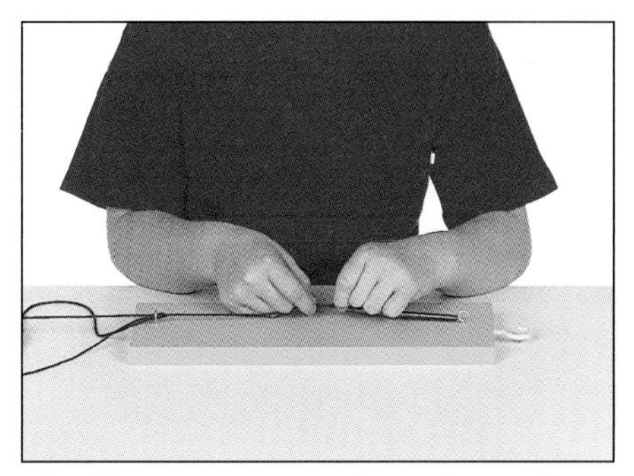

3 HANG THE SPRING from the small screw hook. Pass the free end of the short string from the cradle through the screw eye, and tie it to the free end of the spring. Secure the string with tape if necessary.

4 GLUE A foamcore strip on one side of the spring. Fix a cardboard pointer to the bottom of the spring. Hold the balance by the large hook. Mark a zero on the strip by the pointer when the saucer is empty. Add a series of known weights (in ounces or grams) to the saucer. Mark the pointer position for each weight.

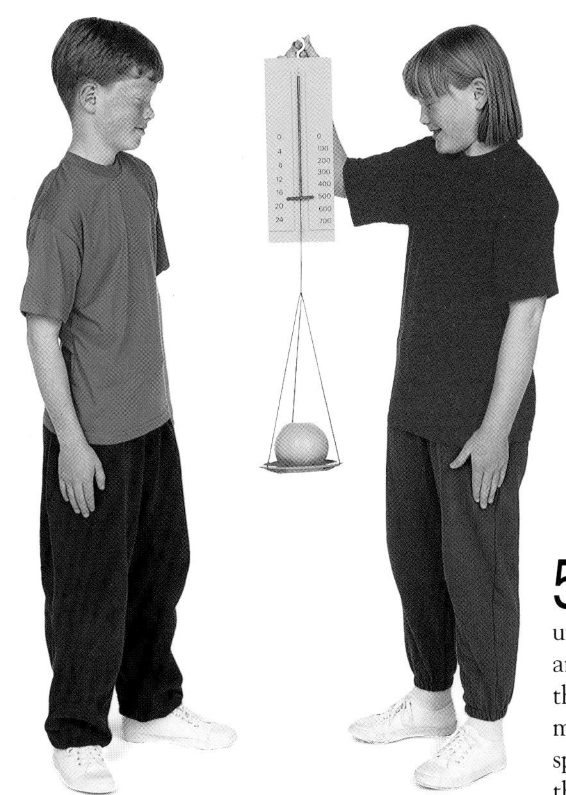

5 THE MARKS FORM a graduated scale of weight units. Weigh objects or amounts by placing them on the saucer. The pointer moves down the scale as the spring stretches and indicates the weight.

Fire extinguishers and aerosol sprays

A FIRE EXTINGUISHER can quickly put out a small fire before it spreads. The extinguisher nozzle is aimed at the fire, and the extinguisher is triggered by a lever at the top. In some extinguishers, pressing the lever causes a jet of water to shoot out of the nozzle. The water smothers the burning material by excluding air that is needed for burning, and cools the fire. Water extinguishes burning materials such as paper or wood but is no use on electrical fires or burning liquids or gases. For these fires, extinguishers containing foam, powder, or gas can be used. Like water, these substances prevent more burning and smother a fire. They are propelled from the extinguisher under their own pressure or by a high-pressure gas. Aerosol cans work in basically the same way as fire extinguishers. They are convenient for spraying many different substances, including paint, deodorant, and insecticide.

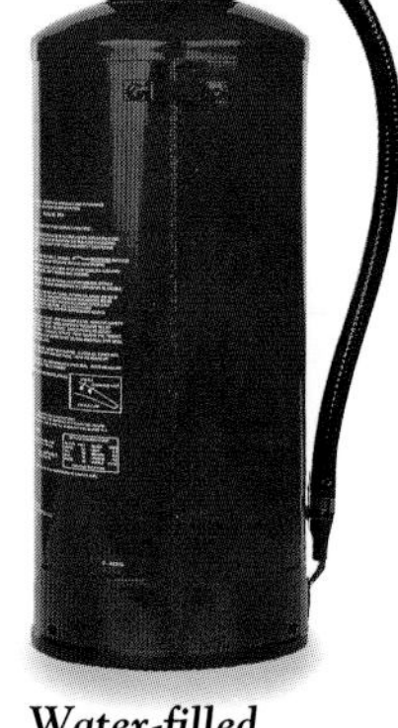

Water-filled fire extinguisher

How a fire extinguisher works

This extinguisher is filled with water and contains a cartridge of high-pressure gas. When the lever on top is pressed, it pushes a pin into the cartridge, releasing gas into the upper part of the cylinder. The gas pushes the water up the discharge tube and out of the nozzle. Other extinguishers contain an extinguishing gas or liquid that is already under pressure, so they do not require a separate gas cartridge.

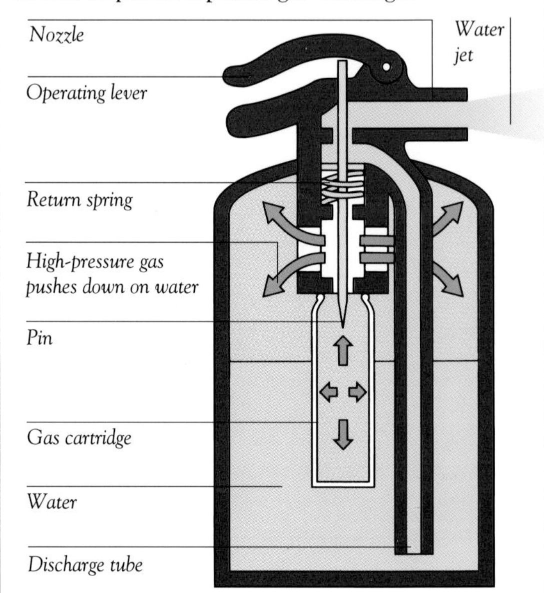

How an aerosol spray works

An aerosol can contains a mixture of the liquid product to be sprayed, such as paint, and a propellant. The propellant is mainly in liquid form but evaporates readily. Its evaporation creates high-pressure vapor above the liquid mixture. When the nozzle is pressed, it opens a valve. The pressure of the vapor can then force the liquid mixture up a discharge tube, through the valve, and out of the nozzle. The liquid propellant in the mixture immediately evaporates, breaking the product up into tiny droplets and forming a spray. Aerosol sprays once used CFCs (chlorofluorocarbons) as propellants. CFCs damage the ozone layer in the atmosphere, so "ozone-friendly" or "CFC-free" aerosols contain a different propellant, usually butane or propane. These propellants are flammable, so aerosol sprays must not be used near flames or hot surfaces.

Ready to spray

Propellant vapor occupies the top half of the can, pushing down on the liquid propellant in the base of the can.

Valve

Return spring

Mixture of liquid product and liquid propellant

Making a spray

Pressing down on the nozzle allows the pressure of the vapor to force the product up the tube and out of the nozzle as a spray.

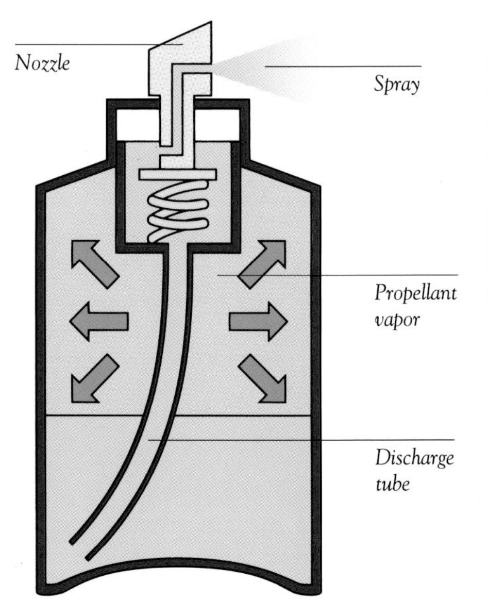

EXPERIMENT

Model fire extinguisher

Adult help is advised for this experiment

Snuff out the flame of a candle with a simple and safe fire extinguisher. It uses two household chemicals to produce carbon dioxide gas. This gas is heavier than air, and it will cover the candle flame and keep out the air so that the candle stops burning. You can also use this extinguisher to produce a foam that will smother the flame. This experiment demonstrates how real fire extinguishers put out a fire. Do not try it with a real fire or any other kind of burning material.

YOU WILL NEED
- *heatproof glass jar*
- *piece of cardboard*
- *modeling clay*
- *baking soda*
- *matches* • *taper*
- *bendable drinking straw* • *bulldog clip*
- *vinegar* • *candle*
- *small plastic soda bottle*

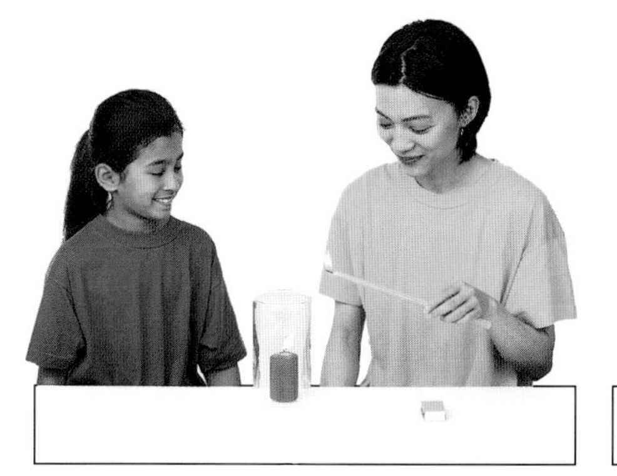

1 PLACE THE CANDLE in the bottom of the heatproof jar. The jar should be about twice as tall as the candle. Ask an adult to help you light the candle with the taper.

2 MOLD SOME modeling clay around the drinking straw, 1 in (2.5 cm) from one end, to make an airtight stopper for the plastic bottle. Clamp the longer part of the straw closed with the clip.

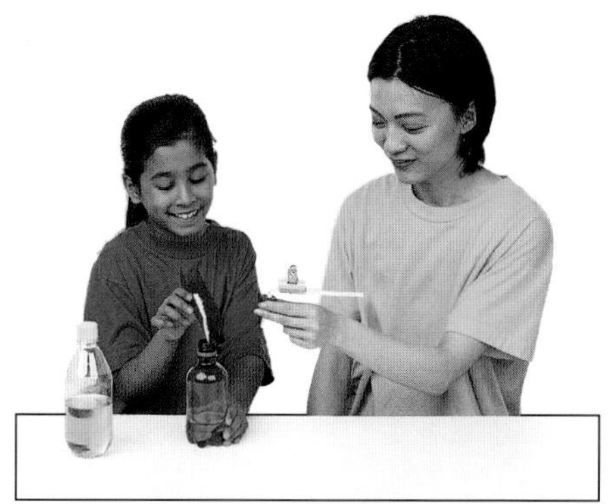

3 POUR 2 IN (5 CM) of vinegar into the bottle. Put a tablespoon of baking soda into folded cardboard, and pour in. Replace the stopper at once, with the shorter end of the straw inside.

4 POINT THE END of the straw into the mouth of the jar and release the clip. The gas produced as the chemicals mix bubbles through the liquid, producing a small amount of foam, which remains in the bottle. The gas rushes out and extinguishes the candle flame.

5 NOW REPEAT THE experiment, but use double the quantities of vinegar and baking soda. This time, enough foam should be produced to reach the bottom of the straw inside the bottle. The foaming liquid rushes out of the straw and extinguishes the flame.

Burglar alarms and smoke detectors

ALARMS PLAY A CRUCIAL ROLE in protecting our homes and even our lives. An alarm basically consists of a sensor connected to a warning device. Burglar alarms have sensors that can detect the opening of a door or window, the pressure of a footstep on the floor, the body heat of an intruder, or even the movement of a person in a room. A fire alarm sounds when it detects the heat from a fire, but a smoke alarm warns about a fire at an earlier stage. It detects the build-up of smoke particles in the air, even if no flames have formed. Warning devices ring bells, sound sirens, and flash lights, and some alarm systems can automatically warn emergency services that something is wrong.

Smoke detector

Many homes have smoke detectors on their ceilings. These detect tiny smoke particles coming from a smoldering object before the smoke builds up and becomes visible. The alarm emits a piercing shriek that is loud enough to rouse sleeping occupants, who might otherwise wake up to find their home ablaze. A chamber in the detector contains a source of radioactivity (too weak to be a hazard) that creates a low electric current in the air of the chamber. Smoke particles interrupt the electric current, triggering the alarm.

EXPERIMENT

Burglar alarm

Make a working burglar alarm that warns you when a door opens. It uses a magnetic switch as a sensor, and a buzzer and LED as warning devices. If the door opens, a magnet on the door moves away from the switch on the doorframe. The switch then triggers the alarm, sounding a buzzer and lighting an LED. The intruder cannot stop the alarm by closing the door; only you can do this using a reset button. *Read pages 10–11 before doing this experiment.*

YOU WILL NEED

• breadboard with base • 9V battery and connector • wire strippers • breadboard wire • normally open magnetic switch with magnet • scissors • double-sided tape • 9V buzzer • LED • NPN transistor (BC441 or equivalent) • 4011B NAND chip • normally open momentary SPST switch • one 220R and two 10K resistors

220R resistors
F33–F36

10K resistors
A10–B10 K23–L23

Wires
A9–B9 A31–B31 C12–C14 D13–D32
G6–E10 E11–E13 E15–E23 F39–L39
F23–G23 K4–L4 K15–L15

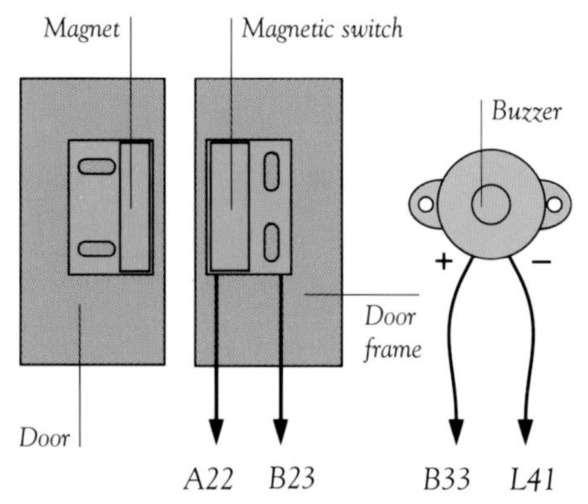

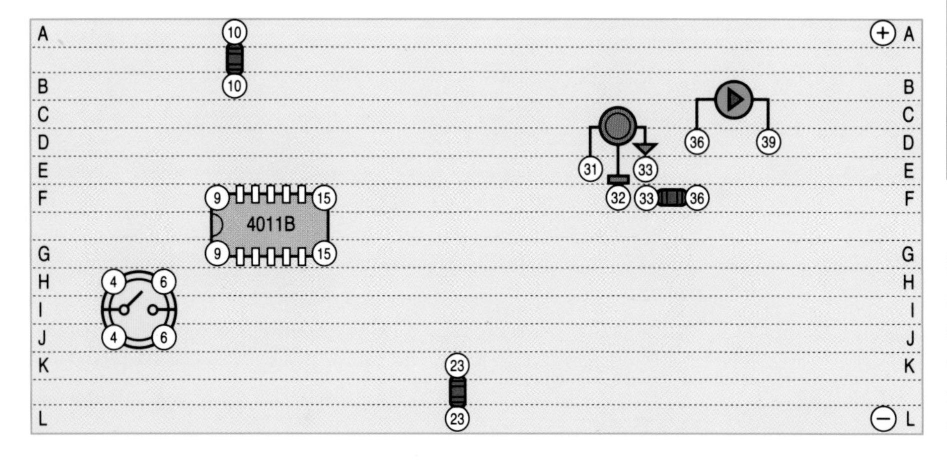

1 ASSEMBLE THE CIRCUIT on the breadboard without connecting the battery. Attach the magnetic switch to a door frame with double-sided tape. Stick the magnet to the door so it is aligned with and very close to the switch when the door is shut. Connect the switch wires back to the circuit board.

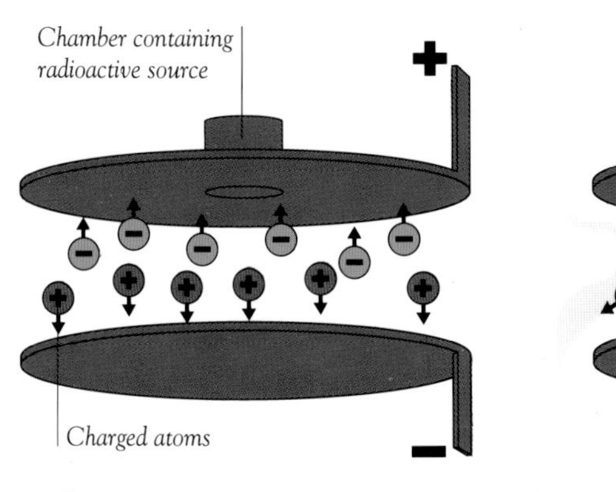

Alarm ready
The radioactive source gives electric charges to atoms of gas in the air. The charged atoms move toward the electrodes and a current flows.

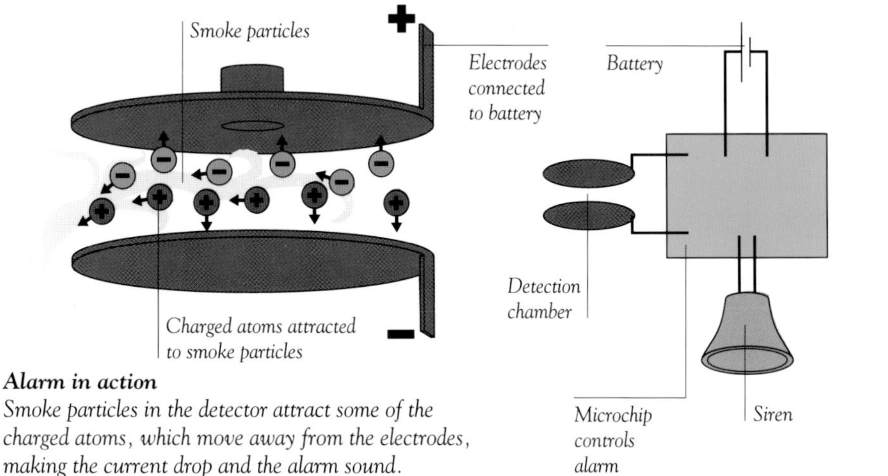

Alarm in action
Smoke particles in the detector attract some of the charged atoms, which move away from the electrodes, making the current drop and the alarm sound.

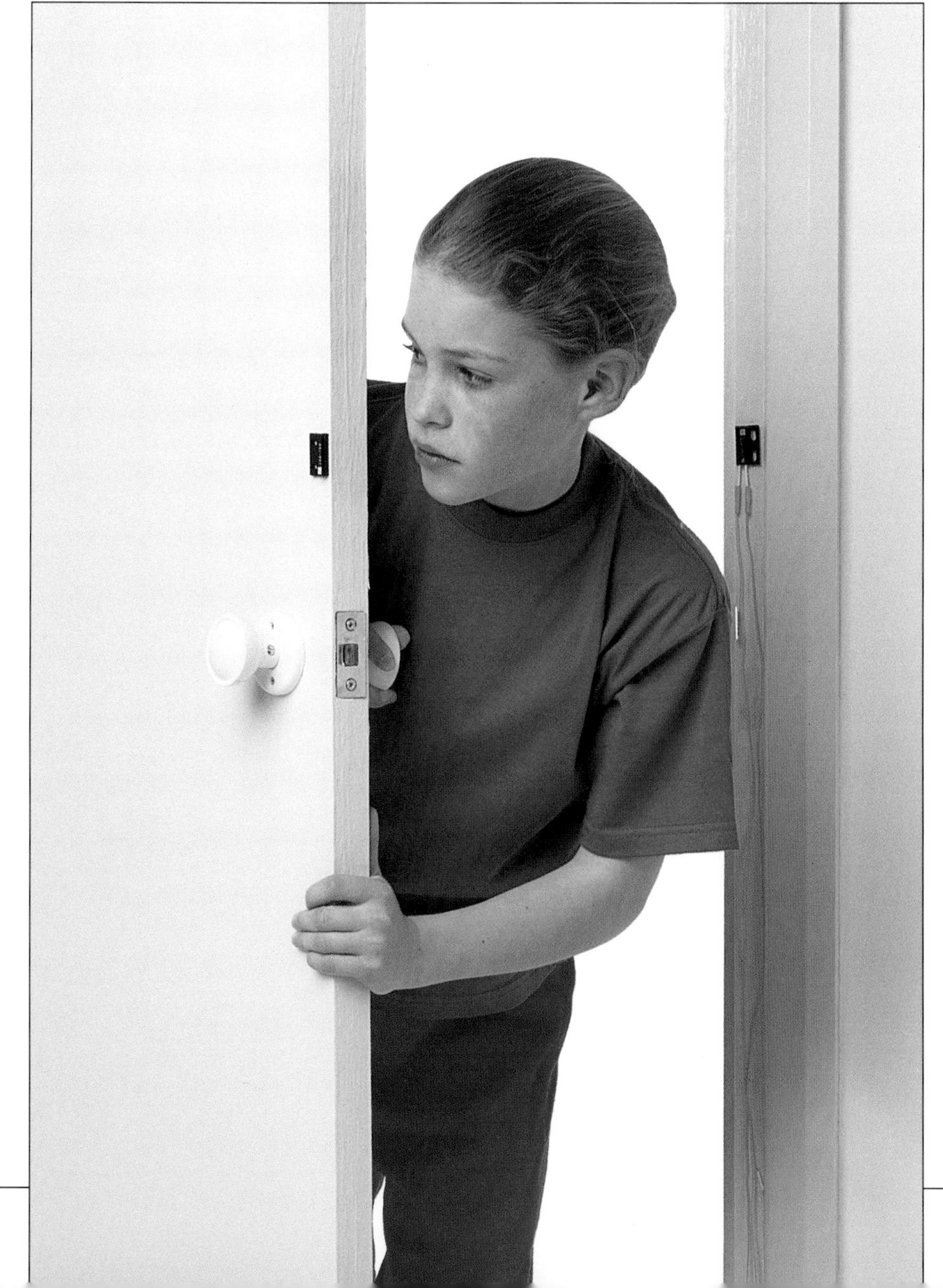

2 SHUT THE DOOR and connect the battery. The magnet closes the switch, arming the circuit. Ask a friend to open the door. This causes the switch to open, sending a signal to the NAND chip to turn on the buzzer and LED. Like a real one, the alarm also sounds if a circuit wire is cut.

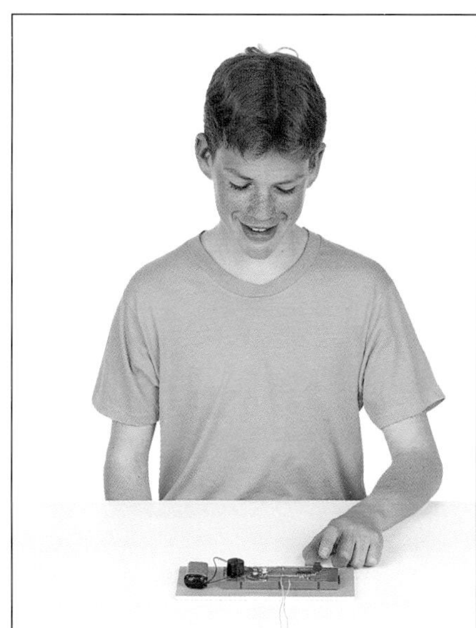

3 SHUTTING THE DOOR does not turn off the alarm. You must shut the door and then press the reset button (the switch on the breadboard), which deactivates the alarm by turning off the buzzer and LED. Pressing the reset button sends a second signal to the chip, which cuts off the current to the buzzer and LED.

Locks

THE NEED TO SECURE valuables led to the development of one of the earliest inventions, the lock, which dates back about 4,000 years to ancient Egypt. Egyptian locks used wooden keys with projections that lifted pins inside the lock. Today's cylinder locks use metal keys that work in the same way. Each key has its own pattern of projections, which pushes up a set of pins in the lock to free a bolt. Losing a key can require replacing the whole lock, a problem avoided by combination locks. Their keys consist of combinations of numbers or letters that the owner memorizes. Entering the correct combination on a set of dials or a keypad frees the bolt. The combination can be altered without changing the whole lock.

Electronic locks

Many locks now use electronic rather than mechanical devices to free door bolts. Some office locks require a code number to be entered on a keypad at the door. The lock matches this number to a number in its memory and operates the bolt if the two numbers are the same. Hotel guests may use a card key that has a magnetic strip. The strip is programmed with a code number recognized only by the electronic lock of the guest's room. Some car locks use remote control units instead of keys. The driver presses a switch on the portable unit, which sends out an ultrasonic code signal that operates the car lock. A door lock can even send out a signal and open itself if an approaching person is carrying the correct unit.

Card with magnetic strip

Cylinder locks

A cylinder lock is often called a Yale lock after its inventor, Linus Yale. This lock uses a flat key with serrations along one edge. The key fits into a cylinder inside the lock. The cylinder has a cam at one end that grips a bolt. Turning the key rotates the cylinder so that the cam pulls back the bolt and the door opens. However, the cylinder cannot rotate unless the right key is inserted. The cylinder and body of the lock have a set of holes that each contain a pair of pins and a spring. The pins block the gap between the cylinder and the body of the lock, preventing the cylinder from rotating. When the key is inserted, its serrated edge pushes up the pins by different amounts. This causes the gap in each pair of pins to line up with the gap between the cylinder and the body of the lock. This frees the cylinder so that it can be turned by the key to pull back the bolt.

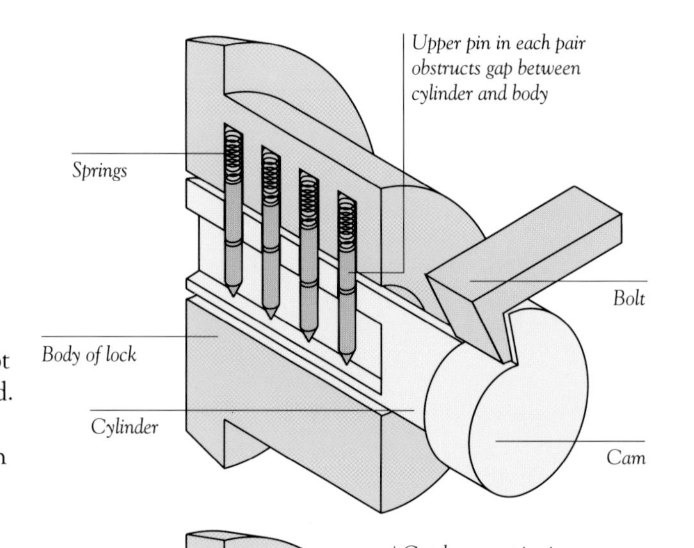

Lock closed
The springs push each pair of pins to the base of the cylinder. The pairs are of different lengths; the upper pins hold the cylinder and lock body together, locking the cylinder and cam so that they cannot turn. The bolt is held in the door frame, which prevents the door from being opened.

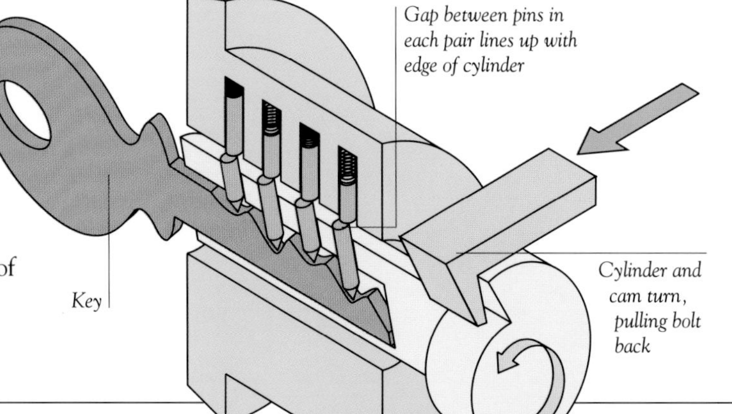

Lock open
Inserting a key raises the pins by different amounts. The correct key causes the gap between the pins in each pair to line up with the gap between the cylinder and the lock body so that the cylinder is freed. Turning the key turns the cylinder and cam, which pulls back the bolt and opens the door.

EXPERIMENT

Combination lock

Build a model safe with a combination lock that uses two numerals, each from a range of 0 to 9. There are 100 possible combinations, but only one of these will open the safe.

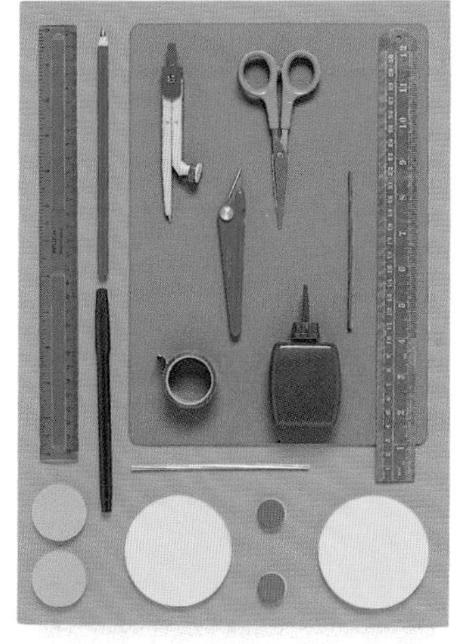

You Will Need

● $\frac{3}{16}$-in (5-mm) *thick foamcore* ● *ruler* ● *pencil* ● *pen* ● *cutting surface* ● *compass* ● *tape* ● *craft knife* ● *scissors* ● *glue* ● *wooden skewer* ● *steel ruler* ● *drinking straw* ● *foamcore disks, two each of diameters 1 in (2 cm), 2 in (4 cm), and 4 in (8 cm)*

1 Cut a 10-in (25-cm) square out of foamcore to make a base. Cut two pairs of rectangles measuring 10 x 5 in (25 x 12 cm) and $9\frac{5}{8}$ x 5 in (24 x 12 cm) and glue them on top of the base to make an open box.

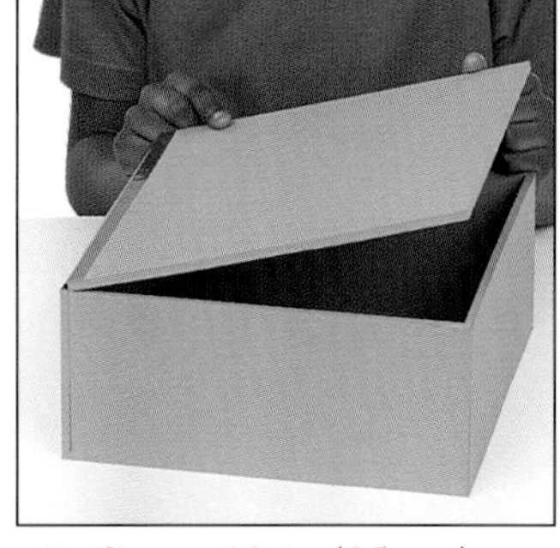

2 Cut a 10-in (25-cm) square foamcore lid. Make a hole in which a straw turns freely along the middle of the lid, 3 in (6 cm) from one edge. Tape the opposite edge to the box so that the lid can be opened and closed.

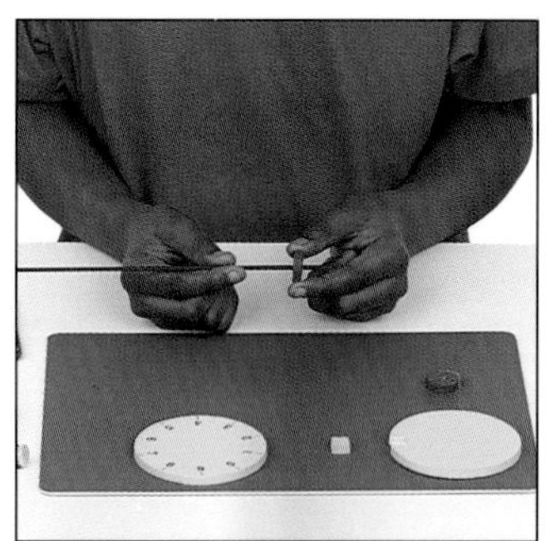

3 Using a skewer, make central holes in the small spacer discs and the large disks. A straw should fit loosely in the spacers and tightly in the large disks. Cut a $\frac{3}{8}$-in (8-mm) square notch into the edge of one large disk.

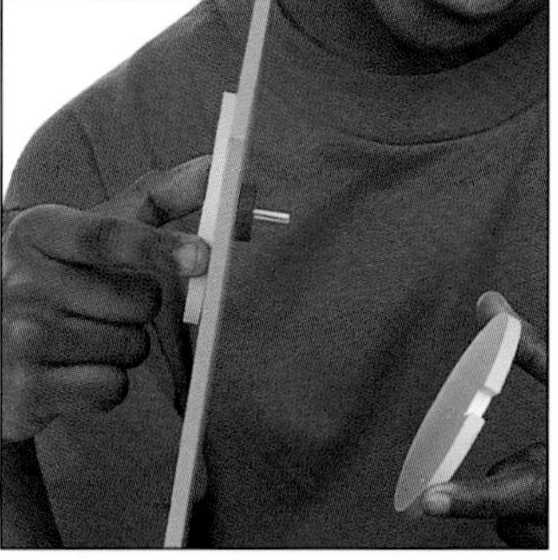

4 Mark the numerals 0 to 9 around the uncut large disk to make a dial. Slide the dial, spacer, and notched disk onto a $\frac{15}{16}$-in (2.5-cm) length of straw passing through the lid as shown below. The large disks should rotate together.

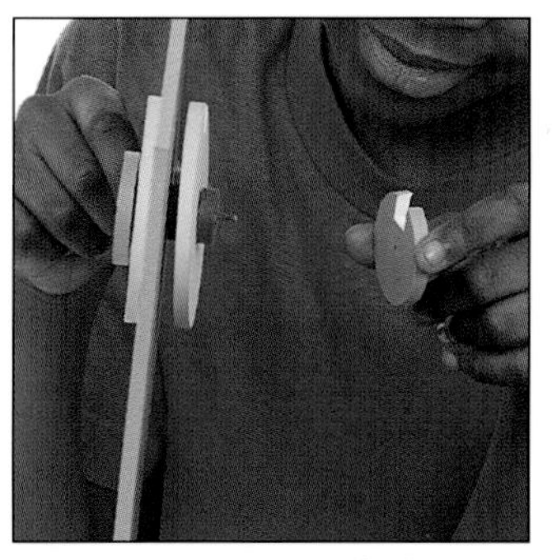

5 Make central holes in the medium disks to fit the skewer tightly. Mark 0 to 9 on one disk. Cut a $\frac{3}{8}$-in (8-mm) square notch in the other disk. Assemble a $1\frac{3}{8}$-in (3.5-cm) skewer, the straw, and disks as shown.

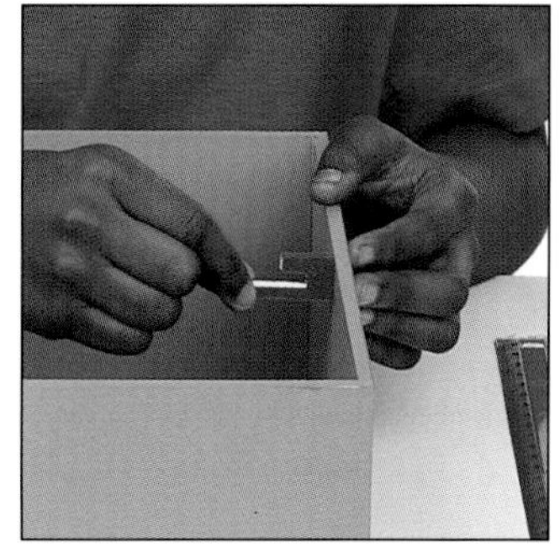

6 Cut a foamcore catch as shown in the diagram. Fix it midway along and flush with the box edge nearest the disks. To set the combination, align each notch with a numeral on its dial. Make sure the notched disks turn with the dials.

Cross-section through discs

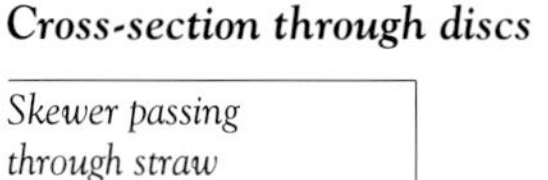

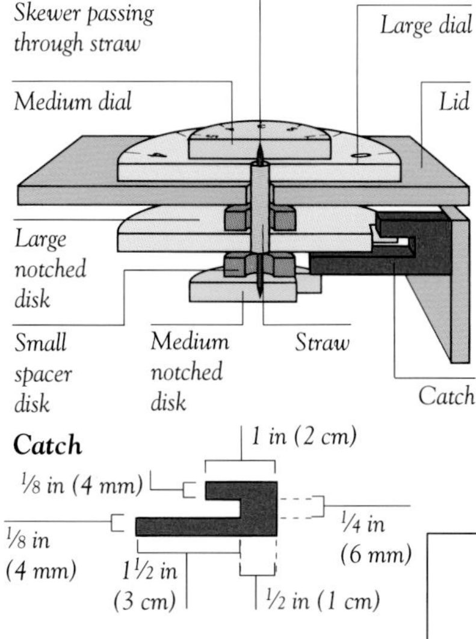

The box opens when the notches line up with the catch. You can set the lock with any number between 00 and 99

7 Mark the position of the catch on the upper side of the safe's lid. Turn the dials so that your chosen combination aligns with this mark. Close the lid, then turn the dials to lock it. Ask a friend to try to open the box by guessing the combination to unlock it.

Clocks

CLOCKS AND WATCHES indicate the time with moving hands or with digital displays that show the time as a set of numbers. Both kinds work in basically the same way. A power source, such as a wound-up spring or a battery, drives the hands or the digital display. A regulator controls the supply of power to the hands or display, so that the hands move or the numbers in the digital display change at a constant rate.

How clocks and watches work

Mechanical clocks and watches are run by weights or springs. Some of them have to be wound up by hand every so often, although automatic (self-winding) mechanical watches use the normal movements of the wrist to wind themselves. An electric clock uses electricity from an outlet, which alternates in voltage at a constant rate and regulates the clock so that it keeps time. Most quartz (electronic) watches and clocks run on battery power. Recently, however, quartz watches that do not need batteries have been developed. These watches use normal wrist movements to drive a tiny generator that produces the small amount of current they require.

Knob to set or change the time

Hands

Pill-size battery

Battery power
A single small battery can run a conventional quartz watch for more than a year because the watch consumes very little current.

EXPERIMENT

Counting the seconds

Adult help is advised for this experiment

The pendulum has been used to regulate clocks since the 17th century. It swings at a constant rate, which can be changed by altering the length of the pendulum. Make a pendulum in which each swing (to and fro) takes exactly 1 second. Use it to measure short periods of time.

YOU WILL NEED

- *wood strip about 24 x 2 x 1/4 in (60 x 5 x 0.5 cm)*
- *ruler*
- *string*
- *about 2 oz (50 g) modeling clay*
- *map pin*
- *adhesive tape*
- *jar lid*
- *scissors*
- *stopwatch*

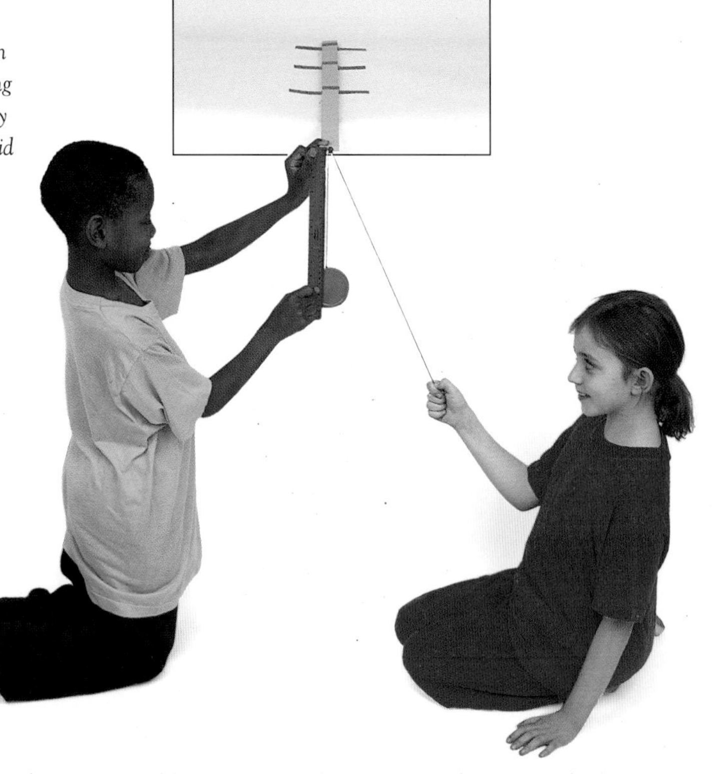

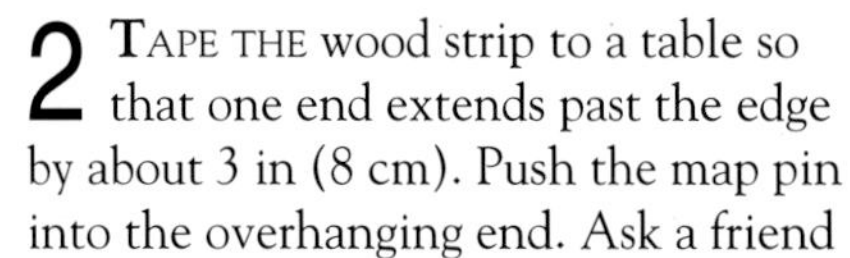

1 FILL THE JAR lid with an even layer of modeling clay. Press one end of a 3-ft (1-m) length of string into the clay, from the center outward, so that the lid is vertical when suspended.

2 TAPE THE wood strip to a table so that one end extends past the edge by about 3 in (8 cm). Push the map pin into the overhanging end. Ask a friend to help you tie the free end of the pendulum string to the pin so that the distance between the center of the jar lid and the pin is 9¾ in (24.8 cm).

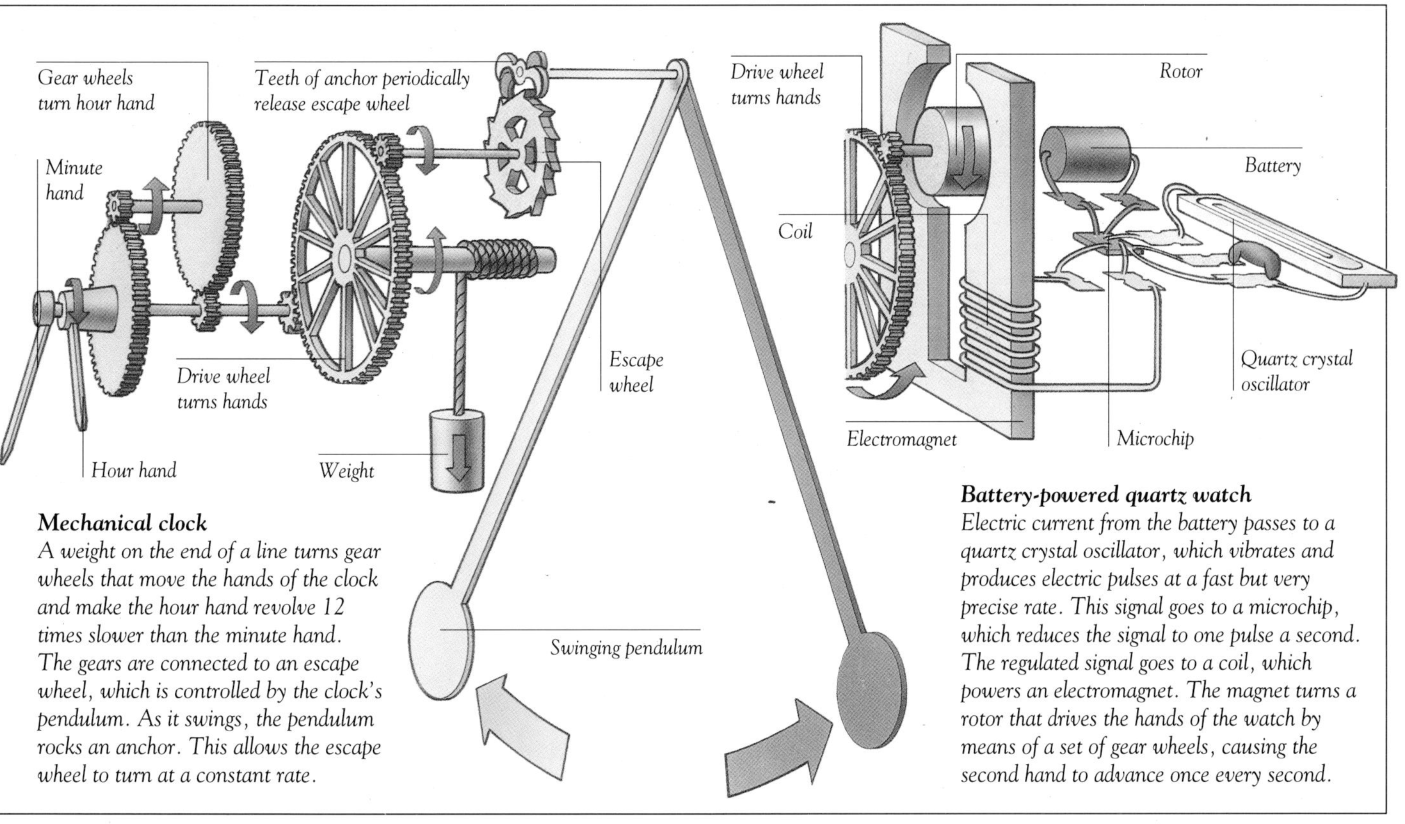

Mechanical clock
A weight on the end of a line turns gear wheels that move the hands of the clock and make the hour hand revolve 12 times slower than the minute hand. The gears are connected to an escape wheel, which is controlled by the clock's pendulum. As it swings, the pendulum rocks an anchor. This allows the escape wheel to turn at a constant rate.

Battery-powered quartz watch
Electric current from the battery passes to a quartz crystal oscillator, which vibrates and produces electric pulses at a fast but very precise rate. This signal goes to a microchip, which reduces the signal to one pulse a second. The regulated signal goes to a coil, which powers an electromagnet. The magnet turns a rotor that drives the hands of the watch by means of a set of gear wheels, causing the second hand to advance once every second.

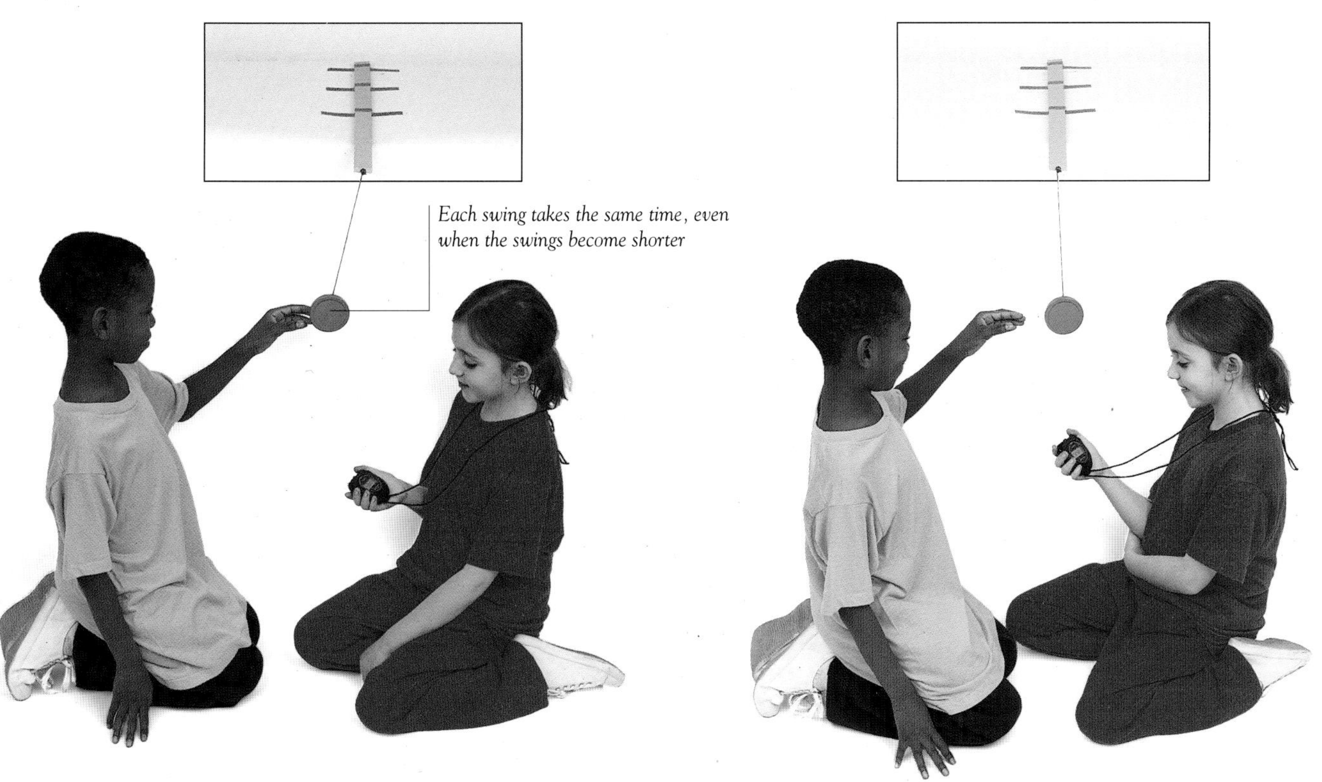

Each swing takes the same time, even when the swings become shorter

3 PULL THE pendulum to one side until it is about 2 in (5 cm) from the vertical. Let it go, making sure it swings from side to side without wobbling back and forth. Once you know that the pendulum can swing smoothly, stop it.

4 PULL THE pendulum out and release it again. As soon as a swing begins, start the stopwatch while your friend counts the number of swings. Check the time that the pendulum takes to make 30 and 60 swings.

Sewing machines

IF YOU ARE ABLE to sew, you can make your own clothes by cutting out pieces of fabric and stitching them together. Sewing is a pleasant pastime that makes it possible to produce individual garments at low cost. However, sewing by hand is slow and demands a lot of practice to do well. A sewing machine is much quicker, and produces perfect stitching in a variety of patterns. Its complex mechanism is usually driven by an electric motor, so that both hands are free to guide the fabric through the machine.

■ DISCOVERY ■

The Singer sewing machine

The American inventor Isaac Singer (1811–1875) built the first successful sewing machine in 1850. The sewing machines of the time were expensive and often unreliable; in only 11 days, Singer built an improved model that was both practical and robust. His sewing machine helped workers in the textile industry as well as becoming sought-after for use in the home.

Sew useful
Singer's design became the basis of the modern sewing machine.

Stitching mechanism

Sewing is a fairly simple action for a pair of well-practiced hands. A needle, carrying a short length of thread, is repeatedly passed through two or more pieces of material and back again. Stitches are formed as the thread is pulled tight and these bind the pieces of material together. The stitching action is much more complex in a sewing machine. The pieces of material are held under a vertical needle in which the eye is near the point. The needle carries one length of thread, which comes from a spool on top of the machine. The needle repeatedly moves up and down as the pieces of material pass below it. Beneath the material, a shuttle comprising a rotating hook and a second spool of thread (called the bobbin) forms stitches that interlock with the upper thread.

1. A new stitch is about to be made. The needle carrying the upper thread descends towards the two pieces of material.

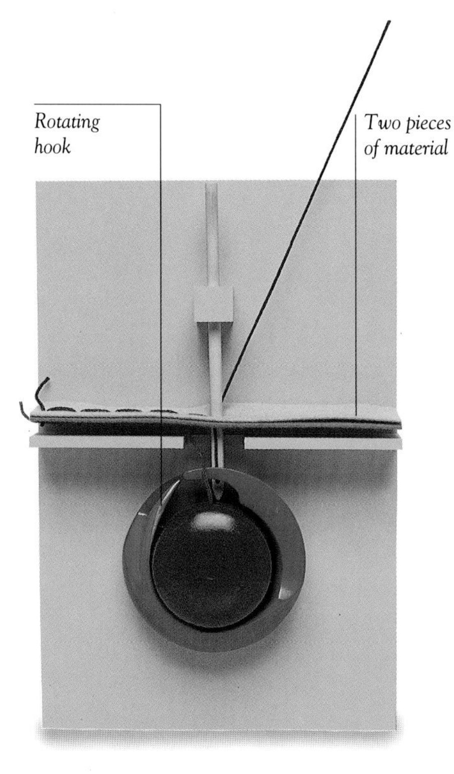

2. The needle pierces the two pieces of material, carrying a loop of the upper thread with it. The hook rotates as the needle moves.

3. The rotating hook passes under the lower thread. The point of the hook then picks up the loop of upper thread as the needle begins to rise.

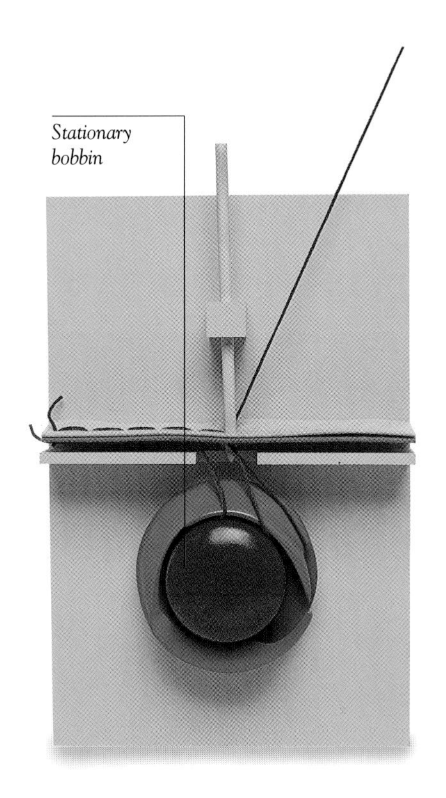

4. The rotating hook pulls the upper thread through the ascending needle. The loop of upper thread becomes enlarged.

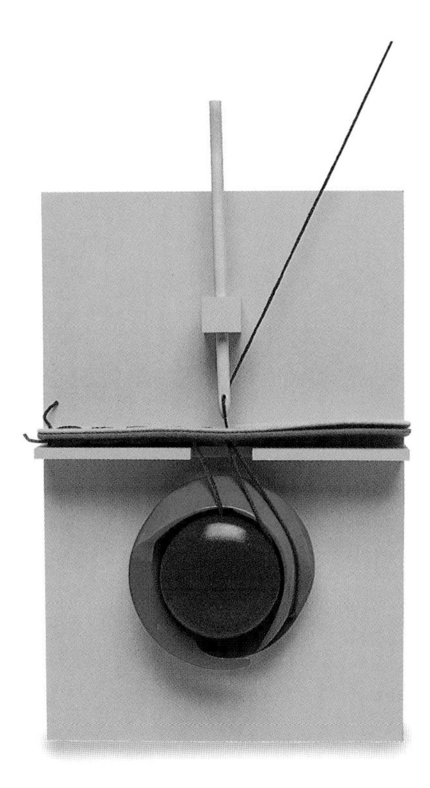

5. The lower thread is still stationary as the needle rises out of the material. The rotating hook loops the upper thread around the bobbin.

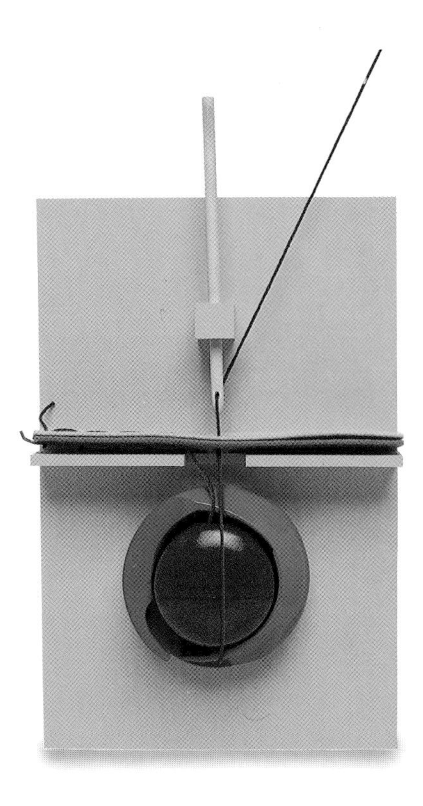

6. One side of the loop of upper thread passes behind the bobbin, and the other side passes in front. The lower thread still does not move.

7. The upper thread passes over the bobbin and loops around the lower thread. The take-up lever on the sewing machine pulls up the upper thread.

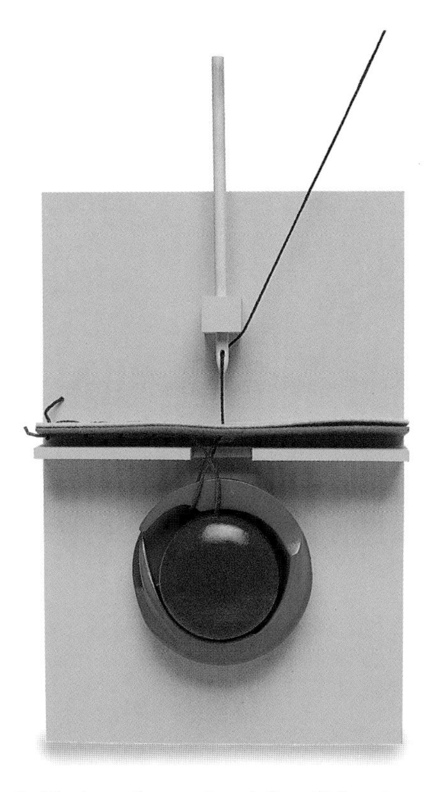

8. The loop of upper thread slips off the point and catches on the other end of the hook. As the loop is pulled up, it closes around the lower thread.

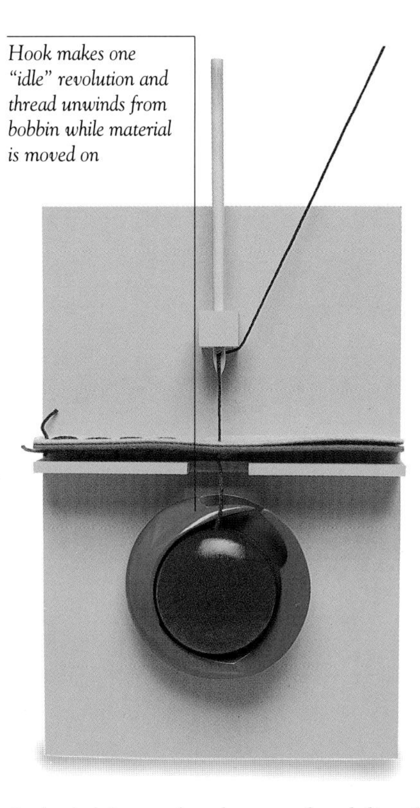

9. A stitch forms when the upper thread slips off the hook and locks around the lower thread. The material can now be moved for the next stitch.

TRANSPORT

On and off the road
Modern highways (left) are designed to enable people to drive at speed and in safety. Bright lights, on vehicles or at the roadside, greatly reduce the hazards of night driving. Cyclists have also benefited from technology. Using a modern mountain bike (above), today's cyclist can ride almost anywhere on land. A wide range of gears minimizes the cyclist's effort.

MODERN TRANSPORT makes most travel very easy. Powerful engines and motors can drive airplanes, trains, and cars at high speed, while sure and safe methods of navigation and traffic control minimize the risks of travel. There is hardly a location in the world that cannot be reached by either public or private transportation. But the sheer number of vehicles brings problems in some places. Congestion can slow traffic to a crawl, and air pollution caused by transport can blight the lives of city dwellers.

FARTHER AND FASTER

EVEN IN AN AGE WHEN TELECOMMUNICATIONS can put us in instant touch with each other and deliver information to us immediately, we still find travel enjoyable or necessary. Various kinds of transport—high-speed or leisurely, long-distance or local—satisfy our different requirements. These vehicles and vessels not only carry people but, perhaps more importantly, also transport freight.

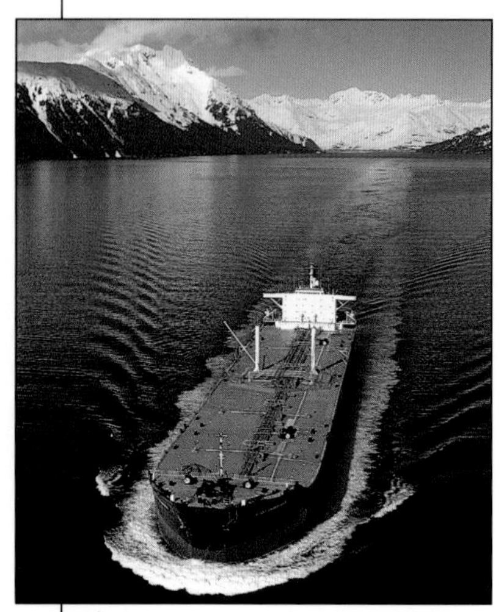

***The world's greatest energy** consumers depend on crude oil, often transported in large tankers from distant oil fields to refineries.*

It was only two centuries ago that transport began to open the world to people. Before then, all transport was wind-blown or powered by humans or animals; journeys that now take hours then took days or even months. New technology made it possible for people to travel quickly. The invention of the steam engine brought self-powered ships, trains, and cars. The development of the gasoline engine not only made the car a practical form of transport, but also made possible the airplane. Ever more powerful engines and the building of road and rail networks have increased the speed of vehicles and the range of journeys. Travel throughout a nation and the whole world beyond is now available to all.

Transport now moves in four environments: land, sea, air and outer space. Road vehicles—cars, buses, trucks, motorcycles, and bicycles—dominate on land, but trains can provide a faster journey. The sea is mainly used by freighters and oil tankers. Ferry boats and cruise liners transport people, and hydrofoils and hovercraft are used for short but speedy crossings. In the skies, airliners, light aircraft, and helicopters provide public and private air transport. Space travel is still reserved for astronauts, but spacecraft are also important for transporting satellites into orbit around the earth.

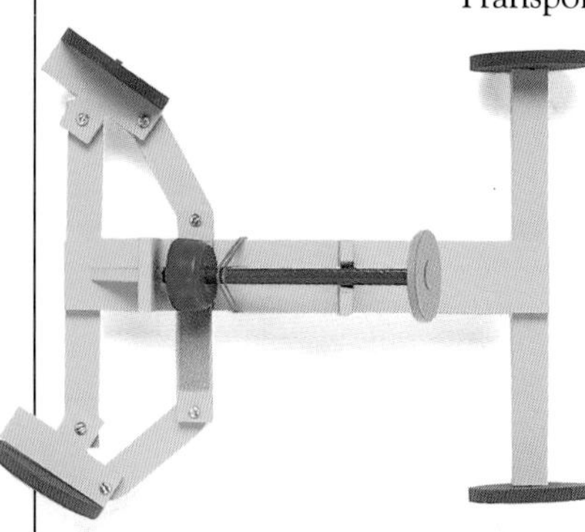

***The steering mechanism of a car** converts a driver's hand movements into movement of the front wheels.*

Methods of moving

Most forms of transport move either by rolling on powered wheels or by reaction. All road vehicles and most trains roll on wheels. Train wheels grip the rails directly, but road vehicles need rubber tires to grip roads. A pattern of grooves and ridges called the tread allows water to escape from under the tires, providing a good grip in rain.

Vehicles that travel through water and air function in a fluid medium, while spacecraft must be able to operate in a vacuum. Such vehicles move by means of reaction: in this process, the vehicle thrusts something backward and, in reaction, the vehicle is pushed forward. The propellers of ships and aircraft thrust water or air backward in order to move forward. Aircraft jet engines and rocket engines both create forward motion by thrusting hot exhaust gases backward into air or space.

Using magnetism is a third way of moving. Maglev (short for magnetic levitation) trains have electric motors called linear motors, which use magnetic attraction and repulsion to propel the train along a track. These trains also use magnetism to lift the train just above the track. Freed from friction caused by contact with the track, Maglev trains can travel very fast.

***Air flowing around the wings of an** aircraft produces lift. This upward force supports the aircraft in the air, provided that the wing angle is not too steep.*

Aircraft can also move vertically; this kind of motion is produced by an upward force called lift. The wings of an airplane produce lift, as do the rotor of a helicopter and the large envelope of a balloon. This upward force does not cease to act when an airplane levels out, or when a helicopter or balloon hovers in the air. In these cases, the lift is equivalent to the vehicle's weight, so that no overall force acts to move it either up or down.

Motive power

Like all other machines, the various forms of transport use three main sources of power. Muscle power propels bicycles, many of which now have a large range of gears that enable the rider to climb steep hills easily as well as speed along on the flat. But most road vehicles burn fuel as a source of power, as do many trains, all powered aircraft, and most spacecraft. Fuel contains a lot of energy, which is why it is the main source of power for most forms of transport. The third source is electricity. Delivered from power stations at high voltage along wires or rails, electricity can drive trains at high speed. Batteries and solar cells also provide electricity. Batteries can be used to power cars, but such vehicles have a limited range. Solar energy has also powered experimental cars and aircraft; it does not give enough power for practical purposes, but has the great advantage of being a clean source that produces no pollution. Batteries and solar energy may develop in the future and come to drive everyday transport.

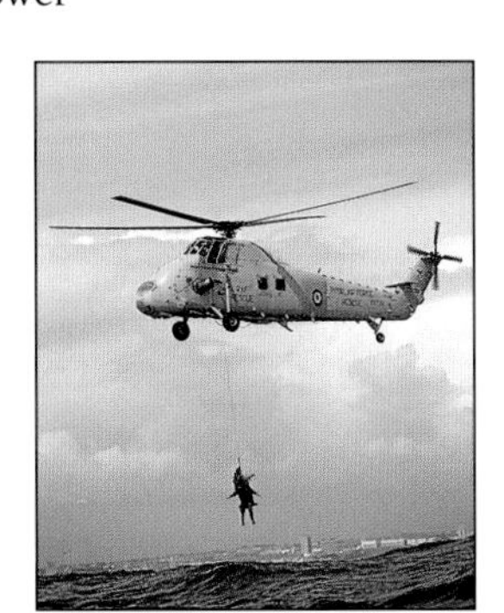

A helicopter's ability *to hover makes it suitable for rescuing people from remote or dangerous places.*

An efficient vehicle operates on a minimum amount of energy, and saves energy by consuming less fuel or moving faster than a less efficient vehicle. Streamlining a vehicle's shape reduces air or water resistance so that the vehicle does not have to use as much energy to push through the air or water around it. Hydrofoils and hovercraft reduce water resistance by raising their hulls just above the surface.

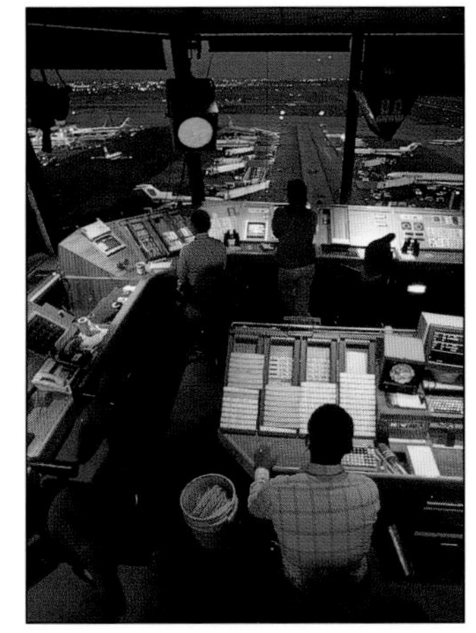

Radar screens in an air traffic *control center display information that gives controllers the positions of all nearby aircraft.*

Efficiency can also be enhanced by recycling energy. For example, some of the energy expelled through the exhaust system of a car engine can be used to drive a turbocharger, which puts energy back into the engine.

Navigation

No type of transport is much use unless it can get you to your destination, preferably by the shortest or quickest route. Computer navigation systems can now give car drivers directions for their journeys, and can even route the car around road work and other obstructions, so some drivers need no longer depend on a map or a navigator.

Ships' crews have long used navigation instruments to find the ship's position. Now, navigation systems that use satellites can pinpoint a ship's location instantly, while radar enables the captain to detect nearby ships and other hazards. The flight of an aircraft may be controlled at certain times by an automatic pilot, which ensures that the aircraft keeps to a predetermined course.

External control is essential for safety, and speeds the movement of traffic. Train signals ensure that only one train can occupy a section of track at any one time, while road networks and traffic lights keep vehicles from crossing each other's path. Air traffic control systems use radar and radio to guide traffic in the sky so that only one aircraft at a time is present in any section of air space. A ground radar bounces signals off all aircraft in the vicinity to find their distance. The signals trigger a device fitted in each aircraft, which sends a radio signal back to the air traffic controllers. The reply signal gives the destination, altitude, and flight number of the aircraft in question. Aircraft also use onboard radar systems to detect storms ahead.

This experimental magnetic *levitation train in Japan is the fastest passenger train in the world. Magnets in the train and in the walls on either side raise the train above the track and propel it forward at speeds of over 300 mph (500 km/h).*

Safety measures

As speeds have increased, safety has become more important. Cars and aircraft now have equipment such as seat belts and air bags that protects passengers in the event of a crash. The vehicles themselves are also designed for safety: in a collision, some cars crumple around the passengers to absorb the energy of impact and leave the passengers unharmed.

Solar power drives this experimental car*, which is taking part in a clean air race in California. Solar cells covering the upper surface of the car generate electricity from sunlight. The electricity powers motors that drive the wheels.*

Trains

TODAY NEARLY ALL TRAINS, apart from a few surviving steam locomotives, are driven by electric motors or diesel engines. Electric trains receive their power from overhead wires or electrified rails on the track. Diesel trains and diesel-electric trains (which contain both diesel engines and electric motors) generate their own power and can run on nonelectrified track. High-speed trains and effective signal systems make rail the fastest and safest form of land travel. Another development, still in the experimental stage, is the magnetic levitation (Maglev) train, which floats above the track on a magnetic field. At speeds of up to 311 mph (500 kph), it promises to outpace even today's high-speed services.

EXPERIMENT

Fast floater

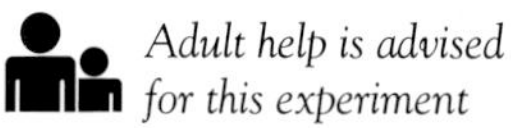

Make a model magnetic levitation train. This type of train uses magnetic fields to lift itself and float above a track. Because no part of it touches the track, friction is minimal, so the train can travel very fast. The model train works by magnetic repulsion. A magnet has two poles: north and south. Two like poles (both north or both south) facing each other repel one another. Here, the magnet in the train is pushed upward (repelled) by the magnets in the track.

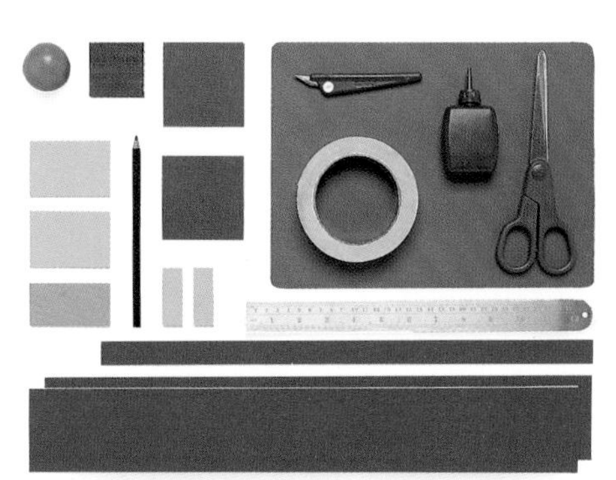

YOU WILL NEED

• modeling clay • 10 rectangular magnets, with poles in the largest faces; let l represent the length of each, w the width, and d the depth • 3/16 in (5 mm) thick foamcore pieces as follows: 2 sides **A**, *$11l$ x $2l$; 2 end panels* **B**, *$2l$ x $2l$; rail* **C**, *$9l$ x w+3/16 in (w+5 mm); 2 spacers* **D**, *l x w; train base* **E**, *$1.5l$ x w+3/4 in (w+2 cm) • pencil • cutting mat • craft knife • glue • scissors • double-sided tape • steel ruler • two heavy cardboard sides* **F**, *$1.5l$ x l*

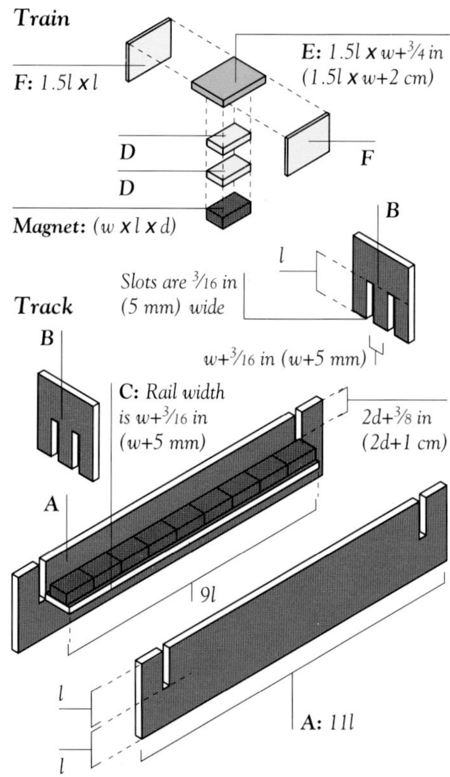

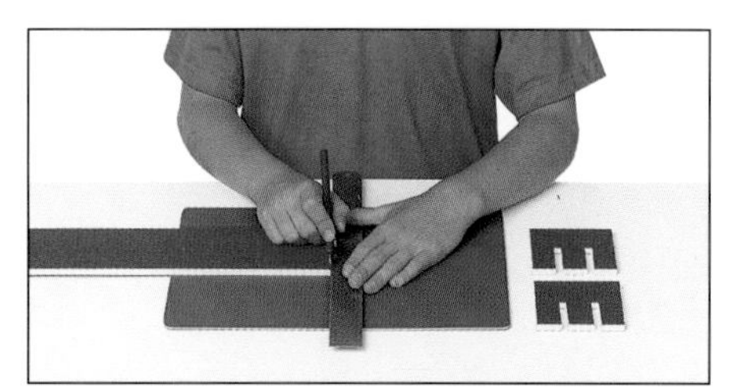

1 CUT TWO slots w+3/16 in (w+5 mm) apart into each panel **B**. Each slot is 5mm (3/16 in) wide and l deep. Cut two more slots $9l$ apart into each side **A**.

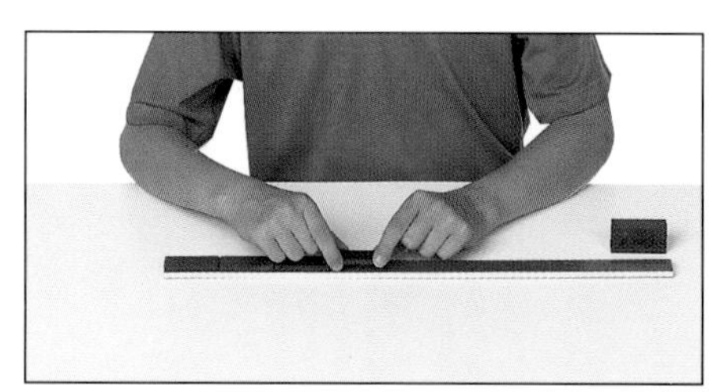

2 USING DOUBLE-SIDED tape, stick nine magnets along the center of the rail **C**, with like poles facing upward. Leave a 1/16 in (2.5 mm) gap on either side.

3 TAPE THE edges of the rail **C** between the slots on the sides **A**, with the magnet faces $2d$+3/8 in ($2d$+1 cm) below the edges of the sides **A**. Slot the two panels **B** onto the sides **A**.

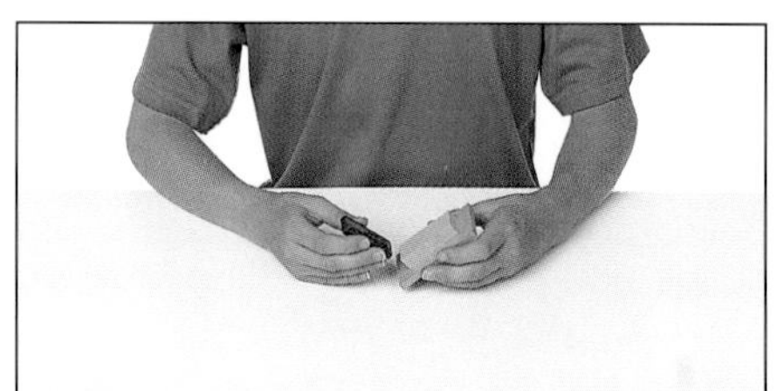

4 STICK THE spacers **D** together. Glue them to the middle of the train base **E**. Find which pole of the last magnet is attracted to the magnet faces on the rail, and stick this face to the spacers.

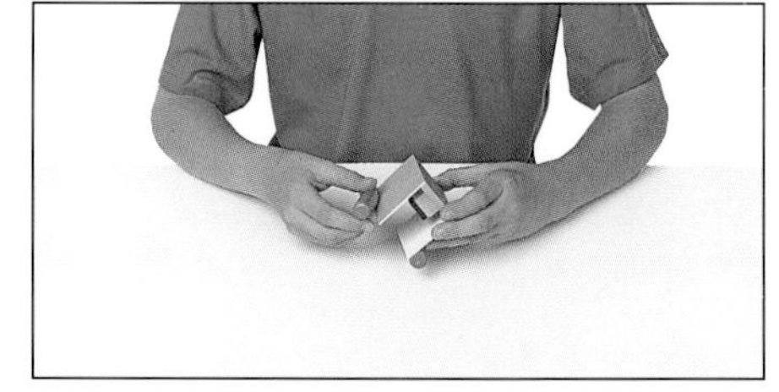

5 STICK THE cardboard sides **F** to the train base so that the magnet hangs between them. Add equal amounts of modeling clay to the bottom of each side, on the outer surfaces.

Maglev trains

These trains use electromagnets, in which electric current passing through a wire coil generates a magnetic field. In the Japanese prototype shown here, fields produced by coils in the train and in walls running alongside levitate the train and also propel it along the track. The train has energy-saving superconducting electromagnets. It is almost as fast as an airplane, but uses only about half as much energy.

Magnetic propulsion
Here, the coils in the walls can be seen.

Japanese Maglev system
Coils in the walls are fed an alternating electric current as the train passes. This current causes each coil to reverse its polarity as each train coil passes it, so that each coil alternately attracts and repels the train coils to propel the train forward. Other coils in the walls interact with coils in the train to lift the train and guide it.

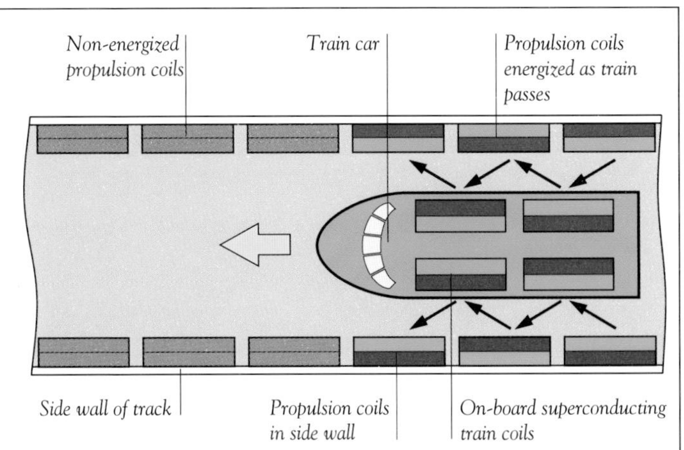

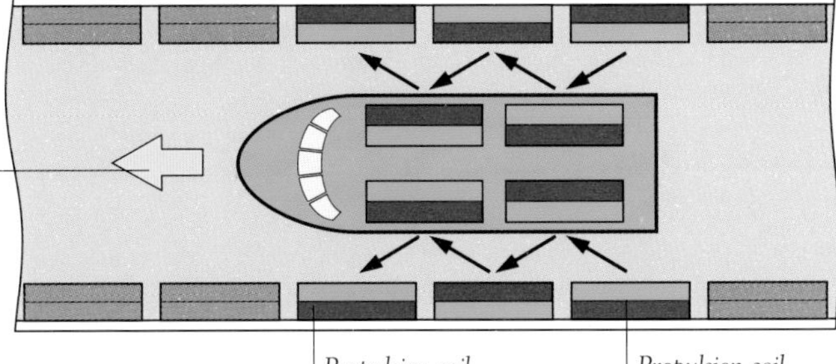

6 Add a train top to the base. Place the train on top of the rail. Add or remove clay from the sides of the train until it levitates stably above the track. Position it at one end of the track and give it a gentle push. Watch how easily it moves.

High-speed trains

The fastest train in regular service is the French TGV (Train à Grande Vitesse). It is an electric train using current supplied from an overhead line to electric motors in power cars at each end of the train. The TGV carries its passengers across France at speeds of up to 186 mph (300 kph), and has achieved a record speed for an electric train of 320 mph (515 kph). In Britain, diesel-electric trains provide high-speed services on non-electrified track. In these trains, powerful diesel engines drive electric generators, which produce current that passes to electric motors in the power cars. The British Intercity 125 is designed to travel at speeds of up to 125 mph (200 kph) on existing non-electrified track, and has achieved a diesel-electric record of 143 mph (230 kph).

Intercity 125
This diesel-electric train travels at high speed on existing track.

TGV (Train à Grande Vitesse)
The electric TGV runs on a new track laid specially for this train.

Cars 1

A CAR IS one of the most complicated machines that a person or a family is likely to possess. The underlying principle governing the car is simple: power is transferred from the engine to the wheels to make the car move. But this transfer of power, together with the need to make driving easy, safe, and comfortable, involves many highly complex systems. In fact, a car is not one machine at all but a whole collection of machines and systems working in concert, each contributing a vital action to make the car perform well.

EXPERIMENT

Steering mechanism

Adult help is advised for this experiment

Build a model of a rack-and-pinion steering mechanism, the steering mechanism used by most cars. When the steering wheel is turned, a pinion (toothed wheel) on the end of the wheel shaft moves a rack to one side or the other. The rack is linked to the wheel hubs and swivels the wheels.

The main systems of a car

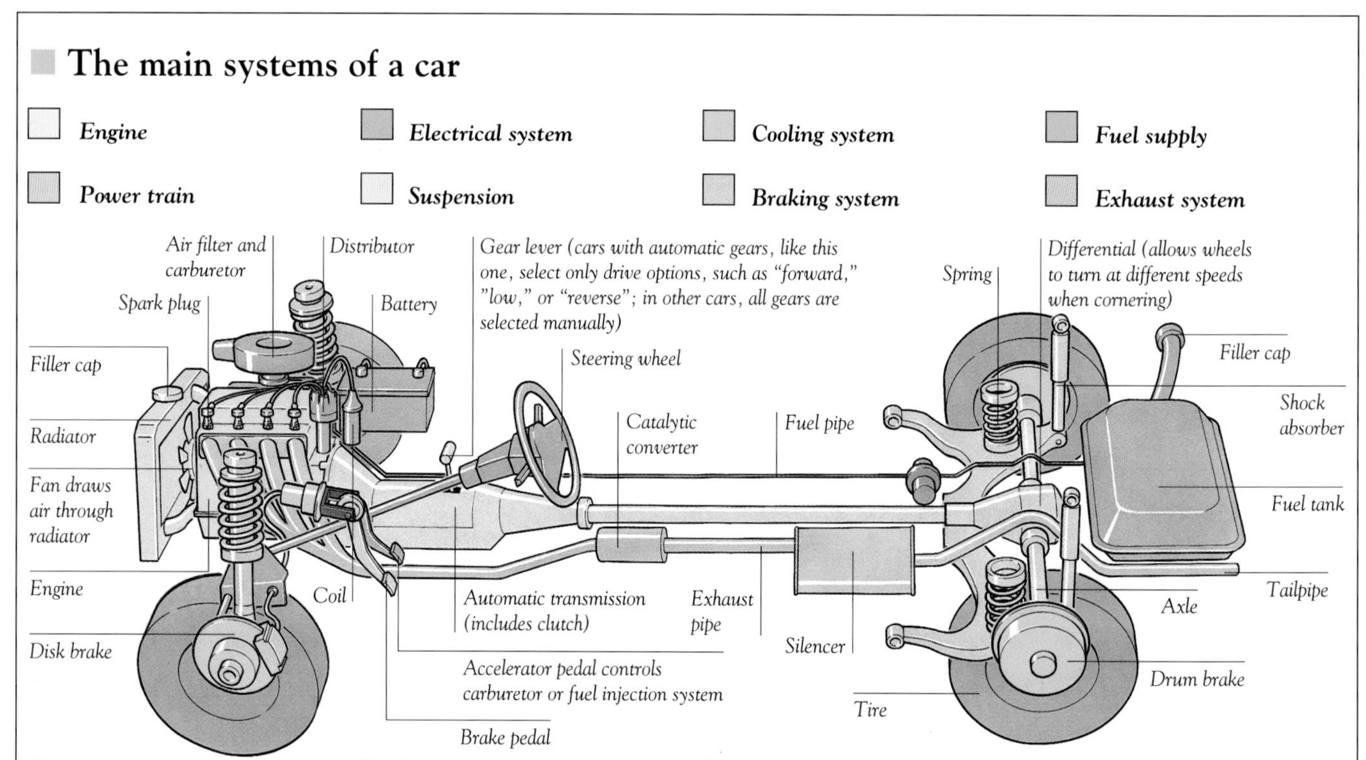

Power train
Shafts transmit the motion of the engine to the wheels via the clutch, transmission, and differential. The power may go to the rear wheels (shown here), the front wheels, or (in a four-wheel drive car) to all four wheels..

Electrical system
A car's electricity comes from a battery, charged by an engine-driven generator. A gasoline-fueled car has an ignition system in which a coil raises the voltage of the current from the battery. The current is then sent by a distributor to the spark plug of each engine cylinder in turn.

Engine
The fuel burns in cylinders inside the engine (pp.18–19), causing pistons to rotate the crankshaft. The crankshaft drives the flywheel and transmission. Oil lubricates the moving parts of the engine.

Exhaust system
Waste gases leave the engine through the exhaust pipe. They then flow through a silencer, which reduces engine noise, and out of the tailpipe. In some cars, the exhaust gases may also flow through a catalytic converter. This removes pollutants from the exhaust fumes by converting them into harmless gases.

Cooling system
Water is pumped around the engine, absorbing heat. It then passes through the radiator, where it loses heat. The hot water is also used by the car's heating system.

Fuel supply
For safety, the fuel tank is usually located away from the engine and passengers. In gasoline-fueled cars, a pump sends fuel to the carburetor (p.102). Here, the fuel is mixed with air, and then sprayed into the engine cylinders. All diesels and some gasoline engines use fuel injectors, not carburetors. These measure precisely the amount of fuel required for maximum efficiency.

Braking system
All four wheels have brakes. Disk brakes grip a disk fixed to the wheel axle, while drum brakes grip a drum fixed to the wheel. Disk brakes are stronger than drum brakes, and are often used on front wheels, which need more braking power than rear wheels. Brakes may be power-assisted.

Suspension
Springs and shock absorbers are mounted on the wheel axles to take up jolts and bumps and give a smooth ride. The shock absorbers reduce the vibrations of the springs so that the car does not bounce as it travels along the road.

You Will Need

• *steel ruler* • *ruler* • *cutting surface* • *foamcore strips 1 in (3 cm) and ½ in (1.5 cm) wide* • *foamcore circles, one 1-in (3-cm) and four 2-in (5-cm) diameter* • *screwdriver* • *scissors* • *craft knife* • *drill and 3/16-in (5-mm) bit* • *skewer* • *round file* • *C-clamp* • *glue* • *pencil* • *rubber band* • *⅛-in (3-mm) nuts and bolts* • *thin nails* • *ridged plastic bottle top* • *2⅜ x ½ in (6 x 1.5 cm) sandpaper strip* • *double-sided tape*

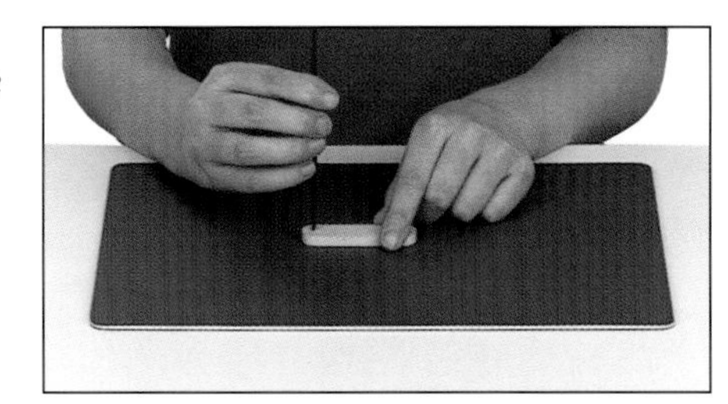

1 Cut out the shapes **G–J** and **M** from ½-in (1.5-cm) foamcore, and the rest from 1-in (3-cm) foamcore, as shown in the diagram below. With a skewer, make holes large enough for a bolt to pass through in pieces **G–L**, where dots are marked. Each hole is ⅛ in (3 mm) from the nearest short edge of the foamcore.

2 Glue the rack guides (**B**, **C**) to the chassis (**A**). Then glue the steering column supports (**D**, **E**, **F**) to the chassis. Glue the front axle (**G**) beneath piece **B**. Ensure the steering rack (**H**) slides easily between its guides. Use nuts and bolts to connect the rack, track rods (**I**, **J**), steering arms (**K**, **L**), and front axle.

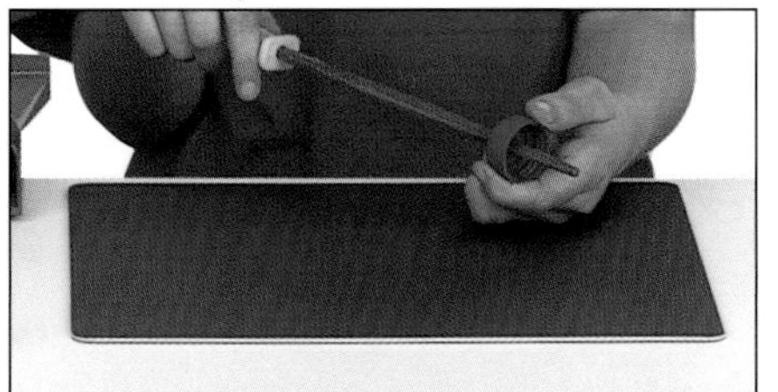

3 Drill a hole through the center of the bottle top. Using a round file, enlarge the hole until a pencil can be fitted tightly through the hole to make a steering column. Glue the rear axle (**M**) beneath the rear of the chassis.

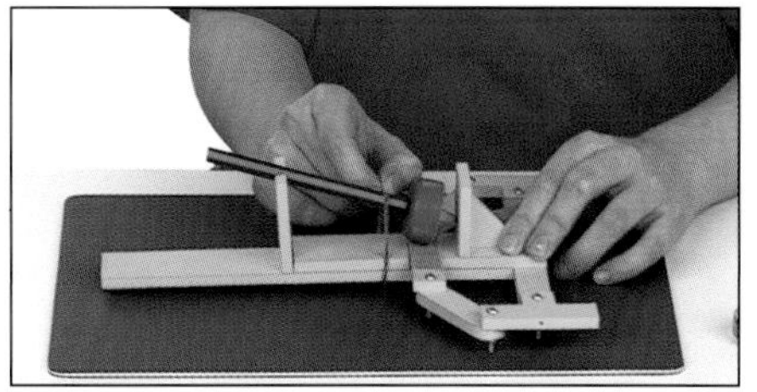

4 Tape the sandpaper to the rack. Place the steering column on its supports so that the bottle top pinion contacts the sandpaper rack, and secure it to the chassis with a rubber band.

5 Nail the 2-in (5-cm) diameter wheels to the ends of the front and rear axles. Attach a 1-in (3-cm) diameter foamcore steering wheel to the steering column. The steering mechanism is now ready to use.

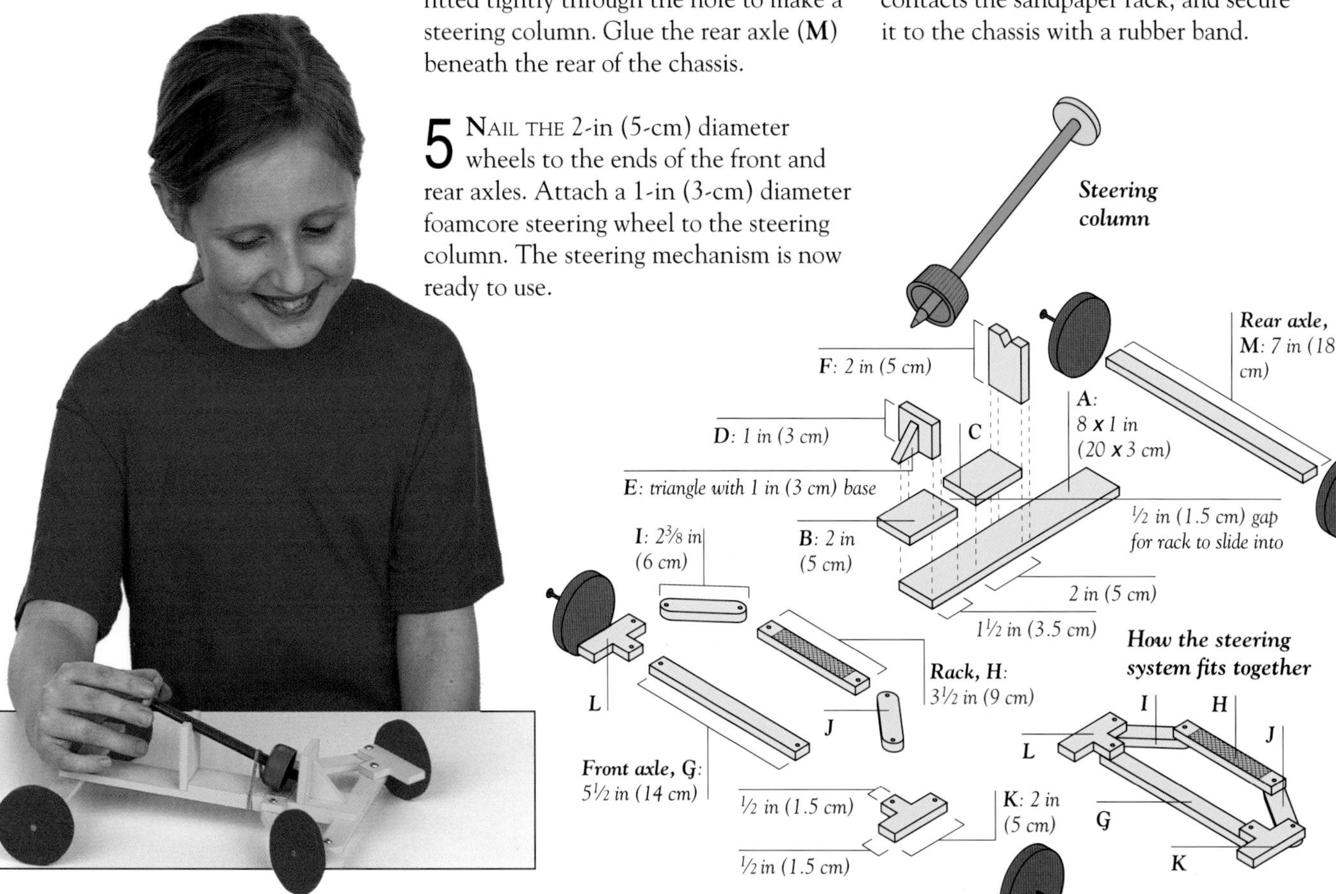

Cars 2

FAST ACCELERATION is a feature that many drivers demand of a car. Pressing down the accelerator pedal feeds more fuel to the engine, through a carburetor or fuel injectors, and the car picks up speed. Other important elements of modern cars are systems to make driving easier and safer. Power brakes and power steering can be power-assisted to reduce the effort required from the driver's feet and hands. Antilock brakes are a safety feature that reduces the chances of skidding during an emergency stop. If the car does crash or come to an abrupt halt, seat belts can prevent the driver and passengers from flying out of their seats.

The first car

The internal combustion engine was invented in 1860 by the Belgian Etienne Lenoir (1822–1900). In 1885, the German Karl Benz (1844–1929) developed a lightweight, practical engine that could reliably power a road vehicle. He also made the first car, a three-wheeler that used his four-stroke gasoline engine.

1886 Benz Motorwagen

EXPERIMENT

Simple carburetor

Blow air through a bent straw to suck up water, and spray it out through a slit in the straw. The moving air in the straw loses pressure, so the higher atmospheric pressure outside forces water up the straw.

YOU WILL NEED

- *glass* • *pitcher of water* • *cutting surface*
- *paper* • *drinking straw* • *knife* • *bowl* • *tape*

1 MAKE A slit across the middle of the straw. Tape the paper upright in the bowl. Put the straw in a glass of water with the slit facing the paper.

2 BLOW through the straw, bending it at various angles, until a fine spray emerges from the slit.

Feeding the engine

In the carburetor, fuel stored in a chamber goes to an air pipe leading to the engine. The pipe narrows at this point, causing the air flowing through the pipe to speed up and lose pressure. This sucks fuel into the airstream, where it forms a spray and evaporates. The accelerator pedal operates a throttle valve, which rotates to control the amount of fuel-air mixture flowing to the engine.

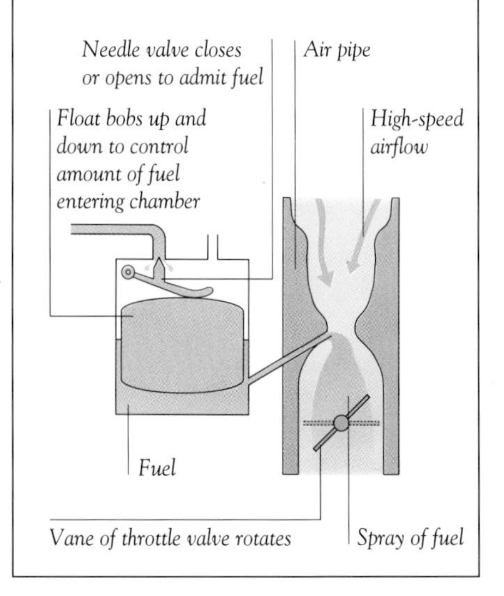

EXPERIMENT

Sitting in safety

Adult help is advised for this experiment

You must always wear a seat belt in a car. The belt has a lock so that it grips you only when the belt begins to move suddenly and quickly, as happens in a crash. However, the belt can extend slowly to allow you to move so that you can reach for things in the car. Make a seat belt that works in this way.

You Will Need

- *wooden base about 20 x 6 in (50 x 15 cm)*
- *scissors* • *screwdriver*
- *round file* • *drill and 1/8-in (3-mm) bit*
- *2-in (5-cm) screws*
- *nuts and 3-in (8-cm) bolts* • *ruler*
- *tape* • *thread spool*
- *rubber band* • *string*
- *vise*

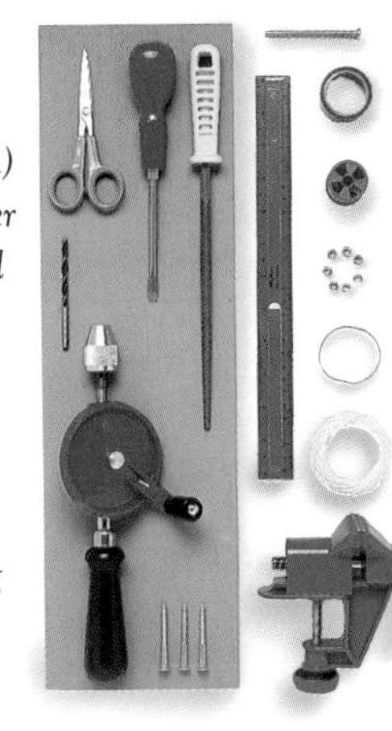

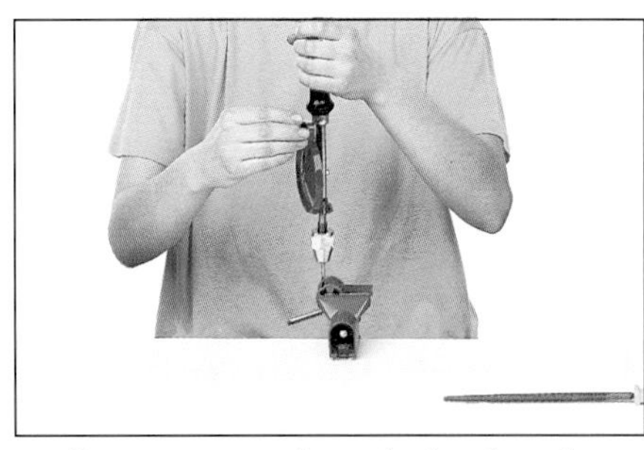

1 Drill a hole through the thread spool, slightly offset from and at 90° to the central shaft. Enlarge the hole with a file until it is slightly wider than the bolt.

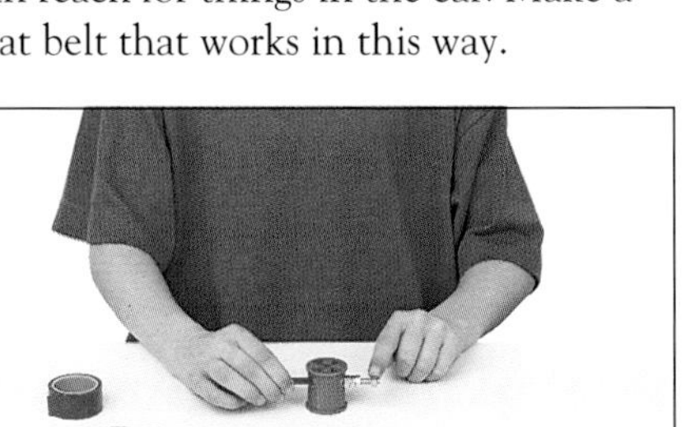

2 Cover the bolt shaft with tape to make it smooth, leaving space for six nuts at the end. Slide it through the hole drilled in the spool. Thread the nuts on to the bolt to weight it at one end.

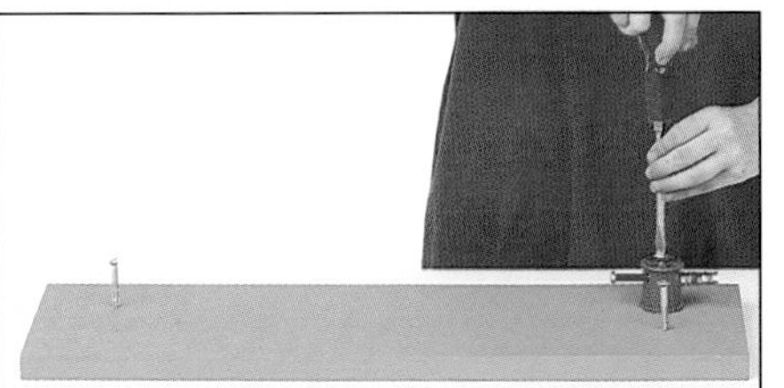

3 Screw the spool to one end of the base so that it turns freely. Add a screw opposite the first at the other end of the base. Add a third screw 2 in (5 cm) from the first to form a right angle.

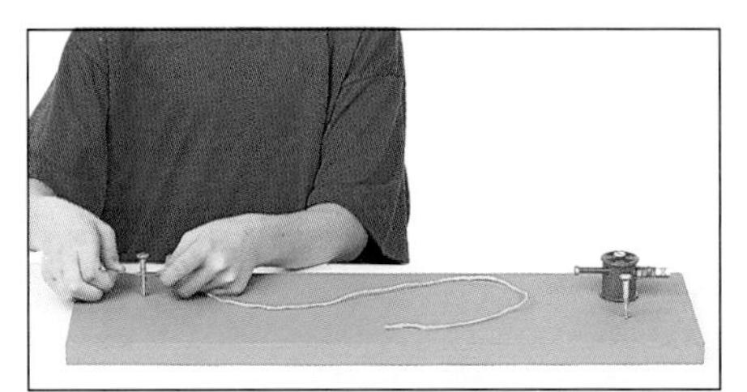

4 Attach a 3-in (8-cm) length of rubber band to the second screw. Tie a piece of string about 20 in (50 cm) long to the loose end of the rubber band. The string is like the belt of a real seat belt.

The spool rotates slowly, allowing the string to be pulled

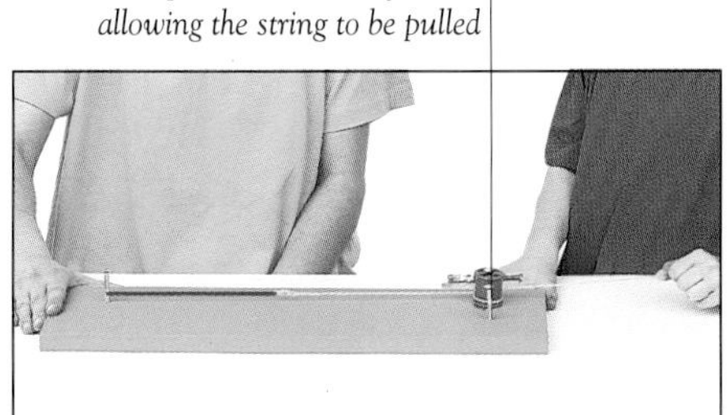

5 Make sure that the bolt protrudes by equal amounts from each side of the spool. Wind the string once around the spool then pull the loose end towards you slowly so that the spool rotates.

6 Relax your grip on the string and let the rubber band pull the string back. Pull the string again, but this time pull it quickly. The string suddenly locks tight, just as a seat belt would in a crash.

The spool rotates quickly, causing the weighted bolt to spring outward and catch on the screw

Ships and boats

PEOPLE ONCE traveled around the world by ship, but jets have made water travel all but obsolete except for short journeys and cruises. However, big vessels are still used to transport large quantities of freight between continents and to deliver huge loads of crude oil to refineries around the world. The speed of such ships is slow, but designers aim to make a ship travel as fast as possible by reducing water resistance on the ship's hull as it cuts through the waves. Hovercraft and hydrofoils, where the hull is raised just above the water surface, achieve greater speeds, but these vessels cannot be built to very large dimensions and are mostly used as ferries crossing short stretches of water. Most ships and boats use propellers (pp.108–109) to drive them through the water, but some have engines that pump out fast-moving jets of water; these water jets drive the vessel forwards in the same way that airplanes are driven by fast-moving jets of air.

EXPERIMENT

Streamlining

Adult help is advised for this experiment

A ship has a streamlined hull to let water flow smoothly around it, without offering much resistance to the ship's motion. Friction between the water and hull creates water turbulence, which slows the ship down. See how water flows around various shapes of modeling clay.

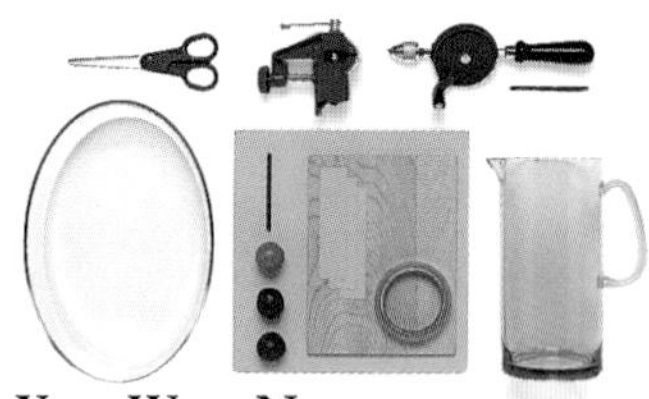

YOU WILL NEED

- *wood board 12 x 12 in (30 x 30 cm)*
- *drill and bit to fit dowel*
- *C-clamp*
- *drilling board*
- *2 in (5 cm) length of dowel*
- *10 oz (300 g) modeling clay*
- *pan wider than board*
- *pitcher of water*
- *scissors*
- *tape*
- *3 cardboard strips 8 x 2 in (20 x 5 cm)*

1 DRILL A centrally-placed hole through the wood board into which the dowel fits tightly. Secure the dowel in the hole.

2 FOLD THE cardboard strips into a star, a square, and a teardrop. Use these as cutters to make three shapes from modeling clay.

Sailing into the wind

When the wind is behind a sailboat, it fills the sails and pushes the boat forwards. A modern sailing yacht, however, can sail at an angle into the wind. This is because the wind blows out each sail so that it is curved. As the air blows over the curved surface, a force is produced at right angles to the wind, in the same way as lift is produced by an aircraft's wings (pp.108–9). The wind force can be considered as two component forces. One makes the boat heel, or tilt, sideways, while the other pushes the boat forward into the wind. The other component pushes the boat sideways, but is counteracted by the keel's weight and the hull's motion through the water.

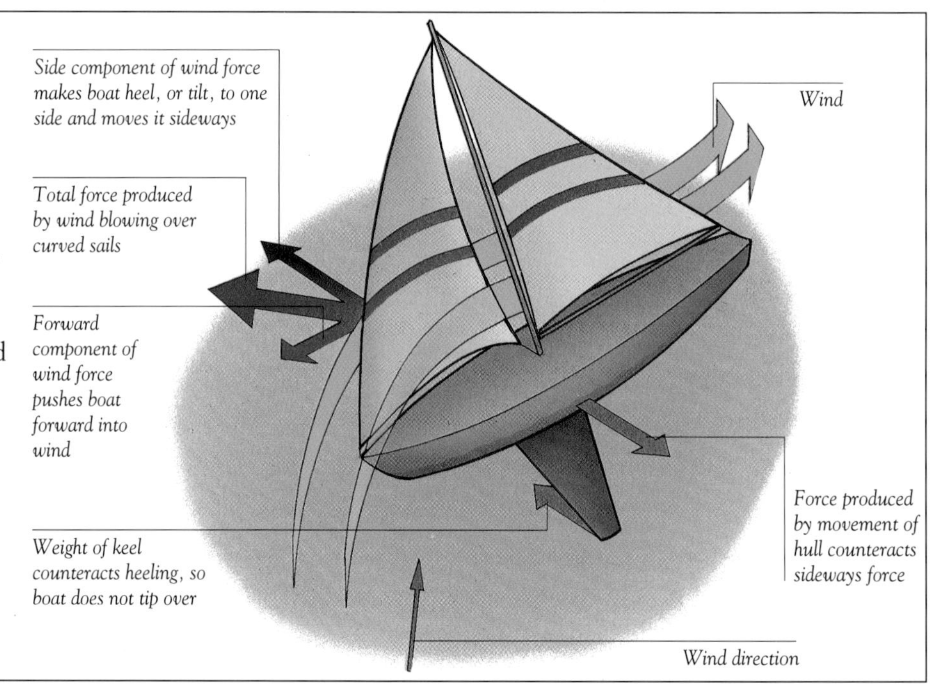

3 STICK THE CLAY star onto the dowel and put the board almost upright in the pan. Pour water over the star and notice how its jagged points produce a very turbulent flow.

4 REPLACE THE STAR with the square. Pour water over it, and compare the flow with that over the star. The flow is smoother, but there is still some turbulence at the corners.

5 NOW REPLACE the square with the teardrop and repeat the experiment. The water flows most smoothly around the teardrop. For streamlining, ships are built with similar shapes and have rounded or pointed ends.

Hydrofoils

At rest a hydrofoil looks like any other boat. The difference is that the hull rests on foils that resemble underwater wings. When the hydrofoil starts to move, water flows over the foils, and they develop an upward force similar to the lift produced by aircraft wings. This force raises the hull out of the water. The hydrofoil can then move at high speed because the water does not drag on the hull. Propellers or water jets beneath the water's surface drive the hydrofoil forward.

Flying on foils
The ends of foils that raise this hydrofoil out of the water can be seen at the front and rear.

Submersibles

DEEP-SEA EXPLORATION and observation is the task of a submersible, which can carry a crew to an ocean floor. With the aid of manipulator arms and underwater robots, the crew can explore the seabed in search of unusual features and strange life forms. Submersibles may also explore shipwrecks and sunken aircraft, often to recover evidence of disasters. A submersible is taken to the scene of operations by its mother ship. The submersible is tethered to its mother ship by an umbilical cable, which delivers power and may carry control signals and video lines. It descends and ascends in the sea using sea water as ballast. On the seabed, the submersible maneuvers with the aid of electrically-powered propellers called thrusters. The crew can talk to the ship and are able to return to the surface in an emergency.

■ DISCOVERY ■

The Turtle

American engineer David Bushnell built the first underwater craft in 1776. Called the *Turtle*, the egg-shaped military vessel carried just one man. Two hand-powered propellers maneuvered it horizontally and vertically, and ballast tanks enabled it to sink and rise.

Probing the depths

Submersibles can operate on the ocean floor, at much greater depths than military submarines. This is because the crew occupies a strong spherical or cylindrical cabin that resists the immense water pressure at such depths. The crew cannot leave the submersible, even in diving suits, so they work with the aid of claw-like manipulators that can hold equipment and collect marine life and samples. Powerful lights illuminate the inky blackness, and the crew can see out through viewports or use video cameras. They may use an underwater robot that can enter and search areas beyond the reach of the submersible, such as the interior of a wreck. The submersible shown here can descend to a depth of 2,300 feet (700 meters) with a single crew member, or to 3,280 feet (1,000 meters) under the remote control of an operator guided by the onboard video cameras. It is used to repair and maintain offshore oil rigs. Buoyancy cylinders (ballast tanks) fill with water or air to make the submersible descend or ascend (p.107).

Video camera
Umbilical cable
Acrylic and fiberglass cabin
Lateral (side-to-side) thruster
Light
Buoyancy cylinder (ballast tank)
OSEL DUPLUS
Forward-reverse thruster
Crash guard
Manipulator
Hydraulic hose for manipulator
Vertical thruster
Anchoring grabber
Air tank for buoyancy cylinders

EXPERIMENT

Simple submersible

Build a submersible using modeling clay and a tube. Make it sink or rise in a bottle of water by squeezing or releasing the bottle. The difference in pressure inside the bottle causes water to enter or leave the tube so that the weight carried by the tube changes. It either becomes heavier and sinks, or becomes lighter and rises. Submersibles use seawater as ballast to sink and rise in a similar way. To sink, the submersible takes seawater into its buoyancy cylinders. This increases its weight so that it sinks. To rise, the submersible uses compressed air to pump out the water and lighten the craft.

You Will Need

• *knitting needle* • *plastic bottle* • *bowl* • *3 in (8 cm) long plastic tube, closed at one end, which fits into bottle* • *small piece of modeling clay*

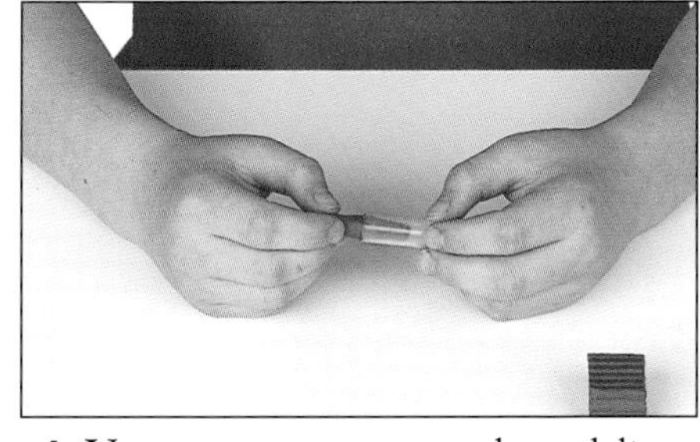

1 Use your fingers to push modeling clay into the open end of the tube until it is about half full.

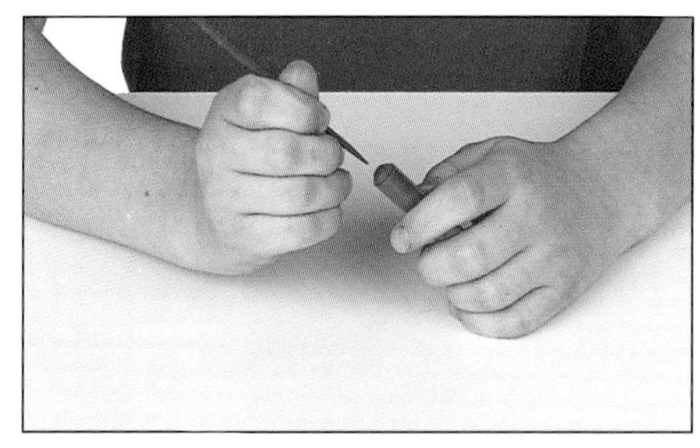

2 Pierce a hole through the clay so that air can pass into the tube. Put the tube in a bowl of water.

3 If the tube floats, add a little more clay. If it sinks, remove some clay. Continue until the top of the tube just breaks the surface.

4 Fill a plastic bottle almost to the top with water. Drop the tube into the bottle and screw the cap on tightly. Squeeze the bottle and note what happens to the "submersible". By varying the pressure applied when squeezing the bottle, you should be able to make the tube sink, rise, and hover.

Squeezing the bottle forces water into the tube and the air bubble inside shrinks; on release, the air bubble expands and forces out water

Aircraft 1

MOST MACHINES THAT FLY through the sky have wings to keep them aloft. As the wings cut swiftly through the air, they develop a strong upward force called lift. This force supports the weight of the aircraft, holding it up in the air. Wings generate more lift when moving quickly through the air than when moving slowly. Slow-moving aircraft, such as gliders, have long wings that stick straight out from the sides of the aircraft, which give lots of lift at slow speeds. Fast airplanes need sweptback wings that cut through the air with ease. These wings produce less lift at slow speeds, however, so airplanes have wing flaps that extend on both takeoff and landing to give more lift.

Modern light aircraft

■ DISCOVERY ■

The first flight

The first known manned flights were by balloon in 1783, but people possibly flew earlier using kites. The first winged aircraft was a glider built by the British inventor George Cayley (1773–1857), and first flown in 1853. It had the basic aircraft form of two wings and a tail. Glider design was then advanced by, among others, the American brothers, Wilbur Wright (1867–1912) and Orville Wright (1871–1948). Their great achievement was to build a petrol engine light enough to be carried in a glider. In this aircraft, called the *Flyer*, Orville made the first powered flight at Kitty Hawk, North Carolina, on 17 December 1903. The flight covered 120 feet (37 meters).

Airfoil

A wing is an airfoil, which means it has a special shape that enables it to fly. The top surface is curved and the underside is flat or almost flat, so the top of the wing is longer than the bottom. As the wing cuts through the air, it deflects the air so that some passes over the wing and some beneath. Due to the wing's shape, the air flowing above the wing moves faster than the air flowing beneath. Because air pressure drops more and more as air moves faster, the air below the wing has a greater pressure than the fast-moving air above and pushes up on the wing to give it lift. Slightly increasing the angle of the wing increases lift, but too great an angle causes the smooth flow of air over the wing to break up and swirl. The air now moves less fast, causing the difference in air pressure to drop. The wing's lift disappears and the aircraft stalls (begins to fall). For simplicity, and to explain the experiment opposite, the diagrams show the air moving rather than the wing.

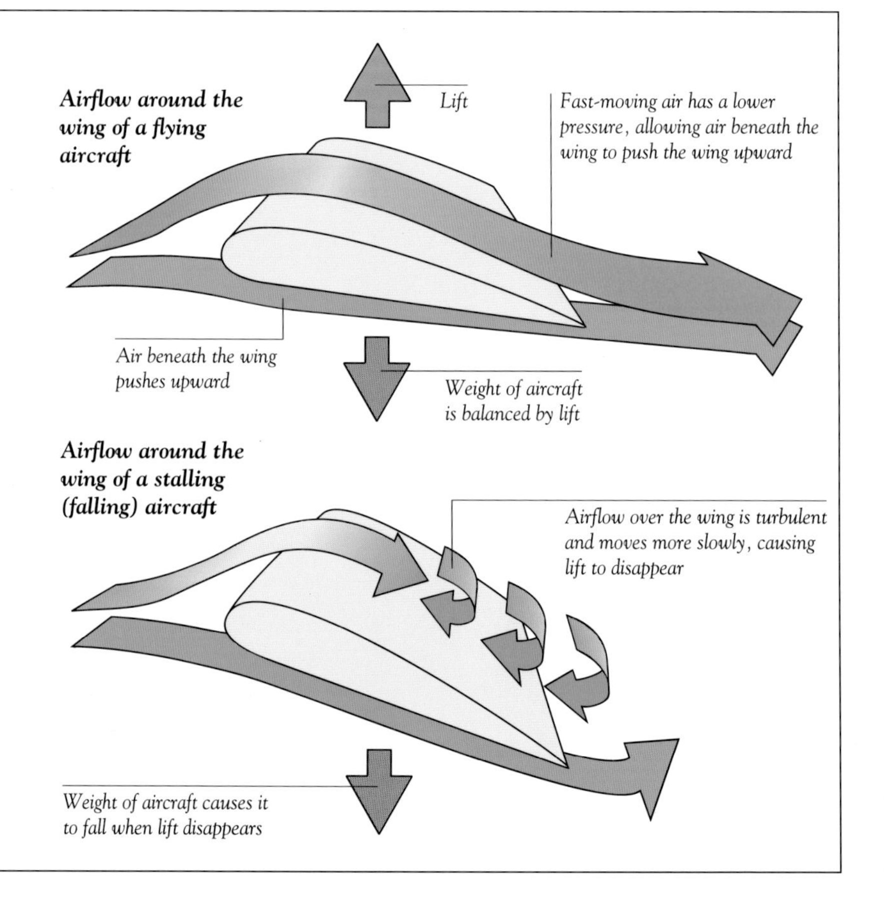

EXPERIMENT

Flying wing

Adult help is advised for this experiment

See how causing air to flow over a paper wing creates a low-pressure zone into which the wing is lifted by air pressure from below. The faster the airflow, the more lift the wing receives.

YOU WILL NEED

● *stiff paper 12* x *6 in (30* x *15 cm)* ● *cutting surface* ● *craft knife* ● *scissors* ● *2 drinking straws* ● *2 knitting needles* ● *hair drier* ● *modeling clay* ● *tape* ● *6 in (15 cm) wide piece of cardboard* ● *ruler*

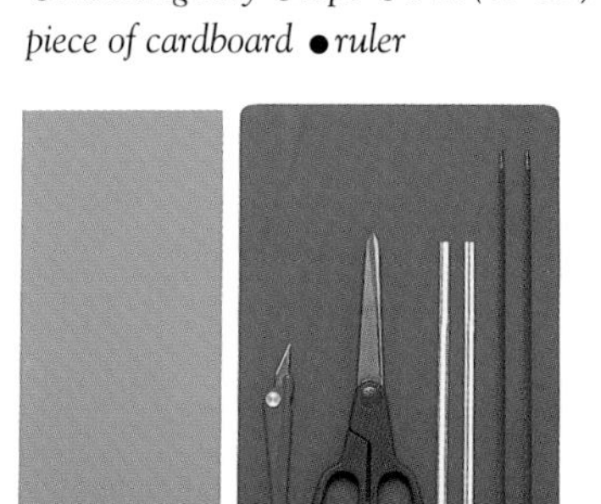

1 FOLD THE PAPER along its width about ½ in (1.5 cm) off centre. Cut holes through the paper 3 in (8 cm) apart, 2 in (5 cm) from the folded edge.

2 TAPE THE TWO ENDS of the paper together to make a wing with a curved upper surface. Push two short lengths of straw through the holes.

3 MAKE TWO slits 3 in (8 cm) apart in a length of straw. Push a knitting needle through each slit until the straw sits at the ends.

4 PUSH THE knitting needles through the straws in the wing (through the curved surface first). Place them 3 in (8 cm) apart in a block of modeling clay.

5 MAKE THE wing fly by blowing air over the top of the wing. Be careful not to push the wing up by blowing air underneath it.

Use the cold air switch if your hair-dryer has one

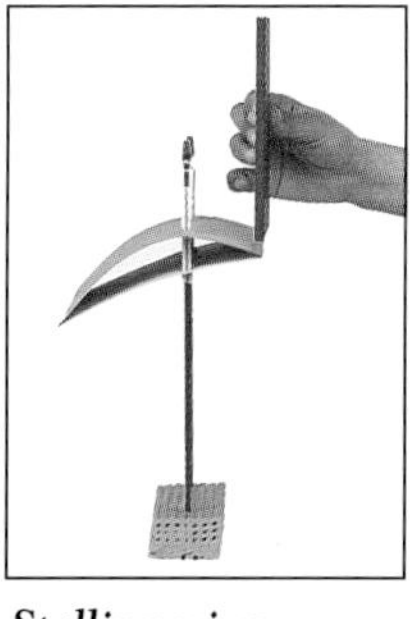

Stalling wing
Show that air flowing over the wing makes it fly by blocking the airflow over the top with the cardboard. The wing should now fall. If it does not, you are blowing air beneath the wing.

Aircraft 2

AN AIRCRAFT'S WINGS hold it up in the sky, but an aircraft also needs a tail in order to fly properly. The vertical fin and horizontal tailplane keep it stable so that it does not slew from side to side or rock up and down. To control the aircraft's height and course, the pilot operates a system of moveable control surfaces—ailerons on the wings, and a rudder and elevators on the tail. Moving these surfaces deflects the airstream passing over the wings or tail, generating forces that change the aircraft's direction. The pilot operates a control column and rudder pedals to move the control surfaces. Shifting the control column backward or forward moves the elevators on the tail so that the aircraft climbs or descends. Rudder pedals move the rudder to turn the aircraft. When turning, the pilot also tilts the control column to one side to move the ailerons on the wings. This banks the aircraft to one side so that it turns smoothly. In small aircraft the pilot's controls are linked mechanically to the control surfaces by cables. In larger aircraft the pilot's commands may be sent electronically by cables or computerized signals to hydraulically powered pistons that move the control surfaces.

EXPERIMENT

Aerial dart

A dart-shaped paper plane is easily made from a sheet of paper and flies well. Make a dart and add flaps to the wings and tail, and see how these surfaces control its flight so that it can change direction. The dart normally flies level. The control surfaces enable it to turn to either side, to climb, to dive, and even to roll as it speeds through the air.

YOU WILL NEED
- *$8\frac{1}{2}$ x 11 in (21 x 27.5 cm) paper sheet*
- *3 paper clips*
- *scissors*

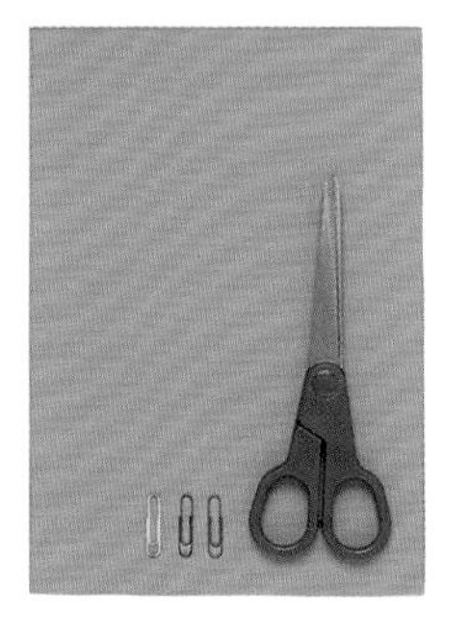

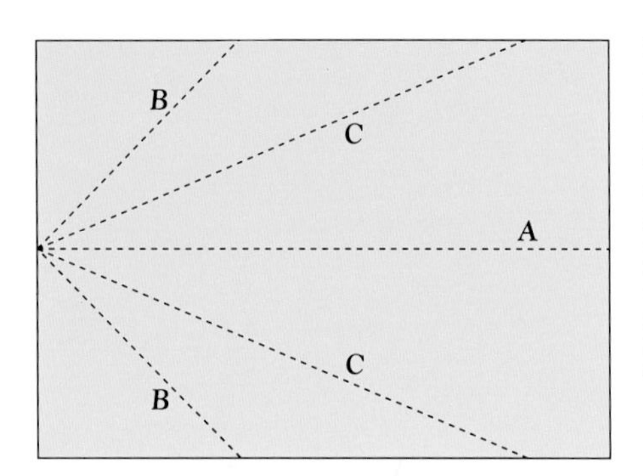

1 MAKE FOLD **A** by creasing upward along the center line of the paper.

2 FOLD TWO corners upward to the center line to make Folds **B**.

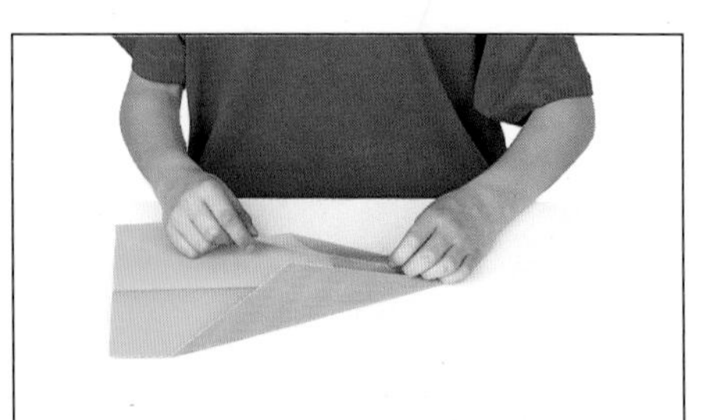

3 FOLD EACH OF THE diagonal sides upward to the center line to make Folds **C**. Next, fold the tip of the nose back inside the paper dart so that the dart's nose is not sharp.

4 FOLD THE sides downward to form two wings 1 in (2.5 cm) above the bottom of the dart. Then cut flaps on the wings and cut a rudder on the tail, as shown in the illustration opposite.

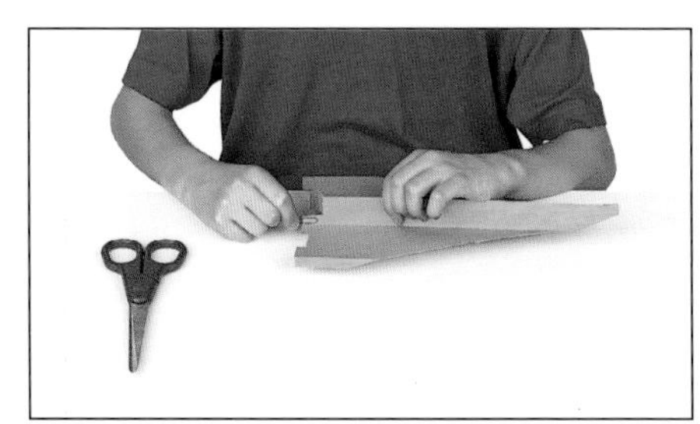

5 ADD ONE or more paper clips to the rear of the dart above the rudder until it is balanced when launched. Fold up a ½ in (1.5 cm) strip at the edge of each wing to stabilize the dart's flight.

Controlling flight

The tail of an aircraft has a vertical fin, containing a rudder, and a horizontal tailplane with two elevators, one on each side. When the position of the rudder or elevators is shifted by the pilot, the pattern of airflow over the aircraft changes and moves the tail. The rudder turns the aircraft to the right or left, while the elevators lower or raise the tail so that the aircraft pitches up or down. The ailerons on the back of the wings usually move in opposite directions, one going up and the other down. This banks or rolls the aircraft, which is necessary to make a smooth turn.

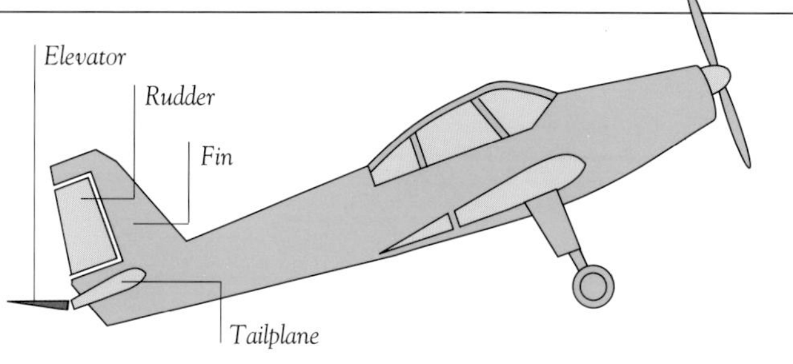

Climbing
Raising the elevators lowers the tail and raises the nose of the aircraft. The angle of the wings increases, giving the aircraft more lift so that it climbs.

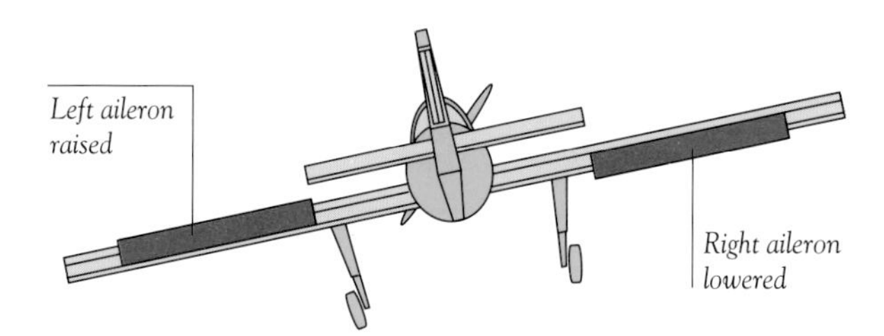

Banking
As shown in the illustration above, raising the left aileron of an aircraft gives the left wing less lift, while lowering the right aileron gives the right wing more lift. The difference in lift on the wings causes the aircraft to bank or roll to the left.

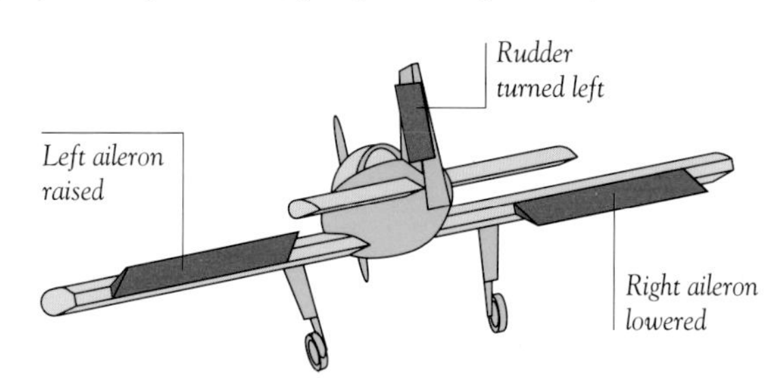

Turning
Turning the rudder to the left swings the tail right and the nose left. As the aircraft turns left, the ailerons also bank the aircraft to the left so that it turns smoothly.

6 LAUNCH THE DART and check that it flies level in a straight line. Change the flight path of the dart by bending the flaps up or down, or put one up and one down. Bend the rudder and one flap to make the dart turn.

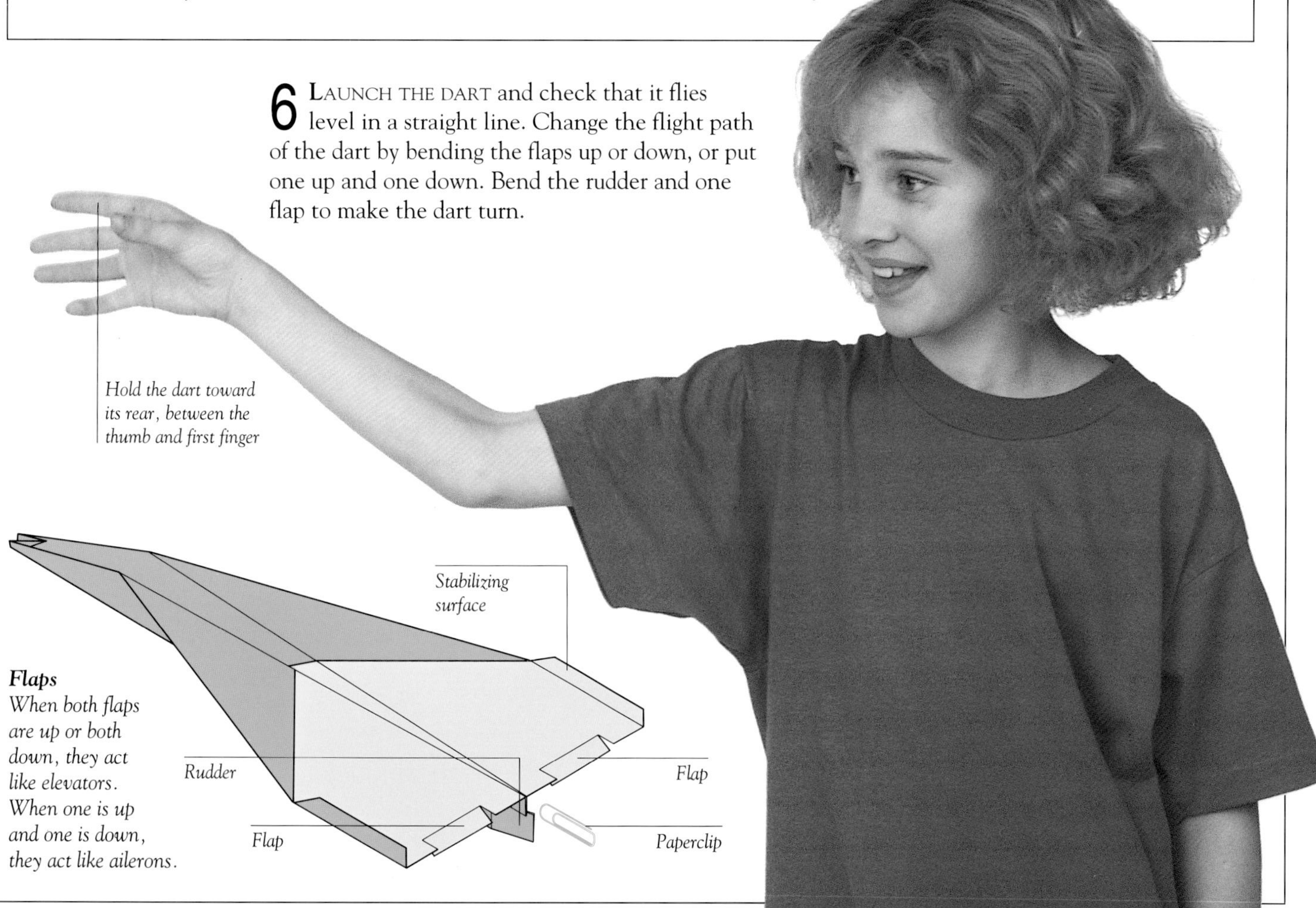

Flaps
When both flaps are up or both down, they act like elevators. When one is up and one is down, they act like ailerons.

Helicopters

HELICOPTERS ARE THE ULTIMATE flying machines, able to fly in any direction and even hover in the air motionless. They can land and take off almost anywhere because they can fly vertically. This makes them especially valuable for rescue work and for transporting freight to locations with no road access. A helicopter has two rotors (sets of rotating blades). Most have a large main rotor on top, and a small rotor on the tail. The main rotor has long, thin blades that act as airfoils (p.108) to lift the helicopter and move it through the air. The tail rotor keeps the body of the helicopter from spinning opposite the motion of the main rotor.

EXPERIMENT

Going into a spin

Show how a helicopter, represented by a fan, would spin out of control with only one rotor. The cabin would rotate in the direction opposite the motion of the rotor blades. The tail rotor exerts a sideways force to prevent this.

YOU WILL NEED

- *portable fan*
- *2 jar lids, one slightly larger than the other*
- *marbles*
- *small piece of modeling clay*

1 PLACE MARBLES in the small lid until it is almost full.

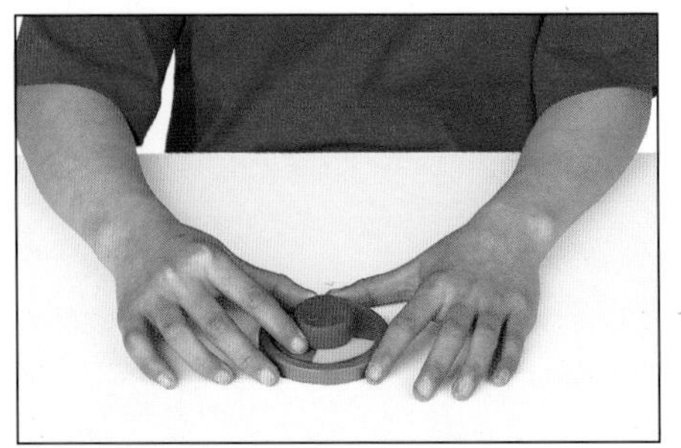

2 STICK A PIECE OF modeling clay on top of the large lid.

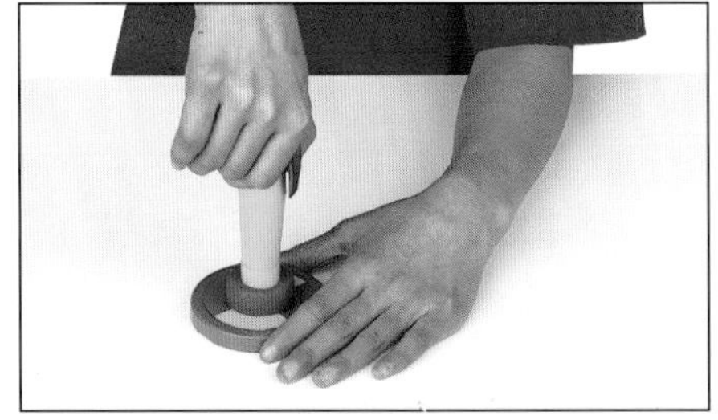

3 STICK THE FAN UPRIGHT into the clay. Place the large lid on top of the small lid and turn the fan on.

Take care to avoid touching the blades when you switch the fan on and off

4 AS THE BLADES rotate in one direction, the large lid and the body of the fan rotate in the other direction.

EXPERIMENT

Flying rotor

Make a model rotor, twirl it, and see how it flies up into the air. The rotor blades on a real helicopter work in the same way to raise and support the helicopter in the air. The blades are shaped like wings. As they whirl around, air flows over the blades to produce an upward force called lift that overcomes the helicopter's weight and raises it into the air. The lift gets less as the model rotor slows, so that it stops rising and begins to descend. A real helicopter can vary the lift to climb, hover, or descend.

YOU WILL NEED

- *steel rule*
- *cutting mat*
- *9 in (22.5 cm) square card*
- *2 colored pencils*
- *6 in (15 cm) dowel that rotates inside spool*
- *craft knife*
- *scissors*
- *thread spool*
- *ruler*
- *cardboard strip about 1 x 26 in (2.5 x 64 cm)*
- *string*
- *double-sided adhesive*

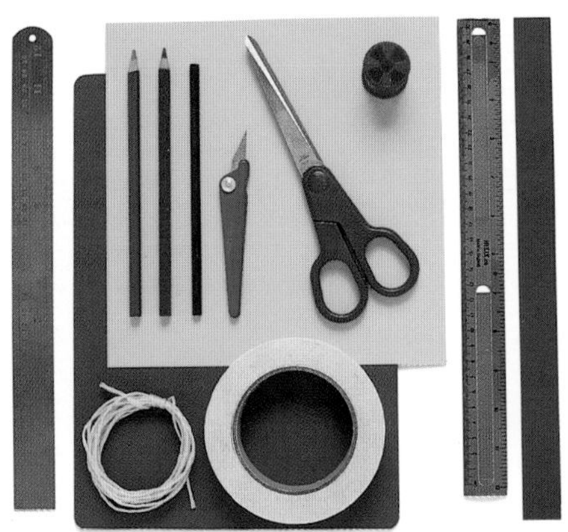

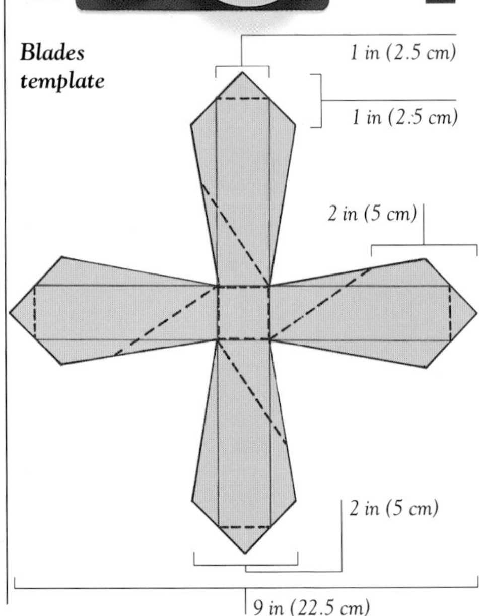

Blades template

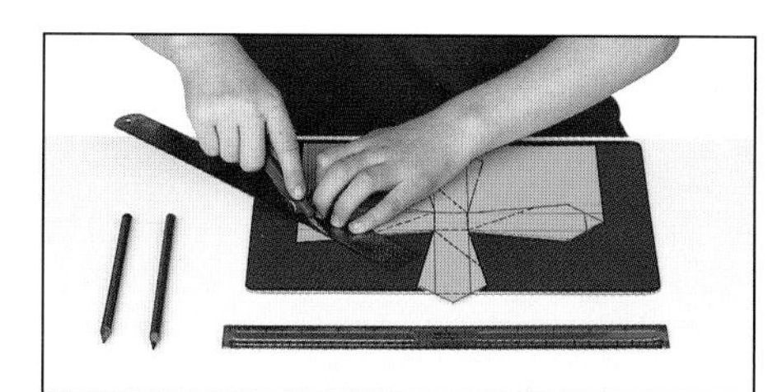

1 USE COLORED PENCILS to copy the blade shapes onto cardboard from the template on the opposite page. Cut out around the solid black lines.

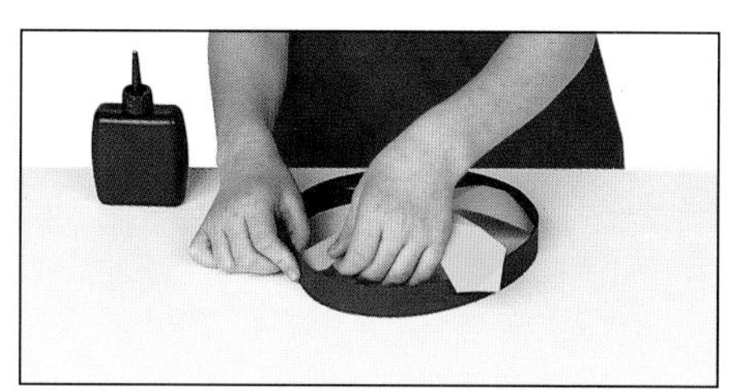

2 FOLD THE BLADES and blade tips down along the dotted lines. Glue the strip to make a circle 8 in (20 cm) across. Glue the blade tips to the circle.

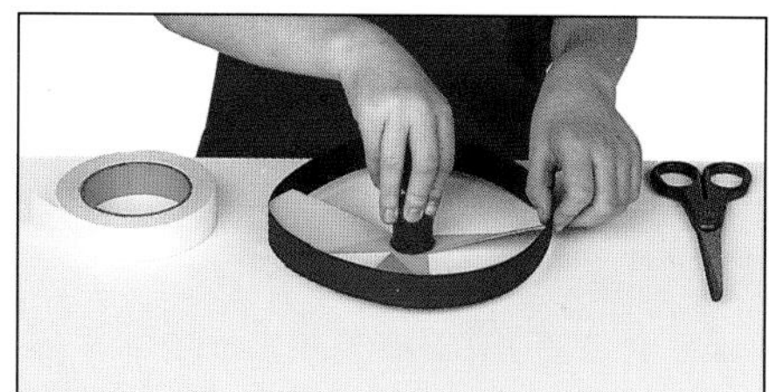

3 STICK THE THREAD SPOOL firmly beneath the center of the rotor with double-sided adhesive tape.

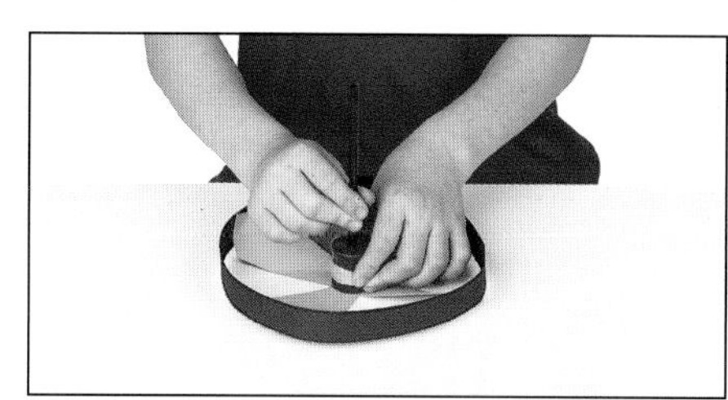

4 WITH THE SPOOL still facing upward, wind about 24 in (60 cm) of string around the spool in a counterclockwise direction. Then insert the pencil into the spool.

5 HOLD THE pencil in one hand with the blades above and pull the string sharply with the other hand to launch the helicopter rotor. Make sure that the pencil is pointing straight up.

How a helicopter flies

The pilot has two rotor controls that both change the angle of the blades. One control increases or decreases the lift of the main rotor so that the helicopter rises, stays at the same level, or descends. The second control makes the rotor and helicopter tilt forward, backward, or sideways, by producing a force called thrust which moves the helicopter in that direction. The pilot also controls the tail rotor. This produces a sideways force to keep the helicopter steady or make it turn.

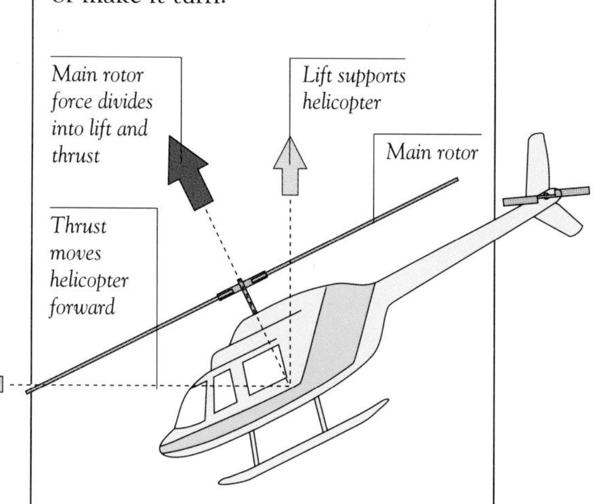

Autopilot

AN AIRPLANE can fly automatically, always heading in the right direction at the right height and compensating for winds that might blow it off course. The airliner's autopilot operates controls to keep the aircraft on a set heading. There is also a guidance system, which is programmed with the route to be taken. The guidance system knows the aircraft's position at any time, and instructs the autopilot to change direction or height as necessary. Guidance systems may receive positional information by radio from satellites or ground stations. Some systems, called inertial guidance systems, are self-contained and, given the airliner's starting position, measure its subsequent movements through the air.

How an autopilot works

An autopilot detects changes in direction and height by means of gyroscopes. These spin rapidly, keeping the shafts of the gyroscope wheels in their original vertical or horizontal orientation. When the aircraft changes direction or angle, the gyroscope frame tilts around the spinning shaft, producing a signal that goes to motors operating the controls.

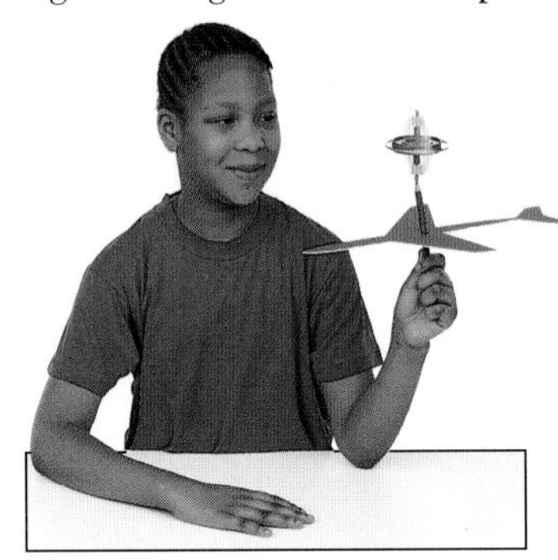

Stable spin
Stick a pencil point through a cardboard airplane shape. Stand a spinning gyroscope on the pencil point. The shaft of the gyroscope is stable and stays vertical.

Changing direction or height
Tilt the airplane, as happens if it turns, dives, or climbs. The gyroscope shaft remains vertical, provided that the wheel continues to spin fast.

EXPERIMENT

Direction finder

Adult help is advised for this experiment

An inertial guidance system uses accelerometers. These detect movements in different directions so that an aircraft can find its position. Make a simple accelerometer that detects four directions of acceleration.

YOU WILL NEED

• glue • poster board • 4.5V battery • insulated wire • cutting surface • paper • screwdriver • foamcore disk of diameter just less than bottle • compass • ruler • steel ruler • 2½ in (6 cm) diameter plastic bottle • piano wire • skewer • pencil • tape • craft knife • pliers • 8 in (20 cm) foamcore disk • four bulb holders with bulbs • aluminum foil • double-sided tape • dense rubber ball about 1½ in (4 cm) across • scissors • wire strippers

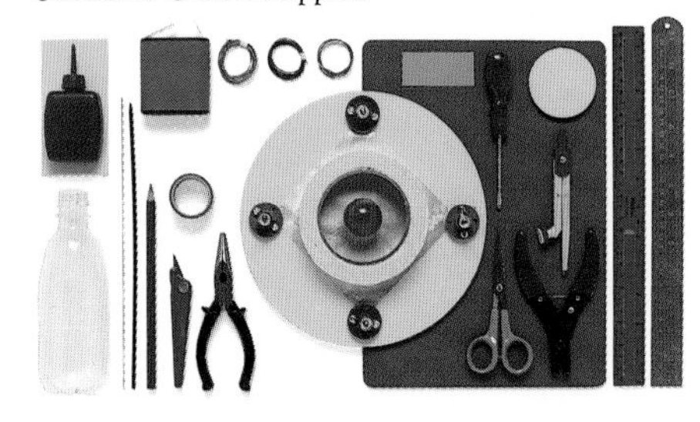

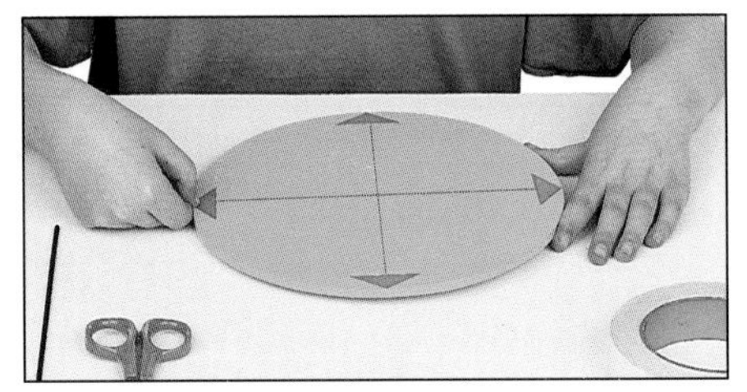

1 GLUE THE SMALL DISK to the center of the large disk. Draw a cross through the center of the large disk on the other side. Using double-sided tape, stick a paper arrow at each end of the cross. Using a skewer, make a hole in the center of the disks and four holes around the cross, 1 in (2.5 cm) from the center.

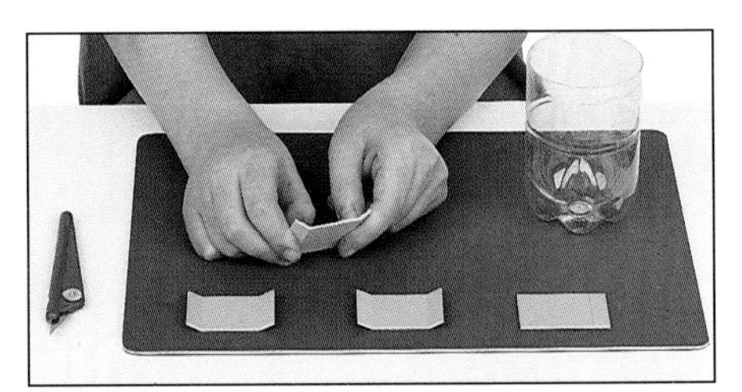

2 CUT THE TOP half from the bottle. Wrap a length of insulated wire around the bottle to measure its circumference. Cut four poster-board contacts about 2 in (5 cm) tall with a width equal to a fifth of the bottle's circumference. Fold a ¼ in (5 mm) strip along the sides of each contact piece.

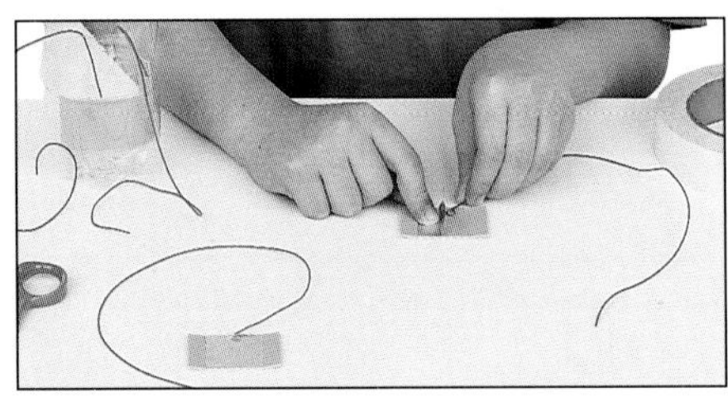

3 COVER ONE SIDE of each contact with double-sided tape, between the folded edges. Stick the end of a 12 in (30 cm) wire, with stripped ends, to each tape patch. Cover all tape with foil. Glue the contacts to the inside of the bottle at equal intervals, foil side inward.

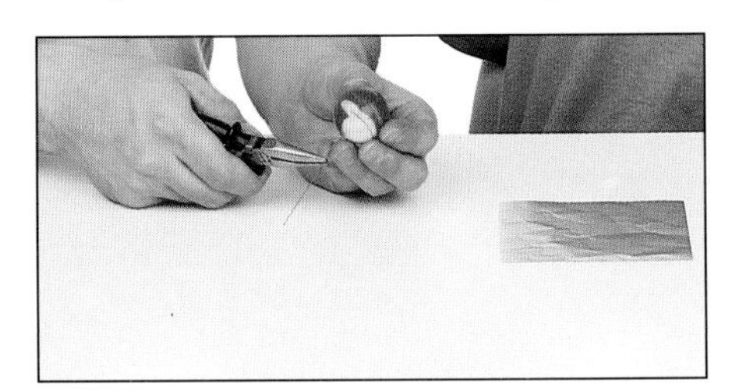

4 CUT A 6 IN (15 CM) length of piano wire. Using pliers to hold the wire, carefully push one end of the wire into the rubber ball. Continue until the ball can be swung from side to side without coming loose from the wire. Wrap the ball and the wire above the ball in foil.

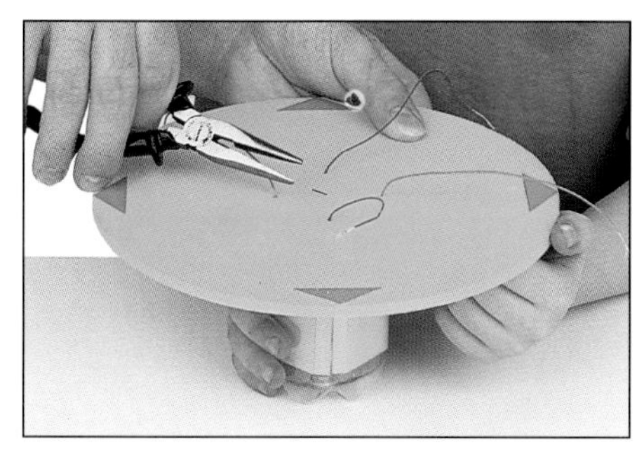

5 PUSH THE PIANO WIRE through the central hole in the small and large disks, small disk first. Bend the wire over the top surface of the disk to secure it. Feed the wires from the four foil contacts through the four remaining holes in the disks. Push the cut-off bottle onto the small disk and secure with tape.

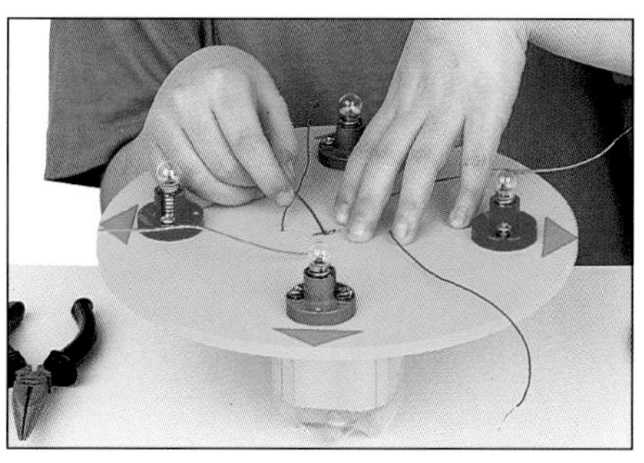

6 USE DOUBLE-SIDED TAPE to stick a bulb holder behind each pointer on the large disk. Cut five lengths of wire about 5 in (12 cm) long and strip both ends of each wire. Wind one end of one wire tightly around the piano wire and secure with tape, as shown in the photograph above.

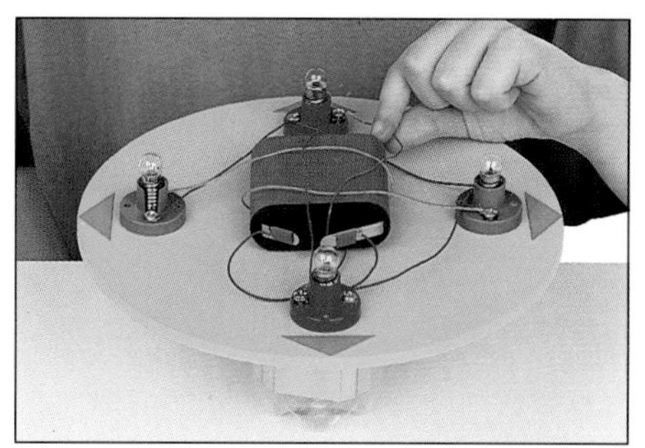

7 STICK THE BATTERY on top of the large disk. Connect the wire leading from the rubber ball to a battery terminal. Secure with tape. Connect each contact wire to the bulb on the opposite edge. Connect the free bulb terminals to the open battery terminal (follow the diagram, below left).

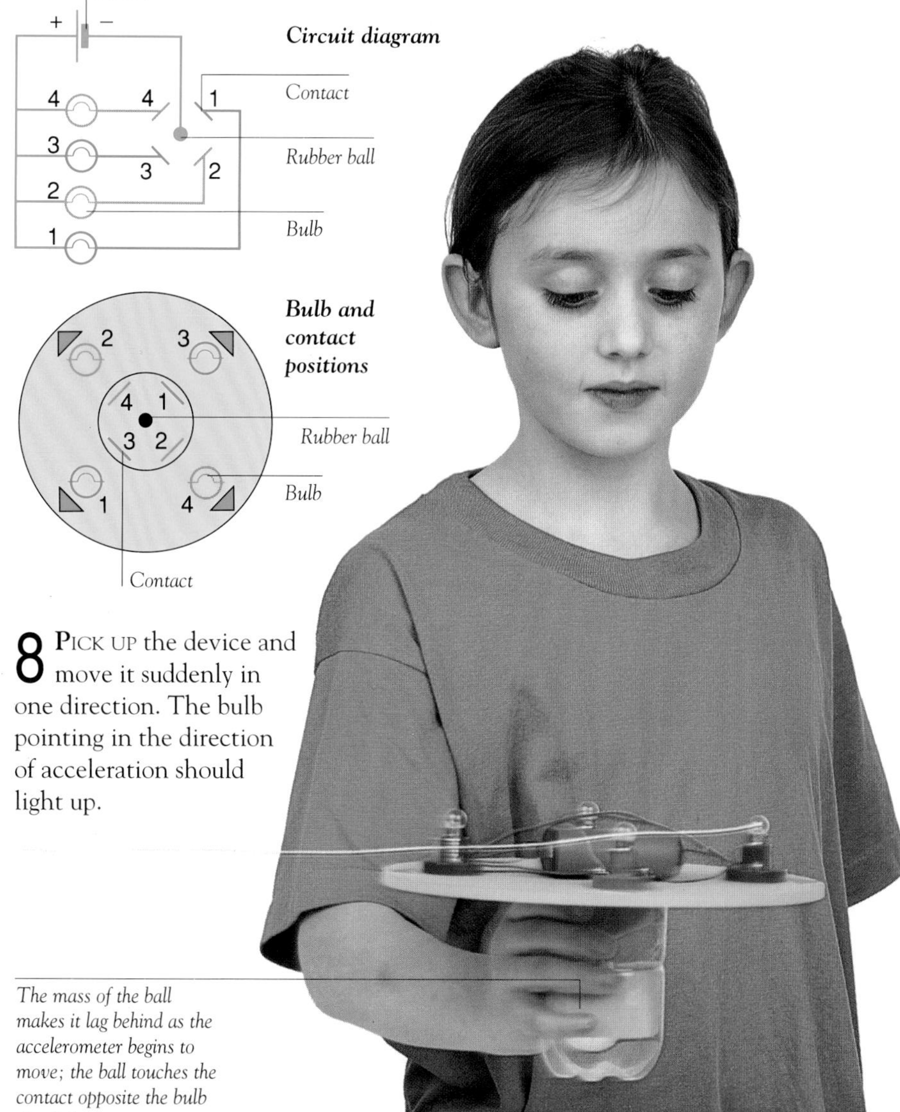

8 PICK UP the device and move it suddenly in one direction. The bulb pointing in the direction of acceleration should light up.

The mass of the ball makes it lag behind as the accelerometer begins to move; the ball touches the contact opposite the bulb that lights up

Accelerometer

An accelerometer in an inertial guidance system has an armature mounted on springs above three coils. An electric input signal in the middle coil causes an output signal to flow from each outer coil. A change of speed, direction, or height in line with the coils produces an acceleration that causes the armature to lag behind and get closer to one of the outer coils. This boosts the output signal from that coil. A computer measures the changing output signals from three accelerometers and continually calculates the aircraft's latitude, longitude, and height.

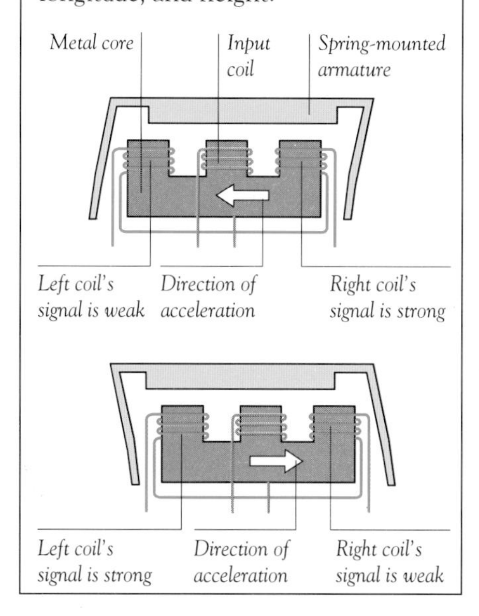

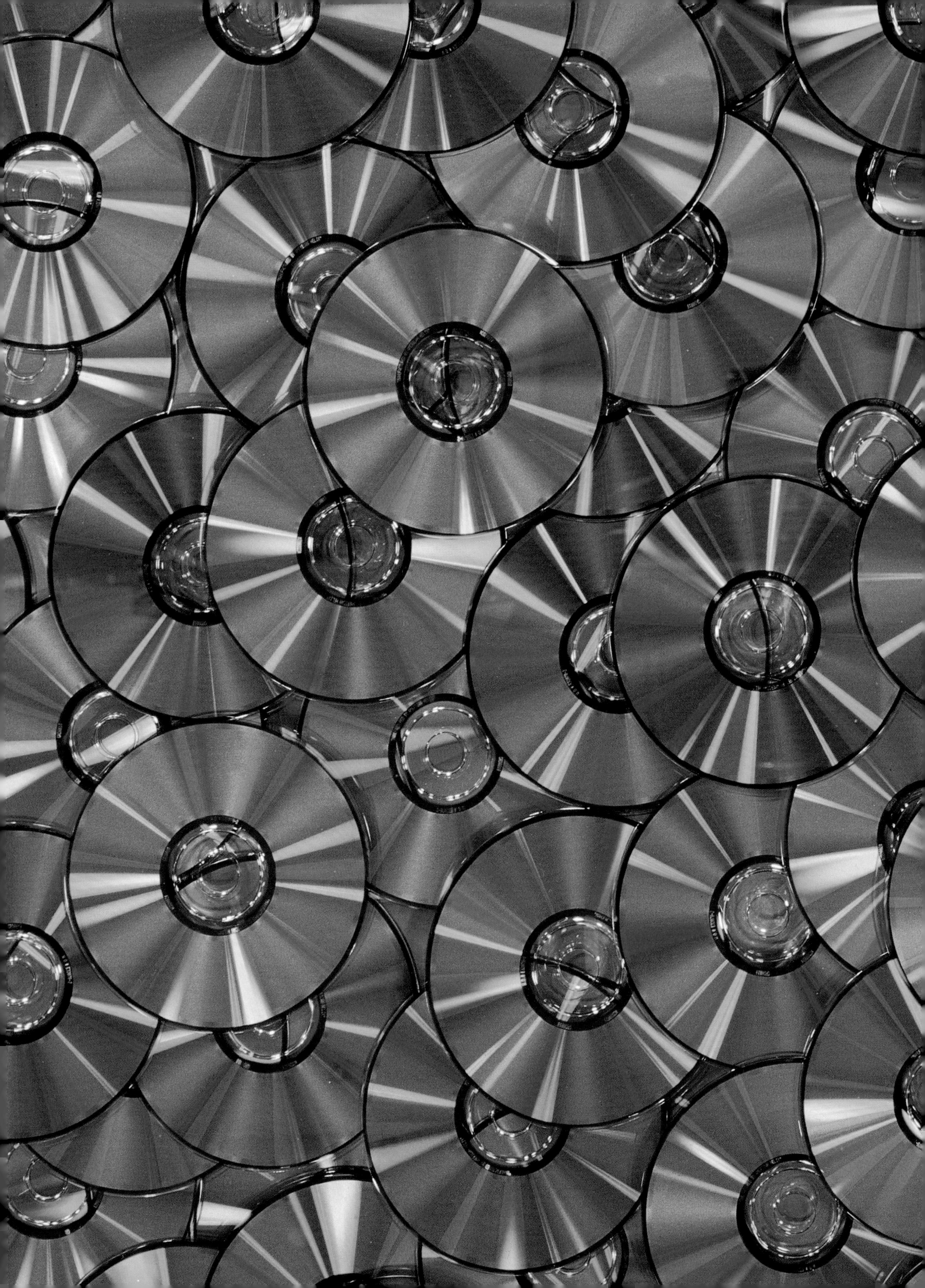

LEISURE

Tradition meets technology
This Irish harp (above) was made in 1820, and its maker began a revival of the old tradition of harp-playing in Ireland. While much music is still made on traditional instruments, the recording of music has greatly advanced with the high-quality sound of compact discs (left). These glow with iridescent colours produced by light reflected by the ultra-thin spiral track on each disc's surface.

OUR LIVES ARE SELDOM FREE from machines, and even our time off frequently involves technology. While we can relax by just thinking or reading, many leisure activities need some kind of artificial assistance. Machines can bring us new ways of enjoying ourselves. Personal stereos, for example, enable us to listen to music anywhere, and computer games provide us with electronic opponents. Technology can also improve traditional pastimes such as sports, where equipment is continually developed to help us perform to the best of our abilities.

PLAY AND PLEASURE

THE PROGRESS OF TECHNOLOGY continues to bring us more and more machines, many of which help us in our work and in the home. By taking over everyday chores and providing swift means of transport, machines leave us with more time to spend at leisure. But few of us leave machines behind when we go out or put our feet up to relax: technology switches on to transform the way that we switch off.

Many people enjoy highly active leisure pursuits, but others prefer to take things easy and relax. All, however, can choose from a far wider range of experiences than the simple pleasures of times past. Some activities are essentially the same, but have been updated by technology. For example, the parlor piano around which friends gathered to sing has given way (for some) to the karaoke machine. Technology has also opened some activities to more people: creating pictures once required painting or drawing skills, but now we can all produce an instant and accurate likeness with a camera. In addition, machines offer us completely new ways to enjoy ourselves. The rides at a theme park, or a flight on a hang glider, bring death-defying thrills that few would have thought possible, let alone enjoyable, a century ago.

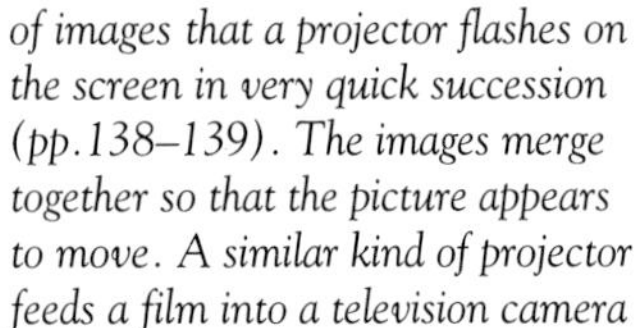

A movie film contains a strip *of images that a projector flashes on the screen in very quick succession (pp.138–139). The images merge together so that the picture appears to move. A similar kind of projector feeds a film into a television camera when a film is shown on television.*

Sport and exercise

Some sports have changed little over time; many ball games are centuries old and are still played in much the same way as when they were first devised. The players' performance has greatly improved, however, partly due to changes in the design and materials of their equipment. Modern tennis rackets are made of superstrong but light reinforced plastics, and can launch a ball at sizzling speed. Rock climbers and mountaineers can climb more difficult routes using improved gear, and are more likely to return to tell the tale. Sailing has also been revolutionized by advances in design and materials, so that today's record-breaking yachts are as much the products of laboratories and computers as of boat builders. Technology has also brought us new water sports such as water-skiing, windsurfing, and parascending. Below the surface, the aqualung has allowed amateur enthusiasts to enter an underwater world that was once visited only by professional divers. The sky, too, has become a new sporting domain for the pilots of parachutes, paragliders, hot-air balloons, hang gliders, ultralights, and other aircraft.

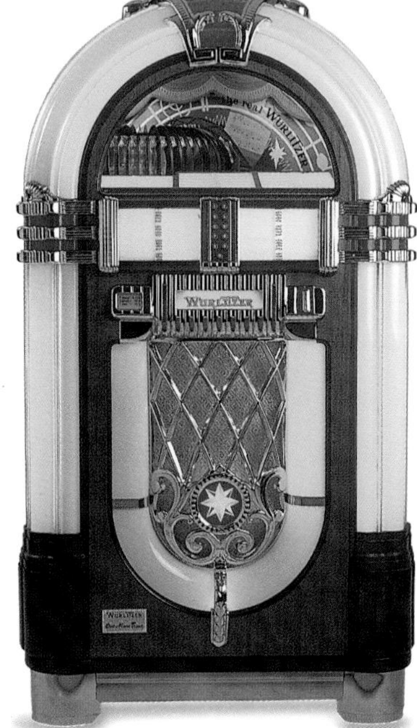

When you insert a coin, and select *a song by pressing a number on a keypad, a jukebox plays your chosen record. This old jukebox contains a big loudspeaker and a large stack of records. Each record is automatically put on a turntable, played by a tone arm holding a needle, then returned to the stack.*

A loudspeaker produces the sound *that comes from a stereo system, radio set, or television set. The earphones used with personal stereos are pairs of tiny loudspeakers (pp.140–141).*

Exercise and training machines have been of great benefit to people involved in sports, as well as providing leisure activities for the less athletic. For example, the bicycle, now often fitted with a wide range of gears, is becoming increasingly popular for exercise, competition, and leisure. Even walking, perhaps the oldest leisure activity of all, has been enhanced by technology, with the development of tough, light boots and good rainwear that make this activity comfortable in almost any terrain and weather.

Arts and crafts

Professional artists strive to attain ever more outstanding results, while amateurs use their skills for personal pleasure. Both ambitions are enhanced by technology. In music, conventional instruments are scientifically designed so that they will respond easily and rapidly to the demands of players. New ways of making music have also been developed, for example with instruments such as the electric guitar, and more recently with synthesizers, electronic instruments that can produce a huge range of sounds. These instruments, along with music programs for computers, have greatly broadened the experience of music-making.

Photography is also solidly based on technology. Recent developments have made the still camera completely automatic, able to produce good photographs while requiring very little expertise from the user. The video camera or camcorder enables people to take moving pictures with ease, and home editing equipment can help make these pictures into good productions.

Modern equipment can enhance hobbies such as gardening, woodwork, and sewing just as it has changed professional people's activities or competitive sports. The results may be as good as those achieved by professionals, and often bring great pleasure to the participants.

Passive pleasures

In addition to improving sports and creative pursuits, technology has given rise to leisure activities that involve no physical effort. Until the twentieth century, people wishing to relax had only a very limited range of pastimes, such as reading, with which to amuse or uplift themselves. Now we can choose to watch television, a video, or a film at the theater; or listen to the radio, a record, a tape, or a compact disc. Computers bring more new ways for us to enjoy our leisure hours. Today we can interact with them, playing games such as chess at any standard from that of beginner to grand master, and enjoying simulations of sports. Computerized leisure activities will become even more complex with the development of virtual reality, in which you can enter a computer-generated world that seems to surround you and to respond to your movements. And in the near future, we may not need to go to the movies, go out to buy or rent tapes or discs, or wait for programs transmitted at specific times. Today's growing information superhighway, which uses cables and satellite links to connect computers to each other, will give access to video, television, music, and a range of other services such as library stacks, newpapers, and magazines. Much of the information available in the superhighway travels along fiber-optic cables, which carry information from huge central data banks in the form of light signals fired from lasers.

An electric guitar has a pickup *that sends out an electric signal when the strings are plucked and vibrate. An amplifier and loudspeaker connected to the pickup turn the signal into sound. Controls on the amplifier can vary the volume and tone of the guitar.*

It has taken several advances in technology to make possible this massive expansion in home leisure services. Perhaps the first was the discovery of electricity, a safe power source for home lighting, which opened dark evenings to a wider range of leisure activities. Many leisure machines in the home are also powered by electricity. They often work by electronics, which can be used to change an image or sounds into an electric signal that can be transmitted or recorded and then converted back into the original form. Most such machines now work by analog technology, a method prone to producing pictures that are fuzzy or sound with background noise. Digital technology improves both transmission and reproduction quality: it is already used in sound reproduction systems, and radio, television, and video will soon use it too. Pictures and sounds will then reach us with even more clarity.

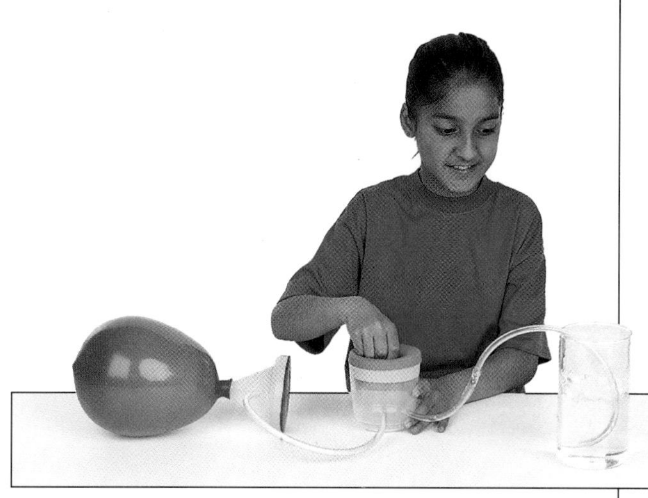

An aqualung enables a diver to breathe underwater without *wearing a diving suit. Air from a supply cylinder reaches the diver's mouth at the right pressure for normal breathing (pp.126–127).*

You can train or exercise in your own home using *gymnasium equipment like the machine in this picture. Specially-designed exercise machines can tone up specific parts of your body, or keep you generally fit.*

Sunglasses

OUR EYES HAVE A natural way of coping with bright light. Their pupils contract to prevent too much light from entering the eyes. However, on sunny days there may be too much bright sunlight for comfort, so people wear sunglasses to reduce the brightness. The simplest sunglasses have thin lenses of dark glass or plastic that absorb much of the light entering them. Polarizing sunglasses also absorb light, but in addition they cut out bright glare, which is produced by sunlight reflecting from shiny surfaces and can be very uncomfortable. Photochromic lenses contain substances that darken in sunlight, and clear as the light dims. Photochromic spectacles are very useful for people who need prescription glasses, because they can be worn all the time and a separate pair of sunglasses is not needed.

Dazzling snow

The summits and top slopes of high mountains often lie above the clouds. The sun usually shines brightly, and white snow reflects much more sunlight than green grass or brown soil. Skiers and mountaineers may need to wear sunglasses in order to see properly.

EXPERIMENT

Light and dark

Adult help is advised for this experiment

Show how sensitively photochromic glasses react to sunlight. The lenses contain clear silver compounds like those used in photographic film. Invisible ultraviolet rays in sunlight make the clear compounds split up to form black silver, and the lenses darken. The rays are less strong when the sun is hidden, and the dark color soon fades as the black silver forms clear compounds again.

YOU WILL NEED
- *cutting surface*
- *photochromic glasses*
- *cardboard* • *tape*
- *craft knife* • *pencil*
- *scissors*

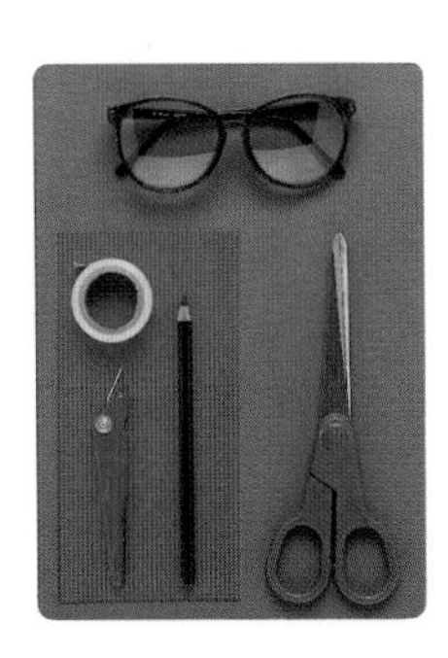

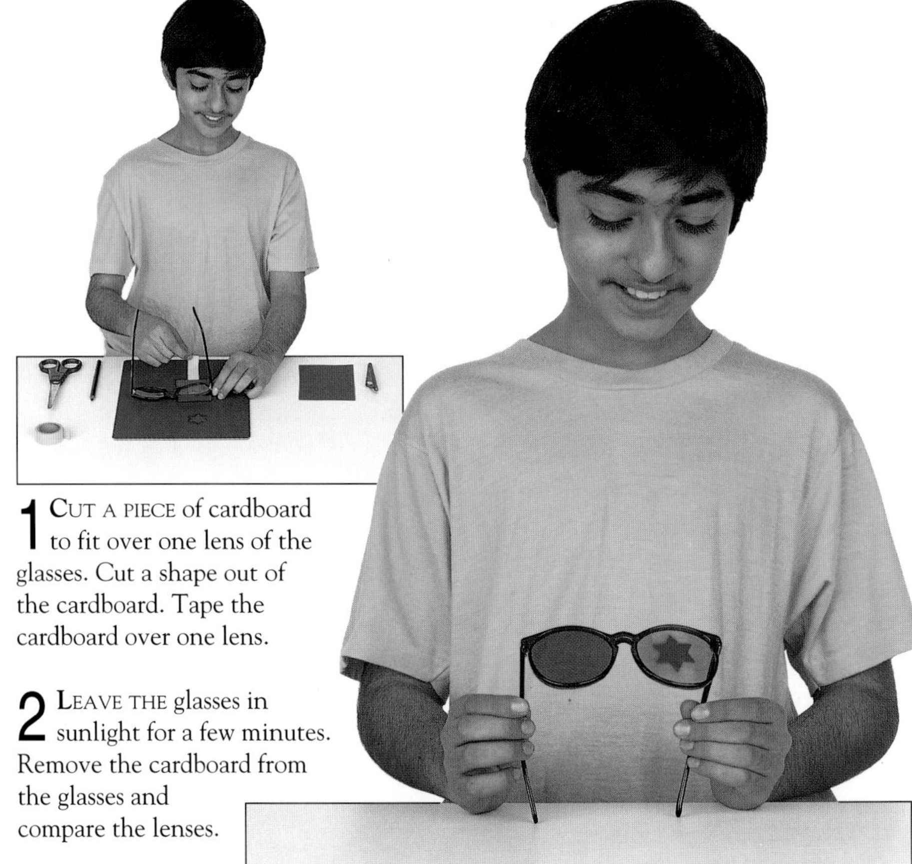

1 CUT A PIECE of cardboard to fit over one lens of the glasses. Cut a shape out of the cardboard. Tape the cardboard over one lens.

2 LEAVE THE glasses in sunlight for a few minutes. Remove the cardboard from the glasses and compare the lenses.

EXPERIMENT

Disappearing trick

Polarizing sunglasses are named for the way they block polarized light. This kind of light is produced when light rays reflect from some surfaces, such as water. Polarized light looks exactly the same as ordinary light to our eyes, but unlike ordinary light it vibrates in only one plane. The glare reflected from water is polarized light, and so it can be blocked by wearing polarizing sunglasses. Show how this happens using light from a bulb and a pair of polarizing sunglasses.

You Will Need

- *polarizing sunglasses*
- *small bright object*
- *large dish*
- *pitcher of water*
- *desk lamp*

1 Place a brightly colored object in the middle of a dish of water. Ask a friend to shine a lamp on the water surface at an angle of about 40°. Look at the dish from the other side so you can see the bulb reflected in the water.

2 Look at the water without sunglasses. Move your head so that the image of the bulb coincides with the object at the bottom of the dish. The bright reflection of the bulb obscures the object.

3 Now look at the water through polarizing sunglasses. The polarized light reflected from the surface of the water is filtered out by the polarizing lenses, allowing the object below the water's surface to be seen.

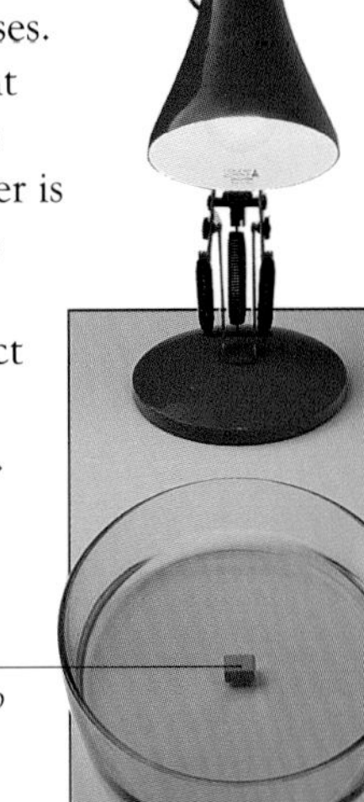

Moving your head up or down may help you to see the object

How polarizing sunglasses work

Light can be thought of as waves of energy. The waves normally vibrate in planes at all angles. Polarized light is light vibrating in only one plane. The light from the bulb shown here (and from the sun) is normal, unpolarized light, but only rays in the horizontal and vertical planes have been illustrated. When the rays strike the horizontal surface of the water, the light becomes polarized. Only the horizontal rays are reflected from the surface. The vertical rays enter the water and are reflected from the bottom of the tank. The lenses of polarizing sunglasses block horizontal rays, causing the surface reflection to disappear. The sunglasses allow vertical rays to pass, making the bottom of the tank clearly visible.

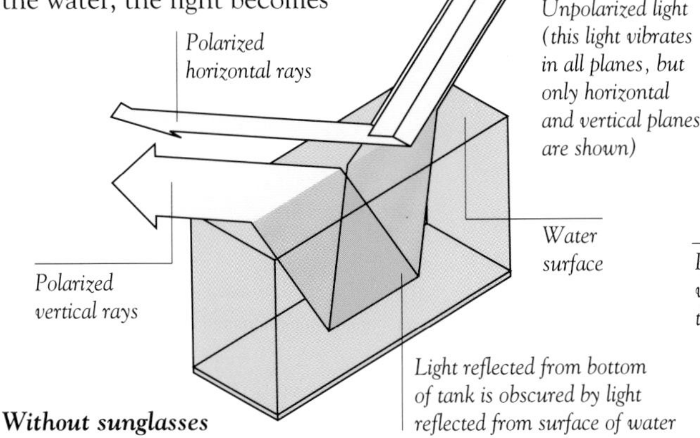

Without sunglasses

Polarized horizontal rays blocked

Polarized vertical rays pass through lens

Polarizing lens

Light reflected from bottom of tank can be seen

With sunglasses

Kites

A DIAMOND OR SQUARE of paper, plastic, or cloth stretched over a light frame makes a simple kite that will fly well. Kites may also be highly elaborate constructions in fantastic shapes. Whatever their shape, all kites fly in basically the same way. The wind pushes the kite up into the sky and because it is tethered to a long line, the kite hovers in the air. In many kites, the surface curves so that air flowing over the kite produces an additional lifting force in the same way as a wing (p.108). Kites may have long tails that help to keep them facing into the wind. A kite may also have control lines that make steering possible, so that it can move in the sky, swooping and climbing at the flier's command. While most people fly kites for pleasure, kites have been used to carry crop-spraying equipment, cameras for aerial photography, and meteorological instruments to collect weather data. New meteorological kites are being developed to reach record altitudes of up to 6 miles (10 kilometers).

How a kite flies

A kite is made to fly at an angle so that it deflects the wind downward, producing a force that pushes the kite up and back. The flier pulls on the line, creating a tension that counteracts the wind force. The two forces almost balance, leaving only a small upward lifting force that overcomes the kite's weight. As the kite rises and its angle lessens, the lift decreases. When the lift equals the kite's weight, the kite hovers.

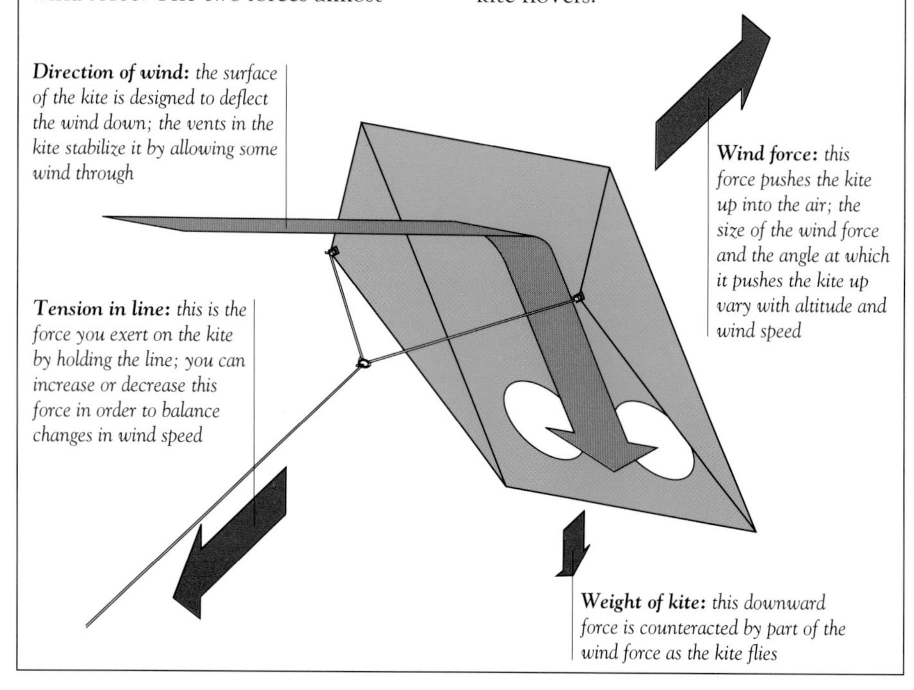

EXPERIMENT

Plastic kite

Adult help is advised for this experiment

Construct a simple kite from strong plastic sheet and wood strip. Have a friend hold the kite facing into the wind as you prepare to launch it (it should fly well in a steady breeze). As the wind blows more strongly or gently, the kite will climb or descend. If the wind drops and the kite begins to fall, you may be able to keep it up in the sky by pulling in the line to increase the pressure of air against the kite.

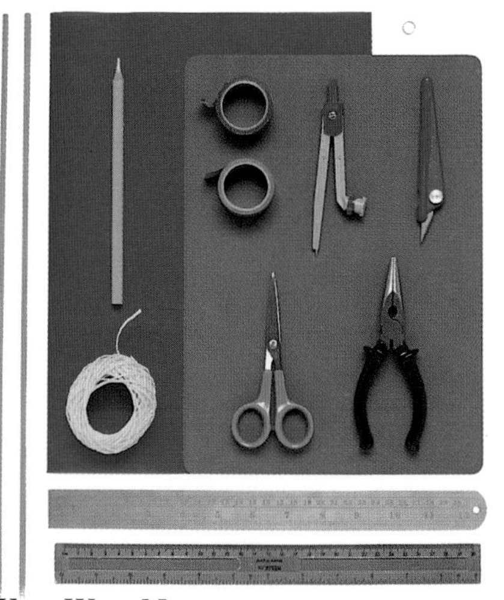

YOU WILL NEED

- *two 22-in (56-cm) lengths 1/8 in (3 mm) square wood strip*
- *sheet of strong plastic 25 x 22 in (64 x 56 cm)*
- *pencil*
- *tape*
- *compass*
- *small metal O-ring*
- *craft knife*
- *kite line*
- *scissors*
- *pliers*
- *cutting mat*
- *stiff wire*
- *steel ruler*
- *ruler*

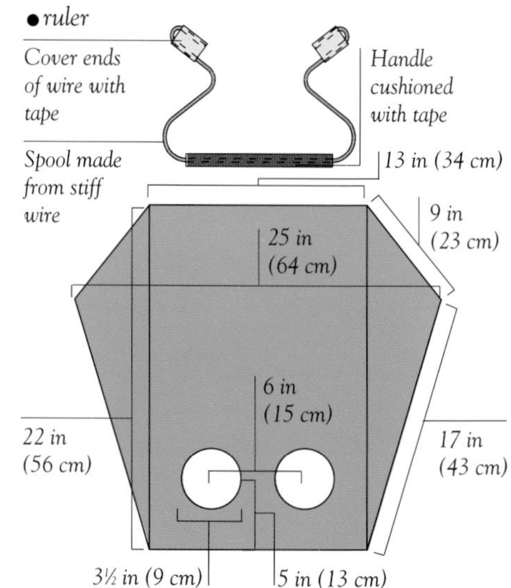

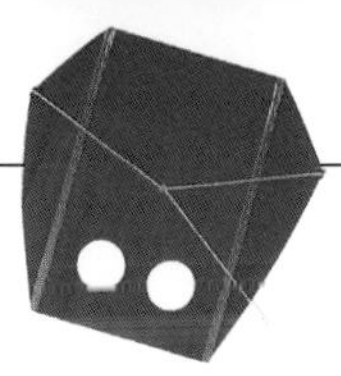

Parascending

You can fly as well as swim at many seaside resorts. Parascending is a sport that enables you to take to the air and soar like a huge bird. You are harnessed to a large kite that resembles a parachute, and then towed on a long line behind a speedboat. As the kite is pulled forward, it soars up into the sky and carries you aloft. The air pushing against the kite lifts it in the same way as the wind blowing on an ordinary kite. The speedboat keeps a constant speed so that the kite flies at a steady height. To end the flight, the speedboat slows and you drop down to land gently in the water.

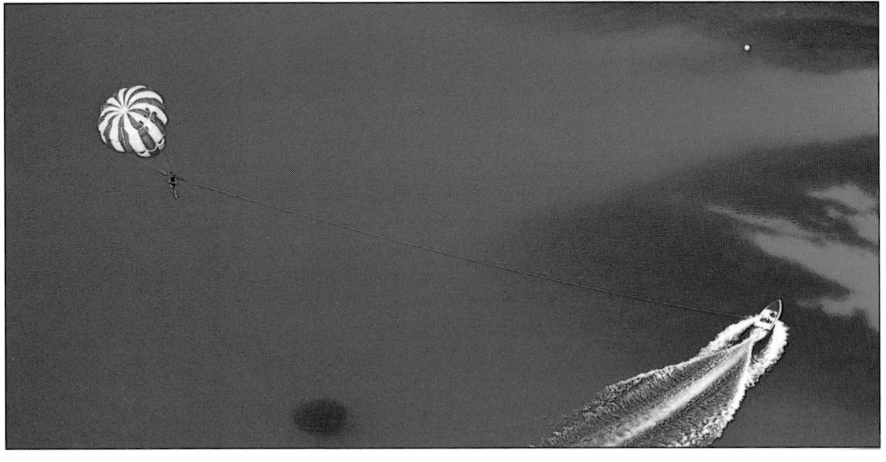

Soaring in safety
The kites used in parascending are similar in shape to a parachute. If the line breaks, the kite will slowly parachute down and land the flier safely in the sea.

1 COPY THE TEMPLATE opposite onto the plastic sheet and cut it out. Stick the two lengths of wood strip to the sheet with tape.

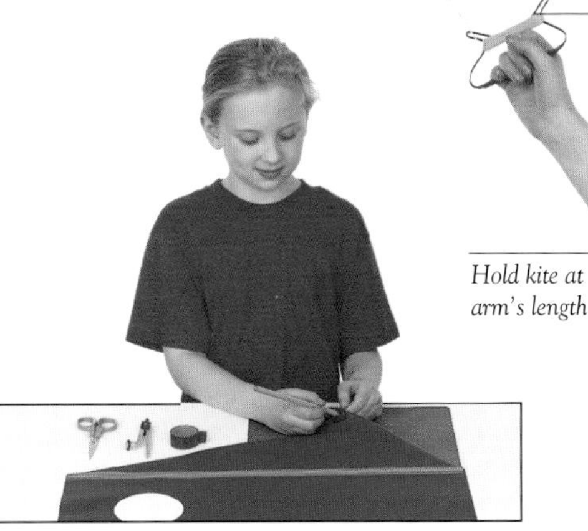

2 REINFORCE THE OUTER corners of the kite with tape. Make a small hole in each corner with a compass, and enlarge with a pencil.

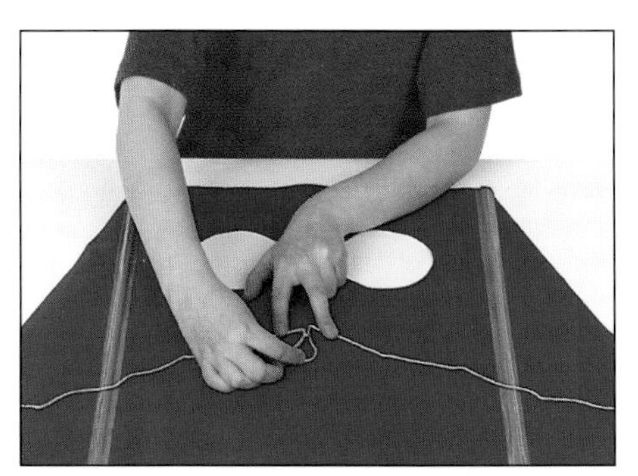

3 TIE THE ENDS OF a 34-in (85-cm) length of kite line securely to the two small holes in the corners to make a bridle. Loop the middle of the line through the O-ring. Pass the loop back over the ring; pull the line taut to secure the line to the ring.

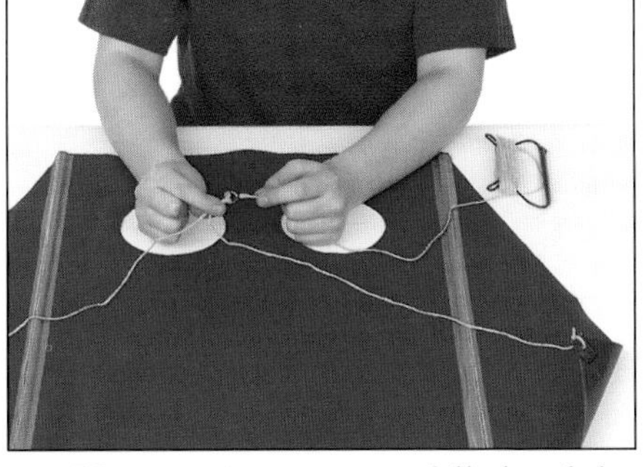

4 USING THE PLIERS, carefully bend the stiff wire into a spool for the line. Cover the ends and the handle with tape. Tie a very long piece of line—at least 80 ft (25 m) long—to the spool and wind it around. Tie the other end to the O-ring.

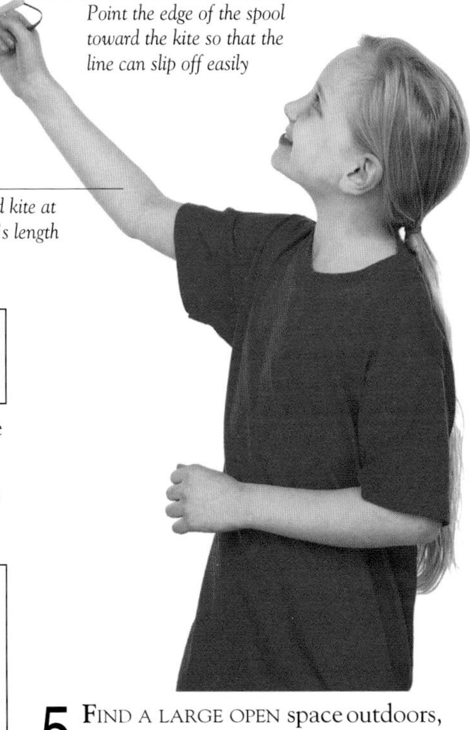

Point the edge of the spool toward the kite so that the line can slip off easily

Hold kite at arm's length

5 FIND A LARGE OPEN space outdoors, away from buildings, trees, and power lines. Choose a day with light to moderate winds. Never fly a kite during a thunderstorm. Stand with your back to the wind, and hold the spool in one hand and the kite in the other. Release the kite into the air, letting out a little line, and watch the wind lift your kite away. If the kite does not fly perfectly straight, retrieve it and move the O-ring along the bridle, away from the direction favored by the kite.

Metal detectors

BURIED TREASURE is the goal of many people who use metal detectors, but they often only find useless metal objects hidden in the ground. The detector can locate certain metals, but cannot tell valuable objects from everyday things. It works by sweeping the detector head, which contains an electric coil, over the ground. A magnetic field produced by the coil penetrates the soil; if the field passes through iron, steel, or certain other metals, a weak electric current is generated in the metal. This current in turn produces its own magnetic field, which passes through the same coil or another coil in the detector head and generates a weak electric signal in this coil. Sensitive electronic components detect the signal and operate a warning light or meter, or sound a warning in earphones, to indicate a find. Metal detectors are also used to detect and reject food contaminated by metal objects.

EXPERIMENT

Searching the sand

Adult help is advised for this experiment

Build a working metal detector that finds iron or steel objects. Such objects disturb a magnetic field from the detector, producing a weak electric signal in a coil around the magnet. The signal is amplified to make a buzzer sound. *Read pages 10–11 before starting this experiment.*

YOU WILL NEED

• large glass or plastic bowl • sand • tape • saw • pen • modeling clay • strong magnet or magnets • scissors • cutting mat • breadboard and base • TL071 operational amplifier (op-amp) • NPN transistor, BC441 or equivalent • 32 SWG wire with thin insulation • insulated wire • two 9V batteries and connecting leads • steel paper clips • buzzer • pliers • wire strippers • large plastic bottle with handle

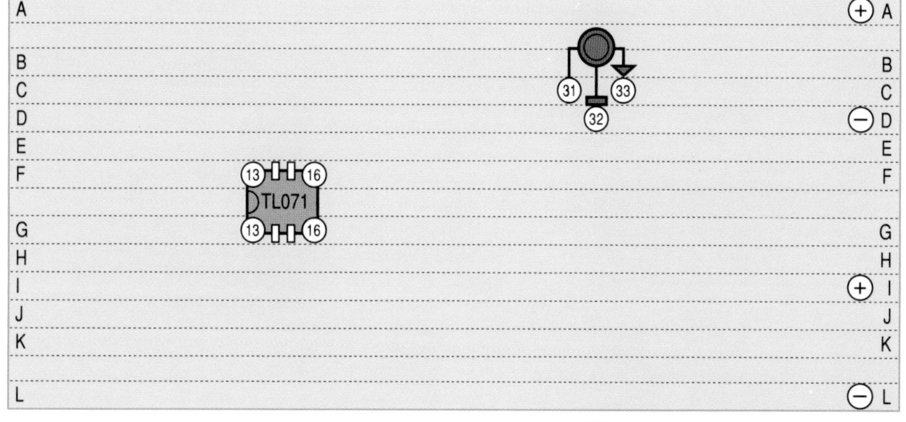

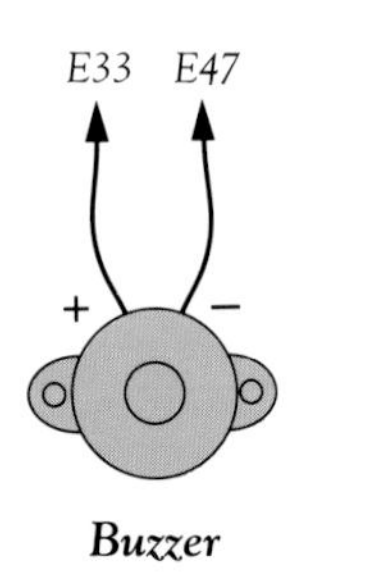

Buzzer

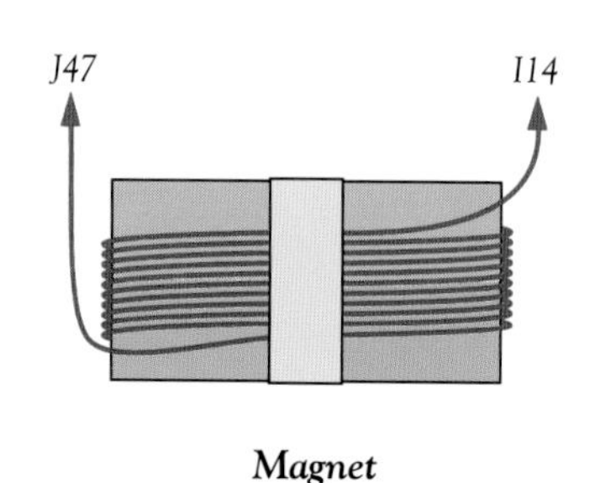

Magnet

Wires

A14–B14 A31–B31 E15–E32
F47–G47 H15–H47 K16–L16

Battery connections

This circuit needs ***two*** *9V batteries in order for the the detector to work. Battery 1 is connected between A47 (+) and D47 (–). Battery 2 is connected between I47 (+) and L47 (–).*

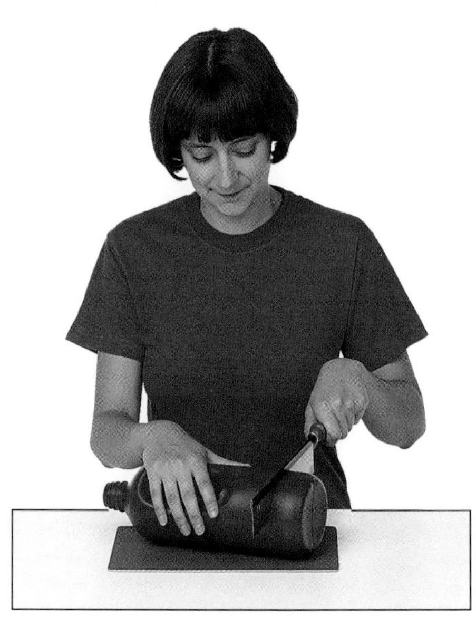

1 ASK AN adult to saw through most of the base of the plastic bottle, about 2 in (5 cm) from the bottom. Leave a flap of plastic to hinge the base to the bottle.

2 Bend the base of the plastic bottle downward to form the head of the metal detector. Stick a cube of modeling clay underneath the hinge so that the base is secured at 90° to the rest of the bottle.

3 Wind thinly insulated wire 200 times around the magnet, leaving pole faces uncovered. Leave 6 in (15 cm) free at each end. Strip the ends. Use ordinary insulated wire to connect the breadboard circuit shown opposite.

4 When the buzzer and coil have been connected to the breadboard, secure the base, breadboard, buzzer, and batteries inside the bottle with modeling clay. Secure the magnet and coil in the bottle with one pole facing downward.

5 Ask a friend to hide some steel paper clips in a bowl of sand, about ½ in (1 cm) below the surface. Move the detector around with the head about ½ in (1 cm) above the sand. The buzzer sounds when a metal object is detected.

Detection by induction

Metal detectors use electromagnetic induction, which is the production of an electric current in a metal object. This occurs when the object is inside a magnetic field and the object and field move relative to each other. This principle is used in many electrical machines, including other kinds of detectors. Ticket and vending machines check the values of coins by detecting the kind and amount of metal they contain. Some traffic lights detect the metal in cars.

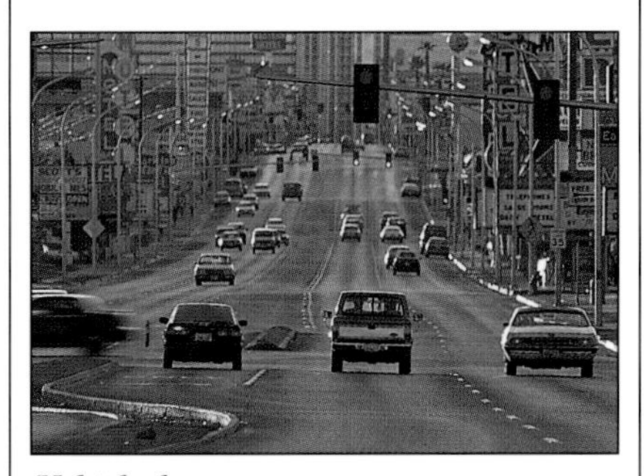

Vehicle detector
Some traffic lights work by detecting and counting vehicles. The lights are connected to an electrified loop of wire in the road surface. Like the coil of a metal detector, this loop detects the metal in every vehicle passing over it and sends a signal to the traffic lights.

Aqualungs

WITH THE RIGHT equipment and training, you can swim underwater almost like a fish. An aqualung or scuba (Self-Contained Underwater Breathing Apparatus) allows you to breathe underwater. Using a mouthpiece, you breathe in air from a cylinder of compressed air on your back. Valves ensure that the pressure of the air you inhale is the same as that of the water, which increases as you go deeper. If the air pressure were different from the water pressure, you would not be able to breathe.

EXPERIMENT

Underwater breathing gear

Adult help is advised for this experiment

An inflated balloon represents the air cylinder, a beaker the diver, and a jar of water the sea. You pull or push part of a balloon stretched over the beaker to represent the diver breathing in or out. Valves open and close so that the air goes from the cylinder to the diver on breathing in, and exhaled air bubbles out through the sea.

YOU WILL NEED

• jar of water • plastic beaker • 4-in (10-cm) length of wide flexible plastic tube • 40-in (1-m) length of narrow flexible plastic tube that fits inside wide tube • drill and bit slightly smaller than narrow plastic tube • tape • scissors • 1 large and 2 small balloons • 12 in (30 cm) of stiff wire • pen • vise • funnel with 4 in (10 cm) diameter mouth • round file • pliers • cork that fits partially into funnel spout • ball bearing that moves freely in wide tube • petroleum jelly • glue • double-sided tape • two 2-in (5-cm) diameter cardboard disks

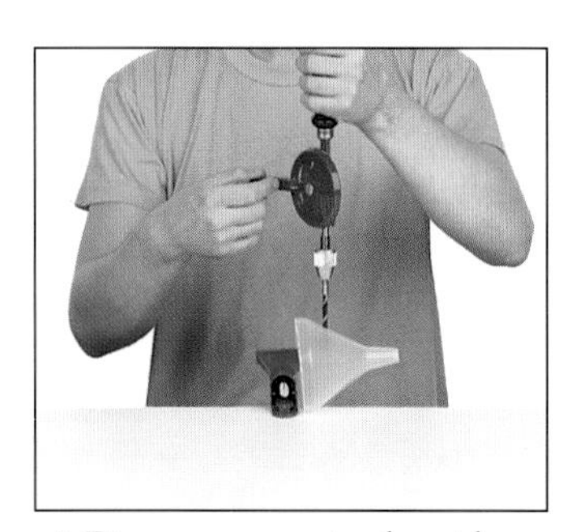

1 DRILL A HOLE in the side of the funnel. Enlarge this hole with a round file until one end of a narrow plastic tube can be squeezed tightly inside it.

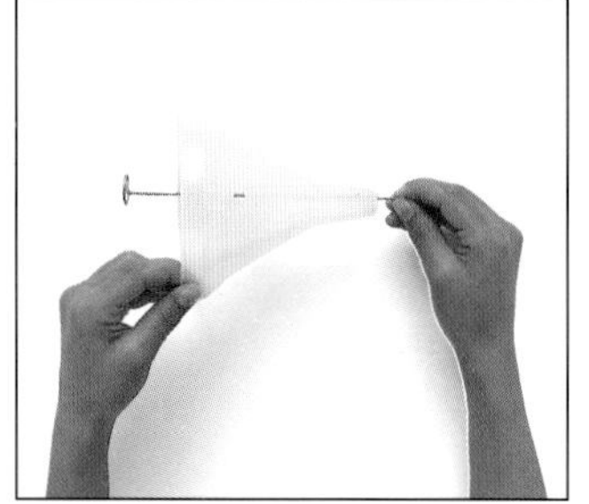

2 USE PLIERS to bend a small coil at one end of the wire, at 90° to its length. Push the other end of the wire through the funnel into the narrow end of the cork.

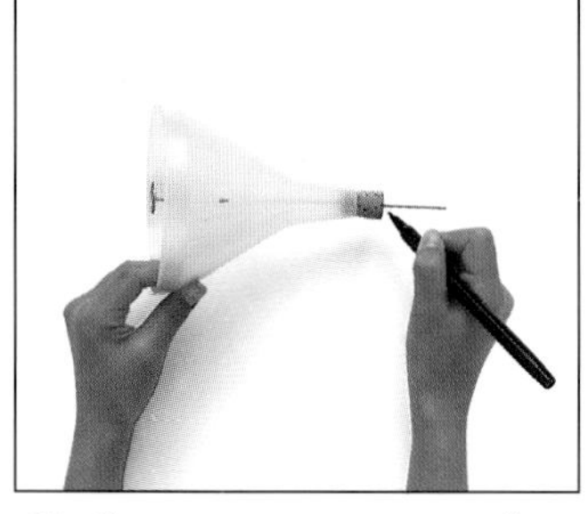

3 PUSH THE CORK onto the wire until it almost fits into the end of the funnel, when the coil is level with the mouth. Mark the wire where it exits the cork.

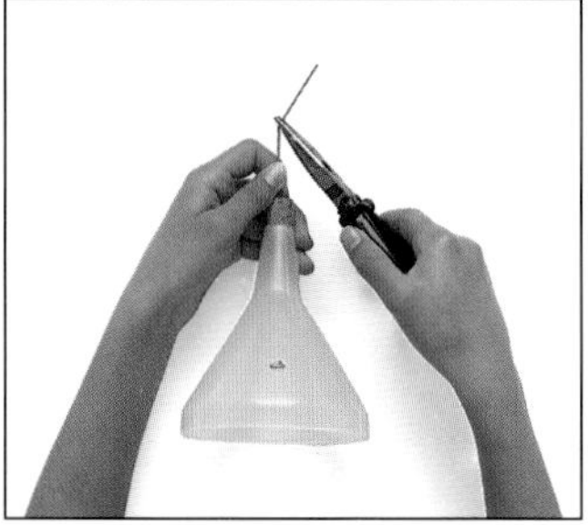

4 MAKE A 90° bend in the wire at the mark. Glue the cork firmly to the bend in the wire. Cut off the excess wire that projects beyond the cork.

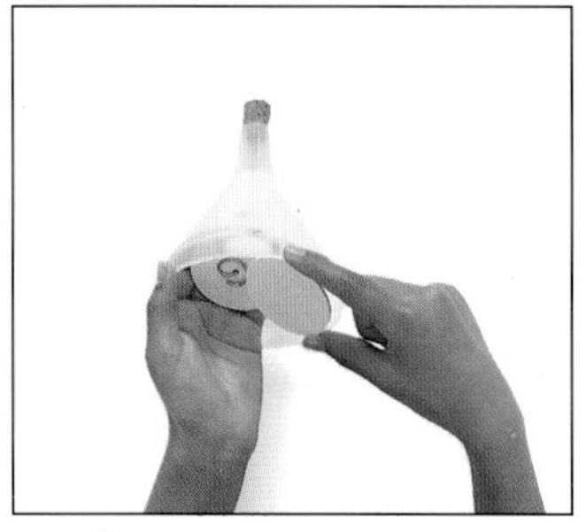

5 CUT A SMALL SLIT in one of the cardboard disks. Push the wire into the slit so that the coiled end of the wire is on top of the disk. Stick the second disk on top of the first with double-sided adhesive tape.

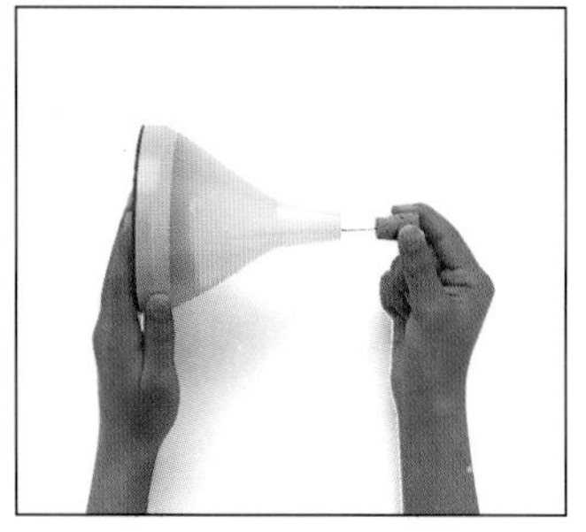

6 STICK DOUBLE-SIDED tape on top of the disks. Cut the neck and half the body from a small balloon. Stretch the remainder tightly over the top of the funnel and fix with tape. Push the disks on to the center of the balloon.

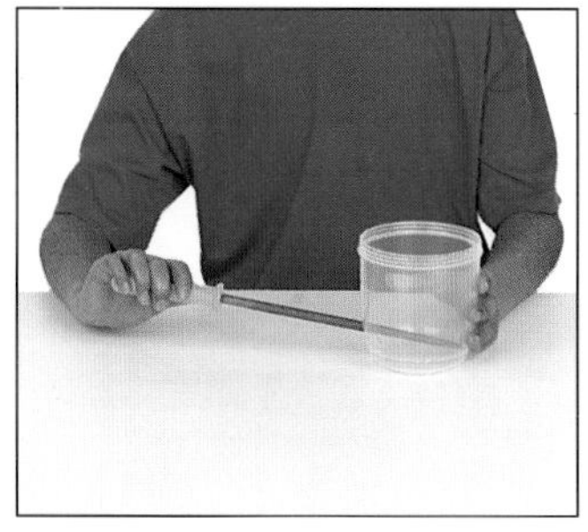

7 USE A CRAFT KNIFE to cut two small crosses 2 in (5 cm) apart near the bottom of the plastic beaker. Enlarge the holes, using the round file, until the narrow tube can be squeezed tightly inside, forming a tight seal.

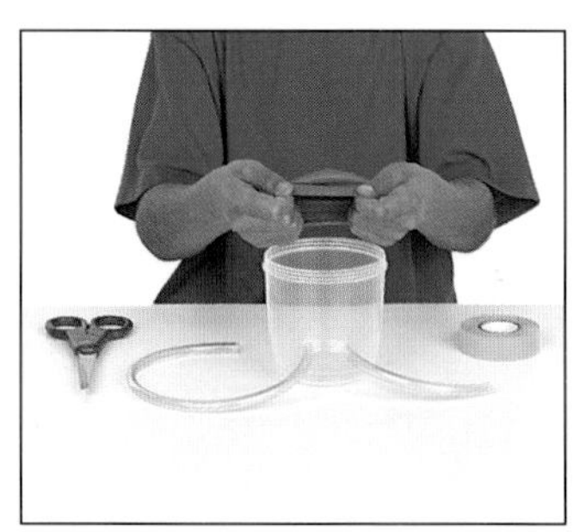

8 KNOT THE NECK of a small balloon. Cut off and discard half the body. Stretch and tape the balloon over the beaker. Push a 6-in (15-cm) narrow outlet tube and a 18-in (45-cm) narrow inlet tube into the beaker holes.

Real aqualung

The diver's mouthpiece is linked to a demand valve, which releases air from the air cylinder when the diver inhales. The surrounding water presses on a flexible diaphragm in the demand valve, giving the inhaled air the same pressure as the water. The air in the diver's lungs is also at this pressure. A reducing valve connects the demand valve to the cylinder of high-pressure air. This valve reduces the pressure of the air fed to the demand valve so that it is just above water pressure.

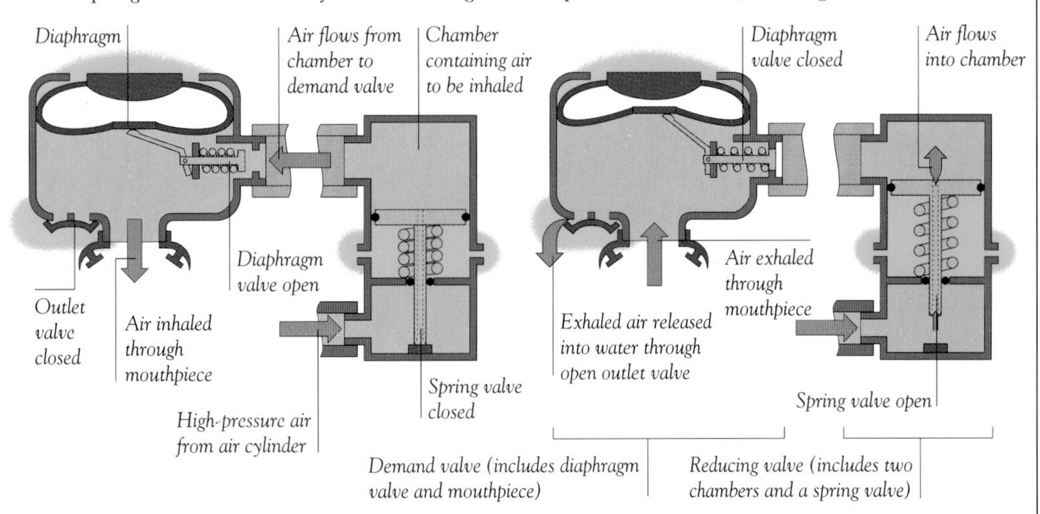

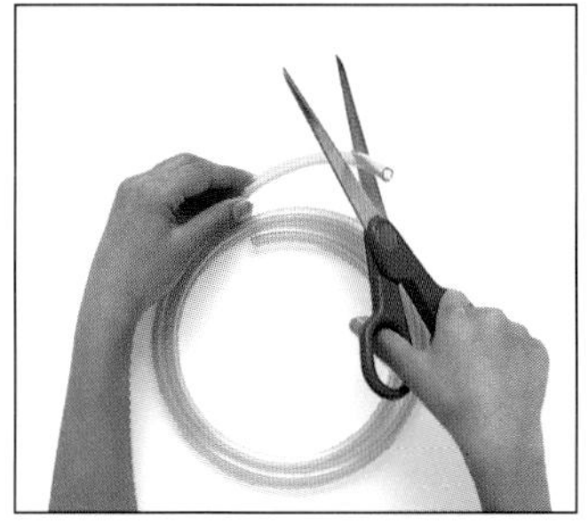

9 CUT A 45° section from the tip of a 16-in (40-cm) narrow tube. Following the instructions shown at right, make a one-way valve to fit on the outlet tube.

10 PUSH THE neck of a large balloon as far as it will go over the funnel spout. Secure with tape if necessary. Grease a short length of narrow tube and insert it into the hole in the funnel.

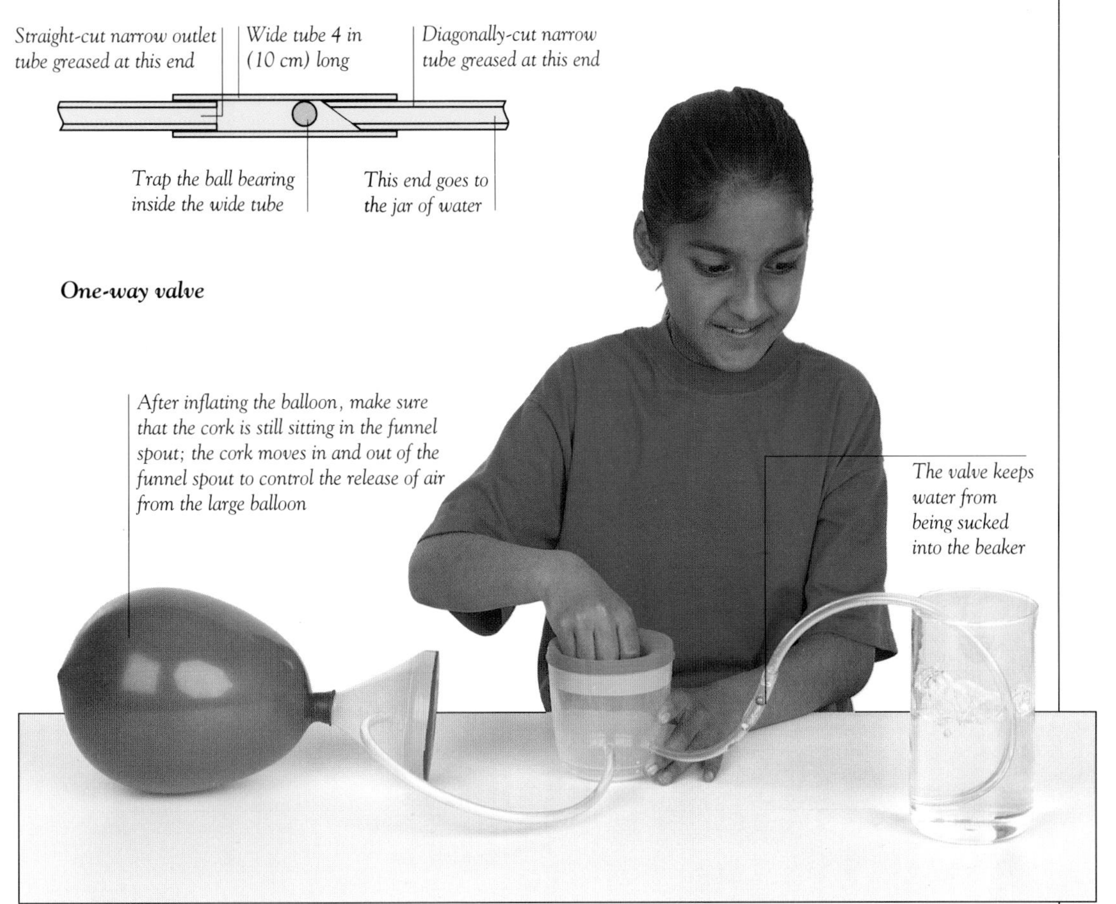

11 DEPRESS the balloon on top of the funnel and blow through the tube to inflate the large balloon, which should then stay inflated. Remove the narrow tube from the funnel. Grease the free end of the beaker's inlet tube and push it into the hole in the funnel. Place the end of the outlet tube in a jar of water. Hold the knot in the balloon on the beaker and repeatedly pull it up and push it down.

Wind instruments

YOU BLOW into a tube, using a mouthpiece or a blow hole, to make a wind instrument sound. A column of air in the tube vibrates and produces a note. The pitch of the note—how high or how low it sounds—depends on the length of the air column and on how hard you blow. There are two kinds of wind instruments. Woodwind instruments have wooden or metal tubes with holes along the sides. Opening and closing the holes with the fingers changes the length of the vibrating air column. Brass instruments have a long curved metal tube without holes.

French horn
This horn contains two coiled metal tubes with a combined length of 30 ft (9 m).

Brass instruments

Blowing a brass instrument, such as a trumpet or a trombone, makes the air in the whole length of the tube vibrate. Changing the pressure of the lips produces only a limited set of notes called harmonics. To play other notes, the length of vibrating air must be changed. In a trombone this is achieved using a sliding tube extension. Other brass instruments have valves, which open to introduce extra lengths of tube. A French horn has three valves, which give six extra notes when used in combination.

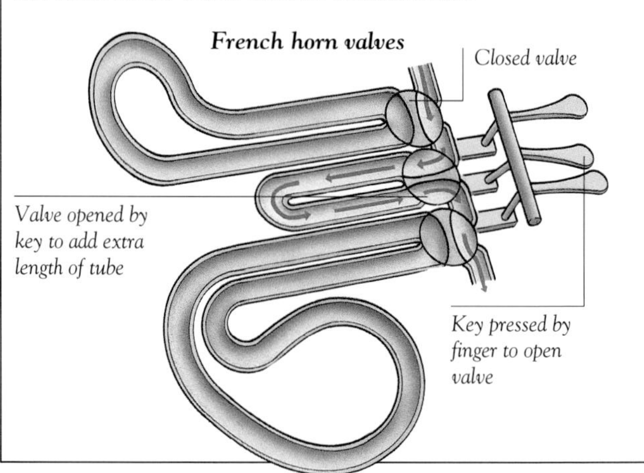

French horn valves

EXPERIMENT

Hose horn

Adult help is advised for this experiment

Take a length of hose and add a funnel for a bell and a faucet connector for a mouthpiece. Press your lips together, and blow through them into the mouthpiece. Relax or tighten your lips, and you should get three or four notes that sound like a fanfare.

YOU WILL NEED

• saw • hose • vise • funnel with end that fits into hose • rubber faucet connector

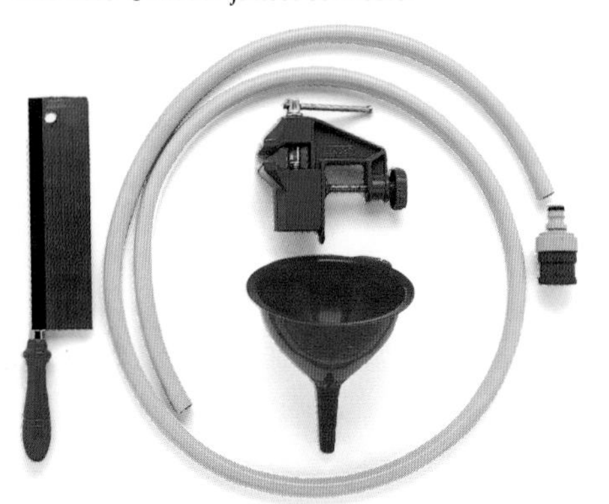

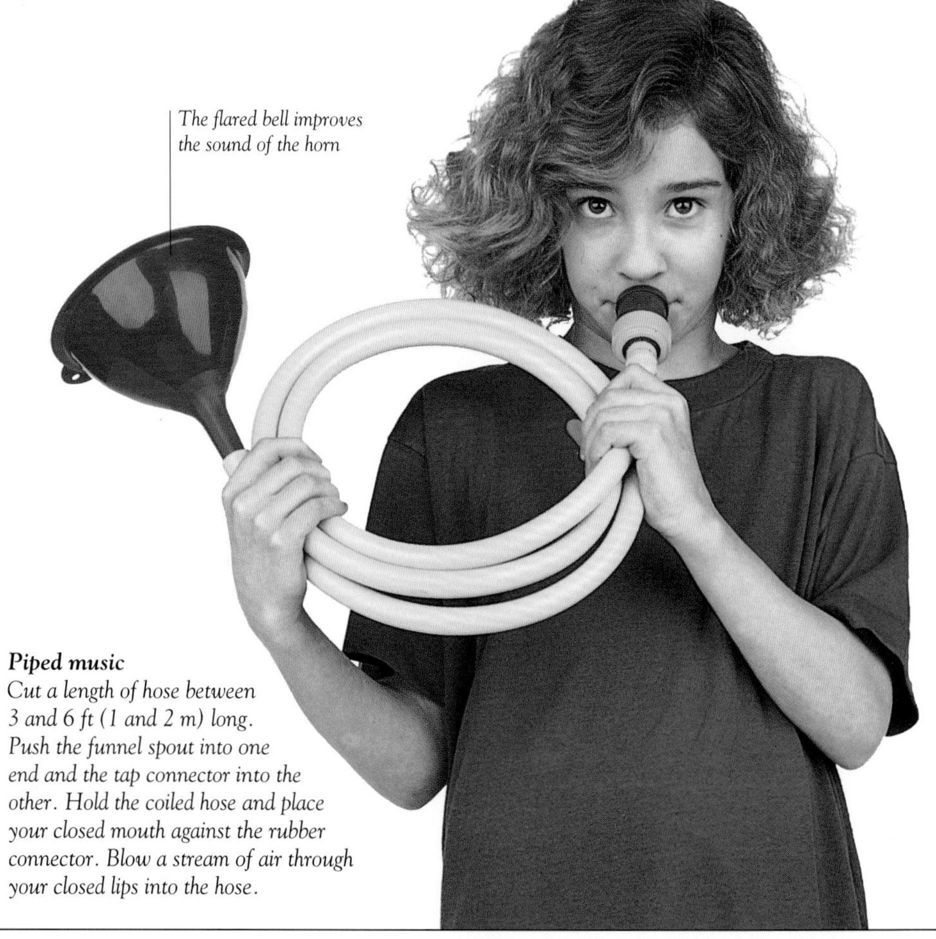

The flared bell improves the sound of the horn

Piped music
Cut a length of hose between 3 and 6 ft (1 and 2 m) long. Push the funnel spout into one end and the tap connector into the other. Hold the coiled hose and place your closed mouth against the rubber connector. Blow a stream of air through your closed lips into the hose.

EXPERIMENT

Blow your own flute

Adult help is advised for this experiment

Make a simple flute that plays the notes of the scale of C major. Purse your lips—as if to say "oo"—just above the mouth hole, and gently blow across the edge of the hole. Play different notes by placing your fingers over the other holes, raising or lowering them to change the length of the vibrating air column inside the flute. Move the cork in or out at the end of the flute to tune it.

YOU WILL NEED

● *ruler* ● *$11\frac{13}{16}$ in (30 cm) rigid plastic pipe, $\frac{3}{4}$ in (2 cm) wide* ● *round file* ● *drill* ● *$\frac{1}{8}$-in (3-mm) and $\frac{3}{16}$-in (5-mm) drill bits* ● *cork* ● *pen* ● *vise*

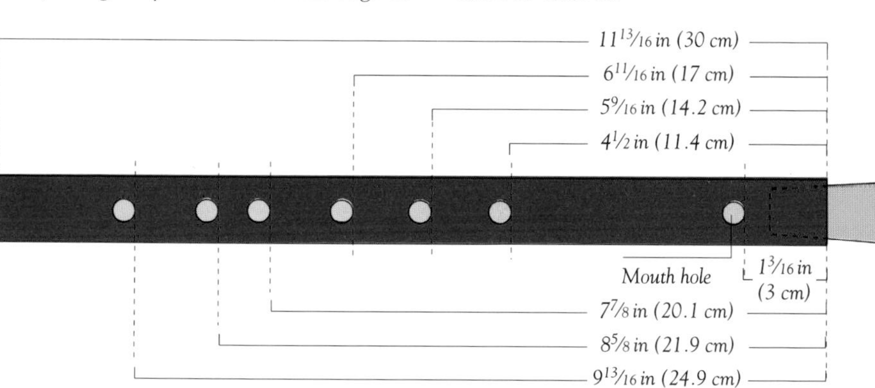

1 MARK SEVEN short lines along the side of the pipe, using the measurements given in the diagram above. Follow the measurements exactly.

2 USING THE $\frac{1}{8}$-in (3-mm) bit, drill a line of holes $\frac{1}{8}$ in (3 mm) from each mark, towards the end of the pipe without the cork.

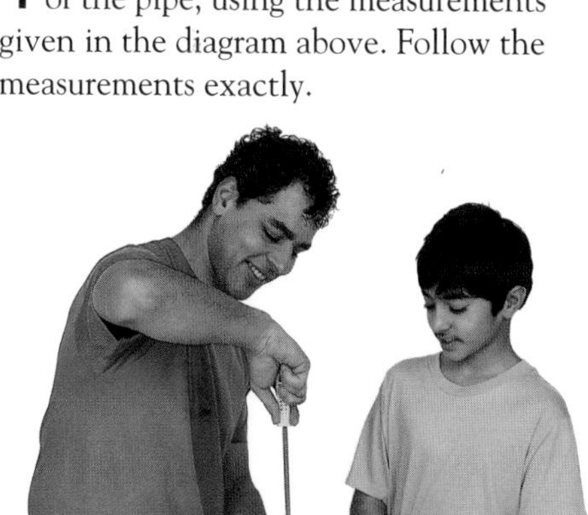

3 USE THE $\frac{3}{16}$-in (5-mm) bit and file to enlarge these holes to $\frac{1}{4}$ in (7 mm) in diameter. The edges of the holes should just touch the marks made earlier.

4 TRIM THE CORK until it fits tightly into the end of the pipe. You should be able to push the cork right up to the edge of the first hole.

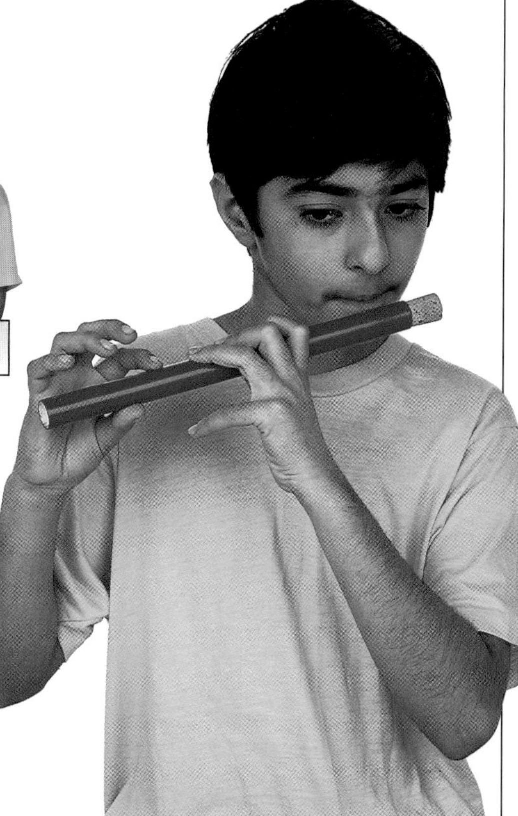

5 PLACE YOUR LIPS at the mouth hole and blow steadily across it to sound a note. It may take some practice to get this right. Cover all the holes and lift your fingers one by one to play seven of the eight notes in a scale of C major.

String instruments

YOU PLAY STRING instruments by plucking a set of taut strings, as on the guitar, or by sliding a bow across the strings, as on the violin. In a piano, the strings are hit by hammers linked to the keyboard. All these actions make the strings vibrate, which in turn causes the body of the instrument to vibrate. The vibrating strings produce only a weak sound, but this is amplified by the vibrating body. The pitch of the note produced depends on the thickness, length, and tension of the string. A thin string sounds higher than a thick one, and shortening or tightening a string makes it sound higher. Guitar and violin players change the length of string that vibrates by pressing the strings against a fingerboard.

How string instruments work

String instruments such as the harp have many strings of different lengths. Others have fewer strings and a fingerboard with thin metal frets spaced regularly along its length. Frets divide the strings into lengths that correspond to specific notes. If a string is played without being pressed against a fret, its whole length from the bridge to the nut vibrates and a note is heard. If the same string is played while pressed against the 12th fret, halfway between the bridge and the nut, a note one octave higher is heard, because the length of vibrating string is halved. Instruments in the violin family have fretless fingerboards. The player learns the exact position of every note.

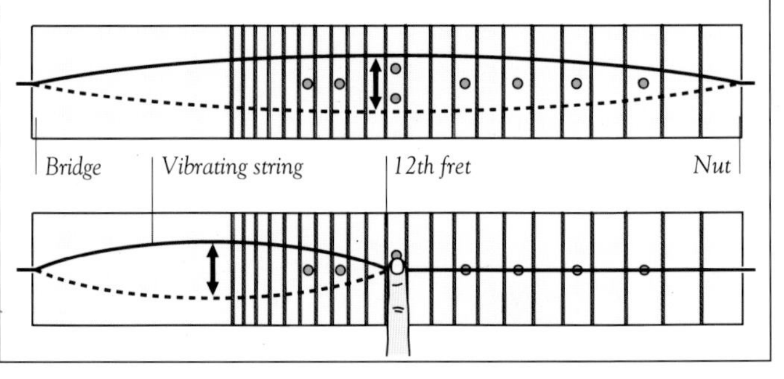

EXPERIMENT

Home-made guitar

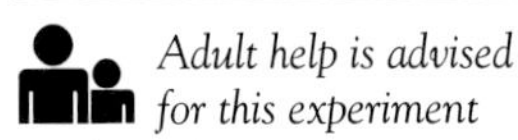

Adult help is advised for this experiment

Make a four-string guitar, and use it to play tunes and chords. Each string is tuned by plucking it to sound a note and turning the screw eye to tighten or loosen the string. Press the second string at the fifth fret. Tune it to the same note as the open (unfretted) first string. Then tune the third string at the fifth fret to the open second string. Repeat this tuning pattern for the fourth string. This is the same tuning as a bass guitar or the lower four strings of a six-string guitar.

YOU WILL NEED

• 2 yd (2 m) of 9/16 x 1¾ in (1.5 x 4.5 cm) wood • round file • thin file • scissors • 18 strips 1¾ in (4.5 cm) long of 1/16 in (2 mm) square hard wood • 6 in (15 cm) strip of 3/8 x 1/8 in (10 x 3 mm) hard wood • pen • vise • coping saw • nylon guitar strings (E, B, and 2 A) • 16 x 24 x 1/8 in (40 x 60 x 0.3 cm) sheet of plywood • cutting mat • drilling board • 4 small screws • screwdriver • 4 screw eyes • glue • paper dots • pencil and pad • calculator • hammer • tenon saw • craft knife • panel pins • drill with 1/16-in (1.5-mm) and 1/8-in (3-mm) bits • sanding block • ruler • long steel ruler

Guitar body

8 in (20 cm)
2 in (5 cm)
C
8 in (20 cm)
D: 6⅞ in (17 cm)
10 in (25 cm)
B: 10⅞ in (27 cm)
1¾ in (4.5 cm)
E: 10 in (25 cm)
F: 12 in (30 cm)
Circular hole 6 in (15 cm) in diameter cut centrally from body front panels
C: 10¼ in (25.5 cm)

A: 30 in (75 cm)

Neck detail

1 cm (3/8 in)
Nut line
3⅝ in (9 cm)
3/8 in (1 cm)
Bridge line
2⅜ in (6 cm)
24 in (60 cm)
1⅛ in (3 cm)
1⅛ in (3 cm)

1 CUT THE neck (**A**) from $\frac{9}{16}$ in (1.5 cm) thick wood as shown opposite. Mark the nut and bridge lines. Calculate $\frac{17}{18}$ of the distance between these lines. Mark the first fret line this distance away from the bridge line.

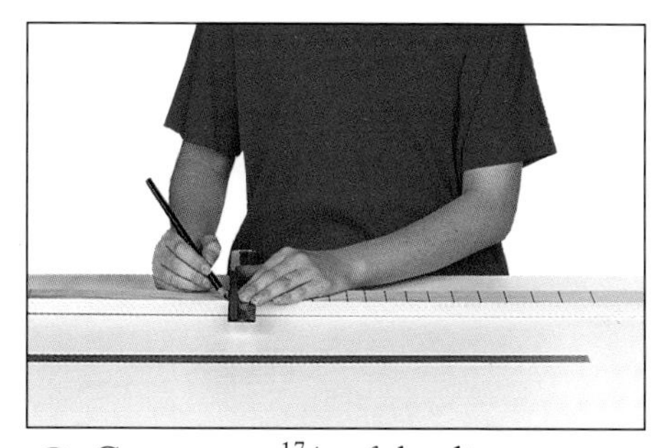

2 CALCULATE $\frac{17}{18}$ of the distance between the bridge line and first fret line. Mark a second fret line this far from the bridge line. Repeat this calculation using the previous fret position until you have marked 18 fret lines.

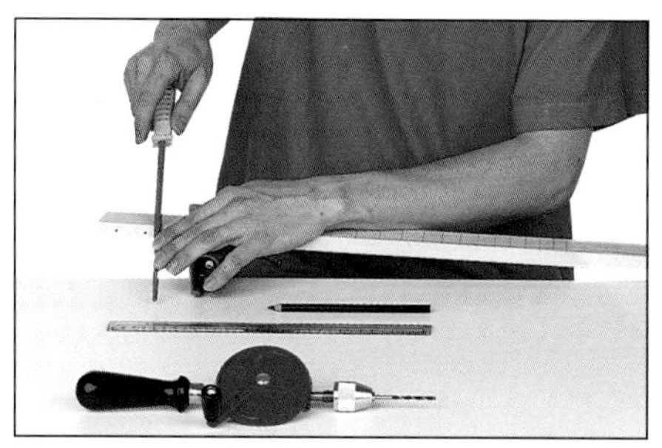

3 USING A $\frac{1}{8}$-in (3-mm) bit, drill four screw holes in the upper surface of the neck behind the nut line, as shown opposite. Drill two holes for screw eyes in each side of the neck. Bevel the edges of the neck near the holes with a file.

4 USING A $\frac{1}{16}$-in (1.5-mm) bit, drill four diagonal holes $\frac{3}{8}$ in (1 cm) apart through the neck near the bridge line. The holes should run from the upper surface of the neck to the end of the neck. They should be parallel and aligned with the holes behind the nut.

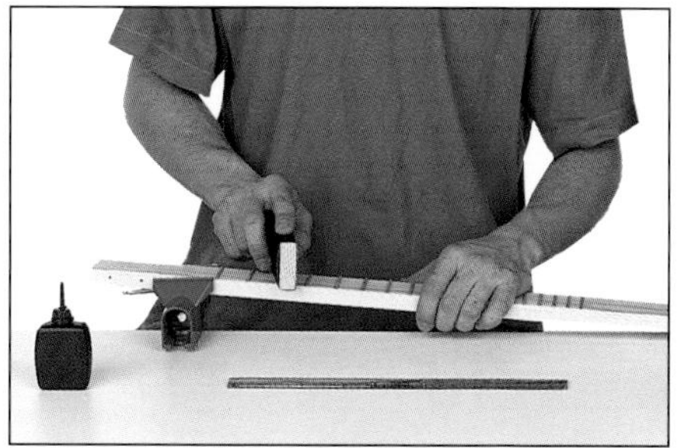

5 GLUE A $\frac{1}{16}$-in (2-mm) square wood strip at each fret mark. Sand them down so that the first fret (near the nut line) is taller than the second, the second taller than the third, and so on. Use a ruler to check that the fret heights diminish evenly towards the bridge line.

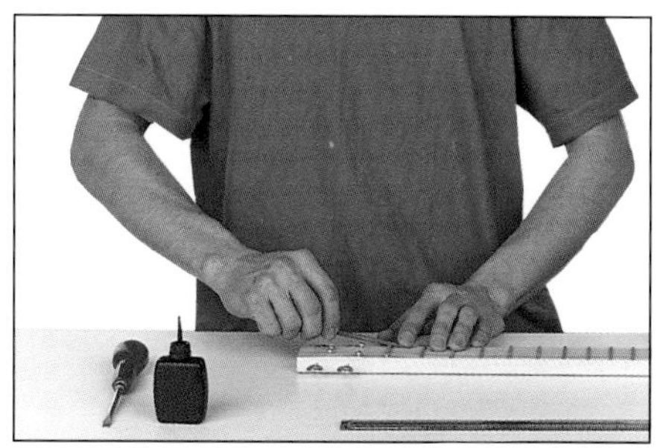

6 ADD SCREWS AND screw eyes to the holes drilled in the neck. Make the nut from a $1\frac{3}{4} \times \frac{1}{8}$-in (4.5-cm x 3-mm) piece of $\frac{1}{8}$-in (3-mm) hard wood and glue it at the nut line. File four small notches $\frac{3}{8}$ in (1 cm) apart in the nut, slightly above the height of the first fret.

7 CUT THE four pieces **B–D** from $\frac{9}{16}$ in (1.5 cm) thick wood as shown opposite. Cut $1\frac{3}{4} \times \frac{7}{16}$ in (4.5 x 1.2 cm) slots for the neck out of **B** and **D**. Cut out the three pieces **E–F** from plywood. Using panel pins and glue, fix pieces **B–E** to the neck. Add the back panel **F**.

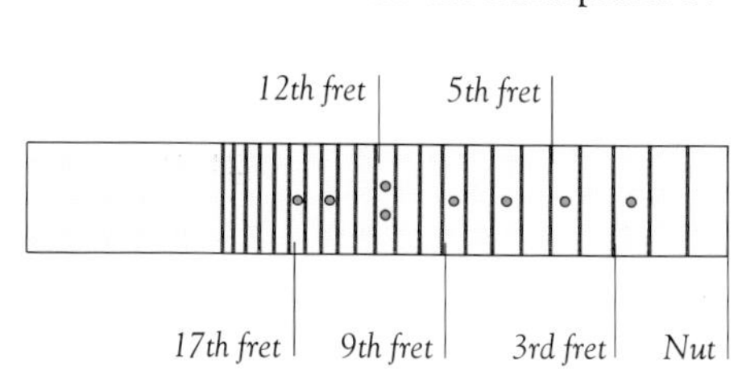

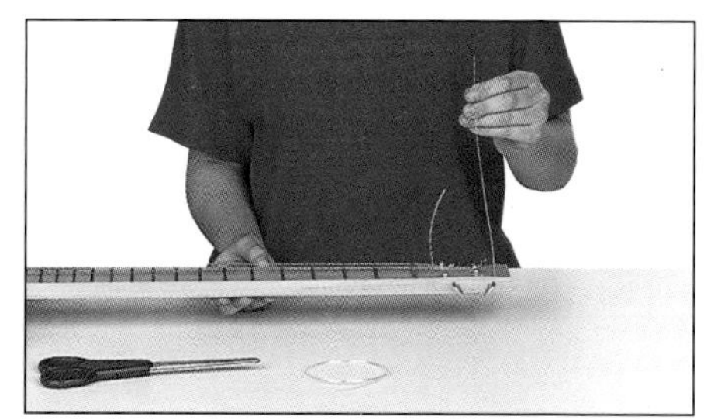

8 KNOT AN END of each string. Pass the strings through the diagonal holes in the neck toward the nut. Pull them taut around the screws, and tie to the screw eyes. Cut a bridge $2\frac{3}{8} \times \frac{5}{16}$ in (60 x 8 mm) from $\frac{1}{8}$ in (3 mm) wood strip and file four small notches $\frac{3}{8}$ in (1 cm) apart.

9 SLIDE THE BRIDGE under the strings at the bridge line. Fit the strings in the nut and bridge notches. Make a taller bridge if the strings buzz against the frets when plucked. Use dots to mark the fret positions (left). Tune the strings, and play.

Finger the frets to get different notes

Pluck the strings singly to play a tune, or strum them to play a chord

Percussion instruments

PERCUSSION INSTRUMENTS are hit or shaken to produce a sound. Some are struck with the hands, but many are beaten with a wooden stick, which may have a soft head. Striking a percussion instrument makes it vibrate. In the case of drums, the air inside the instrument also vibrates. These vibrations make the sound. Often this is just a noise, such as the crash of a cymbal or the rattle of a tambourine. But many percussion instruments have parts that vibrate to give musical notes. A drum has a stretched skin that can be tuned to a specific note, and the xylophone has bars that each produce a specific note.

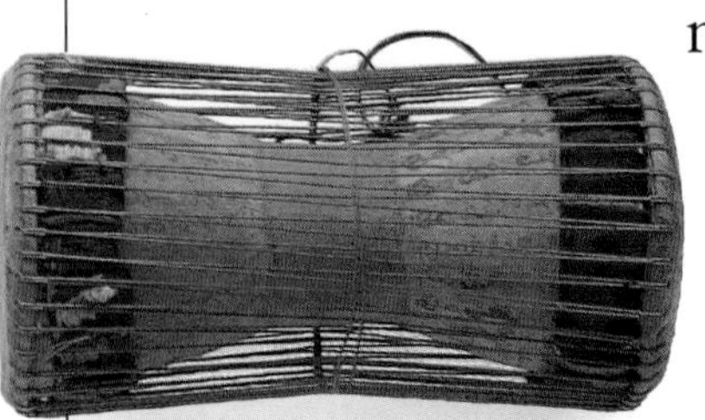

Talking drum
Pressing the cords of this Nigerian drum while beating the skin makes the notes go up and down, imitating the sound of African tonal languges.

EXPERIMENT

Miniature kettledrum

YOU WILL NEED
- *cork with central hole*
- *sandpaper*
- *large sheet of plastic*
- *large rubber band*
- *1-ft (30-cm) length of dowel to fit inside cork*
- *plastic bowl about 1 ft (30 cm) in diameter*

Timpani, or kettledrums, are large bowls with skin stretched across the top. Each one sounds a deep note that can be raised by tightening the skin or lowered by loosening it.

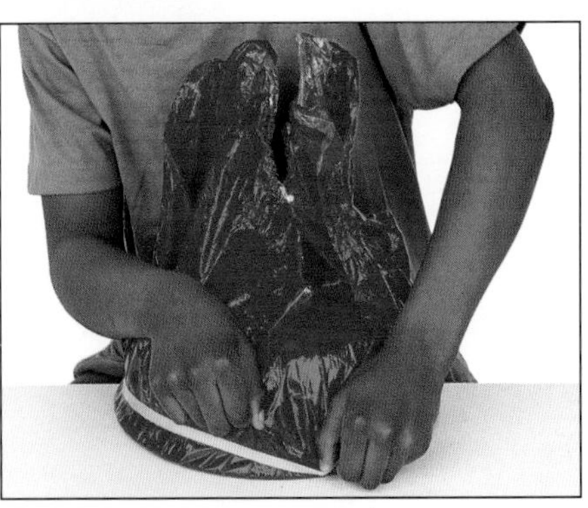

1 STRETCH THE PLASTIC sheet taut over the mouth of the bowl. Secure it with a rubber band just below the lip of the bowl. Gather up the excess plastic under the bowl to make a handle.

2 USING SANDPAPER, smooth down the edge of the cork. Push it onto the end of the dowel to make a drumstick. Strike the plastic to make a sound, and alter the pitch by twisting the handle.

Crash, bang, wallop

Gong
This elaborate gong from Borneo is suspended by a cord and struck in the center. The whole bronze disk vibrates with a deep, long-lasting sound. Cymbals make a sound in the same way.

Tabla
A drum has a skin, usually of plastic or animal hide, stretched over the end of an open cylinder. This small drum from India is played with the hands. The skin is tightened using the thongs at the side.

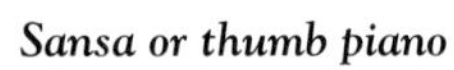

Sansa or thumb piano
This instrument, heard in Africa and South America, is played by twanging the metal tongues with both thumbs. The various lengths of tongues produce different notes.

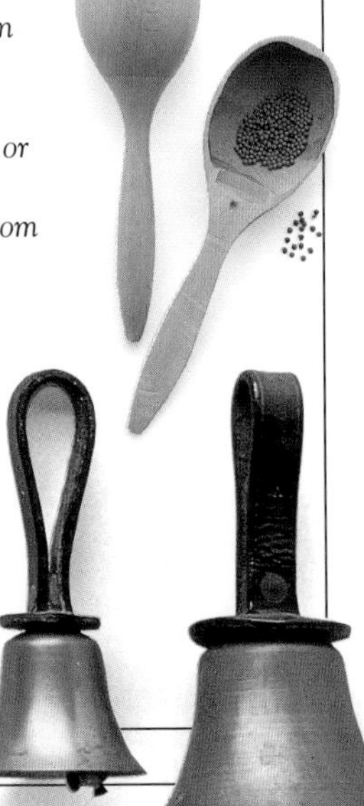

Maracas
Several percussion instruments are shaken to produce a sound. Maracas are pairs of containers with beads or seeds that rattle when shaken. They come from South America.

Bells
A clapper inside a bell swings to and fro, striking the rim to make the bell ring. Sets of bells are tuned to notes of a scale.

EXPERIMENT

Wooden bars and mallets

Adult help is advised for this experiment

The xylophone is heard in orchestras and in the folk music of many countries. Its name comes from Greek and means "wood sound." It has a set of wooden bars that are suspended so that they are free to vibrate when struck. Each bar sounds a different note, depending on its size and weight. This is because a lighter bar vibrates faster than a heavier bar, and therefore sounds a higher-pitched note. Build a xylophone that sounds the eight notes of the major scale, used in many simple tunes. You hold two mallets, using them one after the other to play tunes. You can also play two-note chords by striking two bars at the same time.

You Will Need

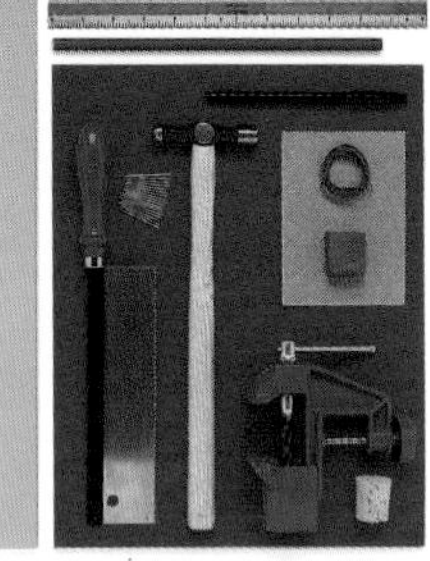

- *1¼ x 5/16 in (30 x 8 mm) wood strip*
- *ruler*
- *two 1-ft (30-cm) lengths of dowel to fit inside corks*
- *wooden base 16 x 12 x 3/8 in (40 x 30 x 1 cm)*
- *pen*
- *tenon saw*
- *1¼-in (3-cm) finishing nails*
- *hammer*
- *sandpaper*
- *rubber bands*
- *vise*
- *modeling clay*
- *2 corks with central holes*

1 Draw two slanting lines on the wood base, as shown in the template below. Hammer nine nails along both lines at intervals of 1¾ in (4.5 cm).

2 Stretch rubber bands along each row of nails. Place them halfway up the nails, making them taut enough to support the wood strips.

3 Make eight bars by sawing the wood strips into the lengths given below left. Place these bars across the rows of pins, on the rubber bands.

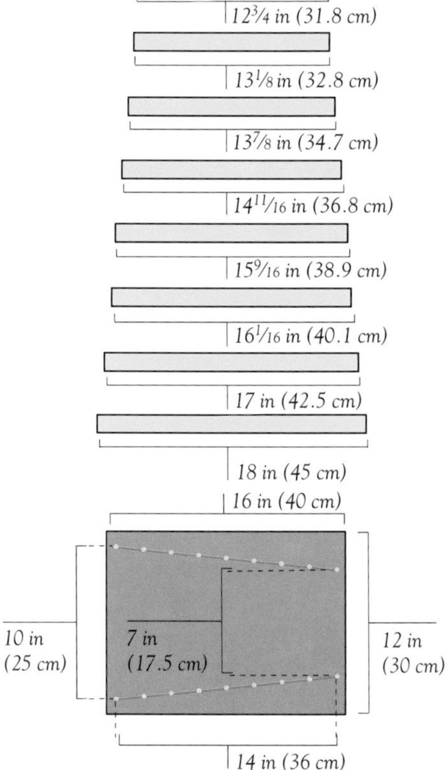

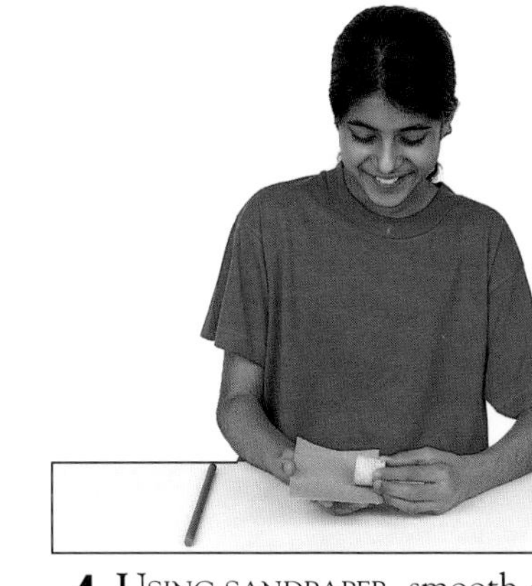

4 Using sandpaper, smooth down the edges of the corks to make rounded tips. Push a cork onto the end of each dowel to make two mallets.

5 Play a scale on the bars with the mallets. They should sound a major scale. If the note sounded by any of the bars is flat (too low), shorten the bar slightly by sanding. If it sounds sharp (too high), add a little modeling clay underneath the ends. Now you can play melodies on the xylophone.

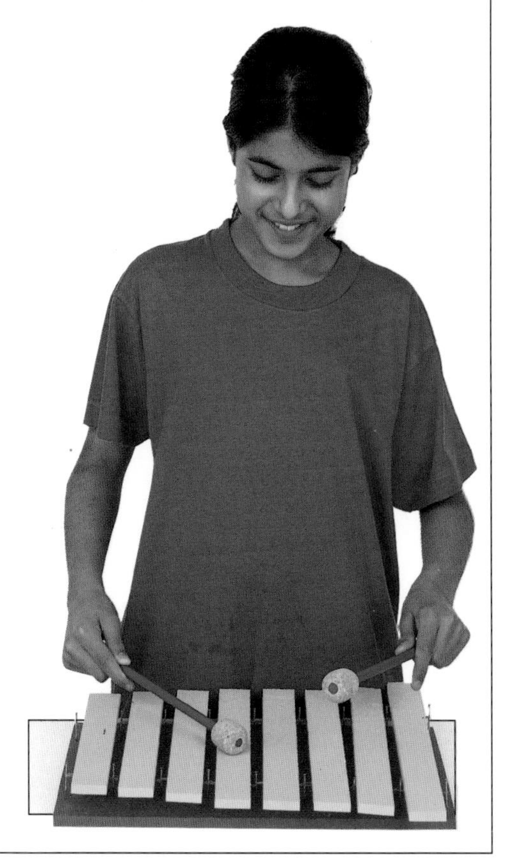

Cameras

WHEN YOU PRESS the button of a camera, a shutter opens briefly behind the lens. Light from the scene in front of the camera enters the lens, which focuses an image of the scene on the film behind the shutter. In the lens, an opening (the aperture) widens or narrows to ensure that the right amount of light reaches the film to produce a good picture. The shutter and aperture often work automatically.

EXPERIMENT

Model camera

Adult help is advised for this experiment

Make a model camera and see how it forms an image on a film. You use a magnifying glass for a lens, and tracing paper for the film. As with a real camera, you move the lens in and out to focus, or sharpen, the image.

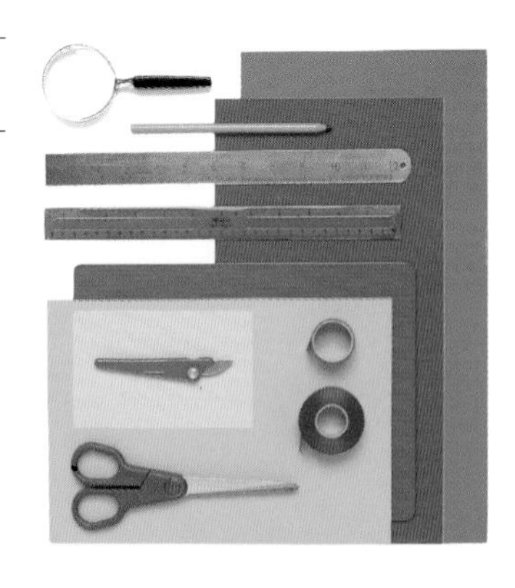

YOU WILL NEED

- *metal-rimmed magnifying lens*
- *pencil*
- *cardboard*
- *steel ruler*
- *ruler*
- *cutting surface*
- *sheet of tracing paper*
- *craft knife*
- *foamcore about 12 x 8 in (30 x 20 cm) square*
- *tape*
- *scissors*

1 FOCUS LIGHT from a bright distant source through the magnifying lens onto the foamcore. When the image is sharp, measure the distance between the lens and board to obtain the lens's focal length.

2 CUT A rectangle from cardboard, as long as the lens's focal length and twice as wide as the lens's circumference. Tape it around the lens, making a focal-length cylinder.

3 CUT A second rectangle 1 in (2.5 cm) longer than the first. Fold a ½-in (1-cm) flap along one longer edge. Divide the rest of the card into quarters parallel to this edge and fold into a box.

4 CUT TWO slots across opposing sides of the box 1 in (2.5 cm) from one end. Slide a tracing paper screen as wide as the box and 1 in (2.5 cm) longer than its width through the slots. Tape taut.

5 CUT A cardboard square with four flaps as a cover for the end of the box. Cut a central square out of the cover slightly smaller than the cylinder. Tape the cover on to the slotted end of the box.

6 MAKE A similar cover for the open end of the box. In this cover, cut a circle slightly wider than the diameter of the card cylinder. The cylinder should be able to move freely through it.

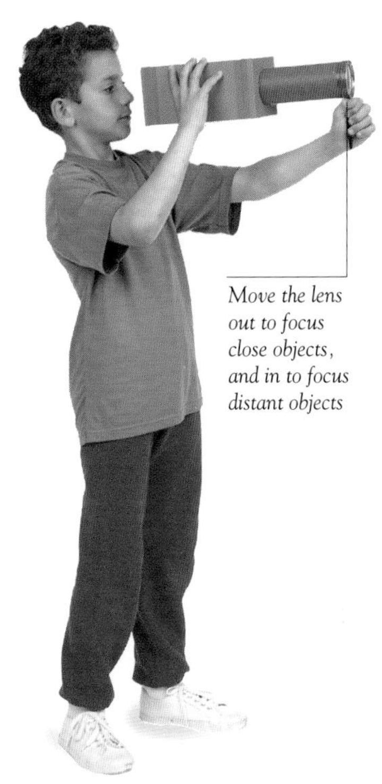

Move the lens out to focus close objects, and in to focus distant objects

7 PLACE THE lens cylinder inside the circular hole in the box. Look into the box through the square hole. Move the lens backward or forward to focus an image on the paper screen.

EXPERIMENT

Camera shutter

Adult help is advised for this experiment

A camera can take still pictures of moving objects because the shutter opens and closes in a fraction of second. Many cameras have a shutter in which a narrow slot moves rapidly across the film. Construct a shutter and see how it freezes movement.

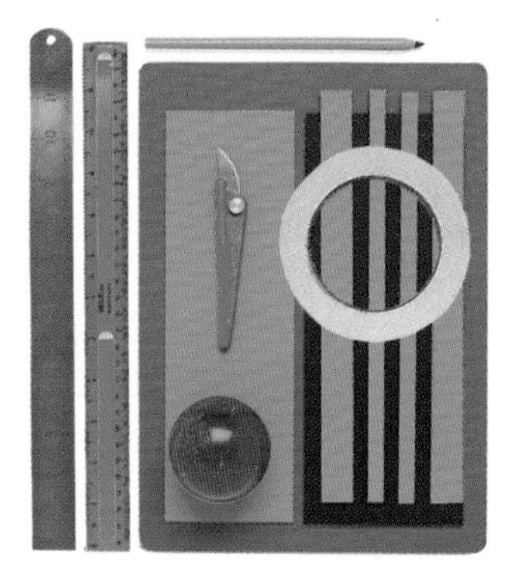

YOU WILL NEED

• *steel ruler* • *ruler* • *pencil* • *cutting surface* • *craft knife* • *rubber ball* • *double-sided tape* • *thick cardboard cut to these dimensions: carrier 10 x 4 in (25 x 10 cm); slider 10 x 3¼ in (25 x 8 cm); 2 strips 10 x ¾ in (25 x 2 cm); 2 strips 10 x ⅜ in (25 x 1 cm)*

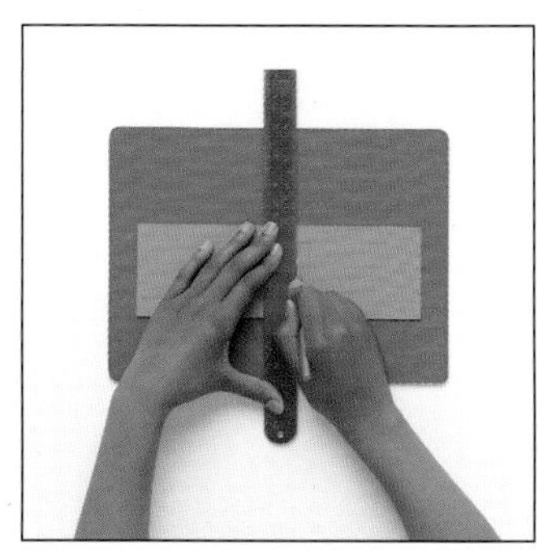

1 MARK A vertical midline on the slider. Cut a vertical slot in the slider, about ¾ in (2 cm) to the left of the midline. The slot should be ⅜ in (1 cm) wide.

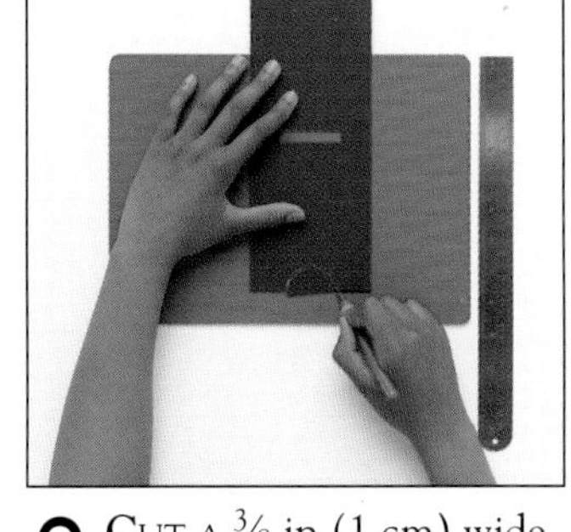

2 CUT A ⅜ in (1 cm) wide vertical slot in the middle of the carrier as shown above. Then turn the carrier and cut a small semicircular hole from the right-hand side.

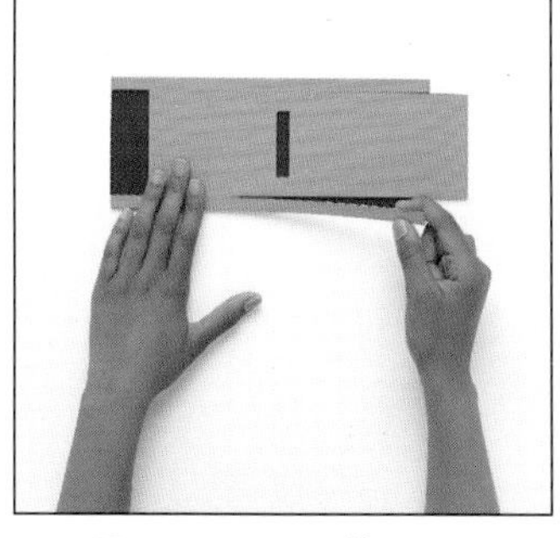

3 TAPE ONE small strip at the bottom of the carrier. Place the slider flush with the strip, but do not tape it. Tape the other small strip above the slider.

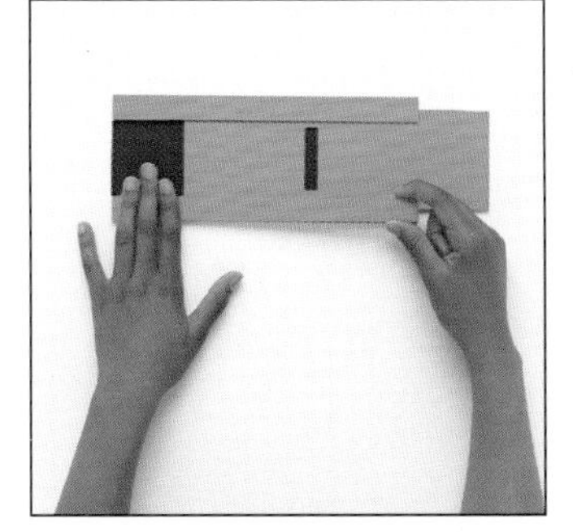

4 TAPE ONE of the two large strips on top of each of the small strips. The edges of these large strips should be flush with the edges of the carrier.

SLR camera

One of the most popular types of camera is the SLR (single-lens reflex) camera. The great advantage of this kind of camera is that the viewfinder shows the actual image formed by the camera lens, so that you see exactly the picture that will appear in the final photograph. Light rays from the scene pass through the lens and are reflected by the mirror into the pentaprism, where they are reflected again so that they pass out through the viewfinder eyepiece. When you press the button, the mirror rises, the shutter operates, and the light strikes the film.

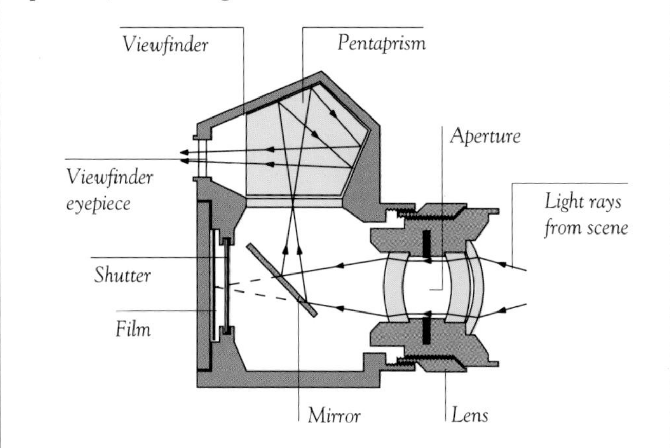

5 ASK A FRIEND to bounce a ball on a table. Look through the slot in the carrier and move the slider back and forth quickly while watching the ball. As the slider slot passes in front of the carrier slot, the ball appears frozen in mid-air. The shutter passes light to your eye for a very short time. The ball hardly moves in this time, so it appears still.

Cinema 1

How does the picture that you see on a cinema screen move? In fact, it only appears to move. A movie film is actually a long strip of still pictures, or frames, that are projected onto the screen one after the other at the rate of 24 pictures a second. As each picture disappears from the screen, it persists in your eyes for a little longer, and you continue to see it until the next picture appears. The still pictures appear to merge together, and because each frame is slightly different from the previous one, the image on the screen appears to move. Television and video also work in this way: the moving picture that you see on the television screen consists of 30 still pictures shown one after the other every second.

EXPERIMENT

Wagon wheel

In a film, the wheels of a fast-moving wagon may seem to stop, or to turn very slowly. As the camera films a wheel, each frame shows a pattern of spokes in a particular position. At certain speeds, the pattern in one frame exactly matches that in the next frame, or the position of the spokes differs only slightly. When the film is shown, we see the spokes as stationary or moving slowly—sometimes backward. Create this illusion using a desk lamp or overhead light, which flashes rapidly so you see a series of still images.

EXPERIMENT

Moving picture show

Adult help is advised for this experiment

Make a simple movie projector and a circular "film" with 16 still images. As you spin the film in the projector, the images move. You view the film through slots and see the images one after the other, just as a real movie projector projects a series of images on to a screen.

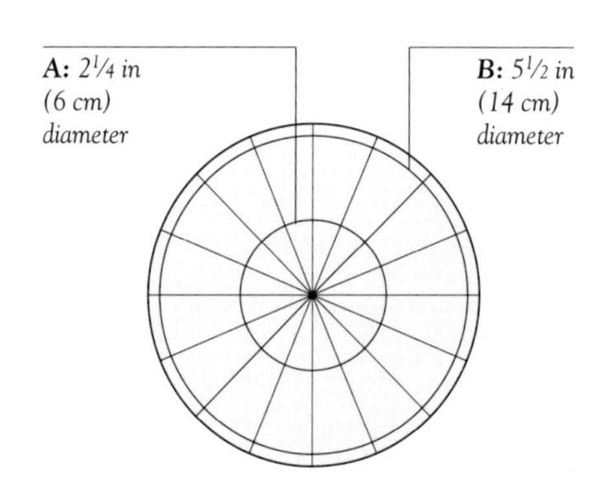

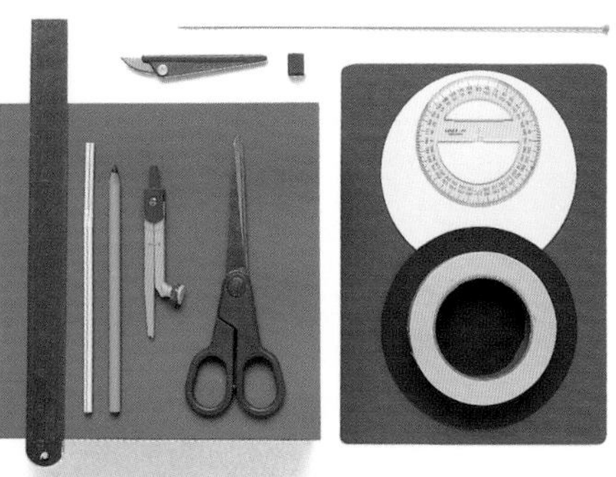

You Will Need

- *steel ruler* • *craft knife* • *knitting needle* • *small piece of modeling clay* • *cardboard box at least 10 in (25 cm) tall, about 8 in (20 cm) deep, with large slots cut from both sides* • *drinking straw* • *pencil* • *compass* • *scissors* • *cutting surface* • *2 cardboard disks $6\frac{1}{4}$ in (16 cm) in diameter, one black and one white* • *protractor* • *double-sided tape*

1 and 16 | **2 and 15** | **3 and 14** | **4 and 13** | **5 and 12** | **6 and 11** | **7 and 10** | **8 and 9**

1 Copy the disk template (left) onto each cardboard disk. Cut slots along each radial line on the black disk. The slots are $\frac{1}{8}$ in (3 mm) wide and extend from circle A to circle B.

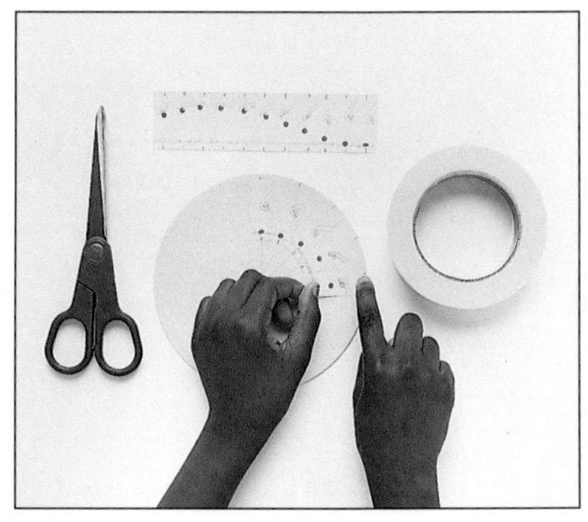

2 Copy or photocopy the eight images (left) twice onto paper. Cut out and stick the images on the white disk in sequence (numbers above images) along the lines between circles A and B.

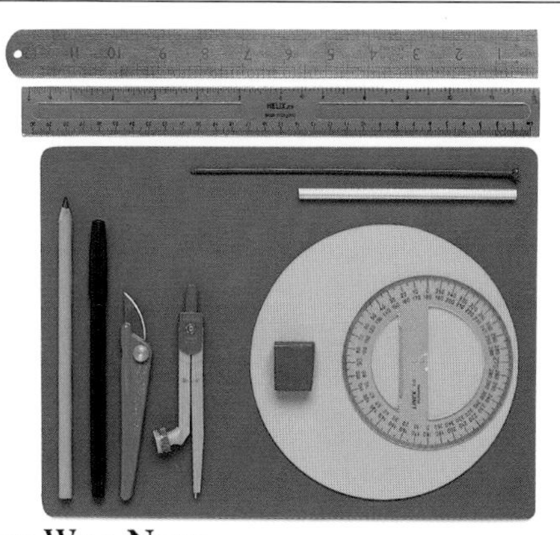

You Will Need

• steel ruler • ruler • cutting mat • knitting needle • drinking straw • pencil • pen • craft knife • compass • cardboard disk 6¼ in (16 cm) in diameter • modeling clay • protractor

1 Copy the template (opposite page) onto a cardboard disk. Draw a stripe on each radial line, between the circles **A** and **B**. Push the pointed end of the knitting needle through the center of the patterned side of the disk. Place the straw on the needle. Cover the point of the knitting needle with modeling clay.

2 Carry out the experiment under artificial light. Hold the straw in one hand and twirl the knitting needle quickly with the other. The stripes on the spinning disk usually appear blurred, but at certain speeds they seem to stop briefly, or to rotate slowly forward or backward.

3 Make a hole 6 in (15 cm) up along a vertical midline in each end of the box. Enlarge with a knitting needle. Cut a slot 1½ in (4 cm) tall and ⅛ in (3 mm) wide, 1⅛ in (3 cm) above one hole.

4 Enlarge the holes in the ends of the box so that the knitting needle rotates. Push the needle through the box and disks in the order shown above, placing a straw spacer between the disks.

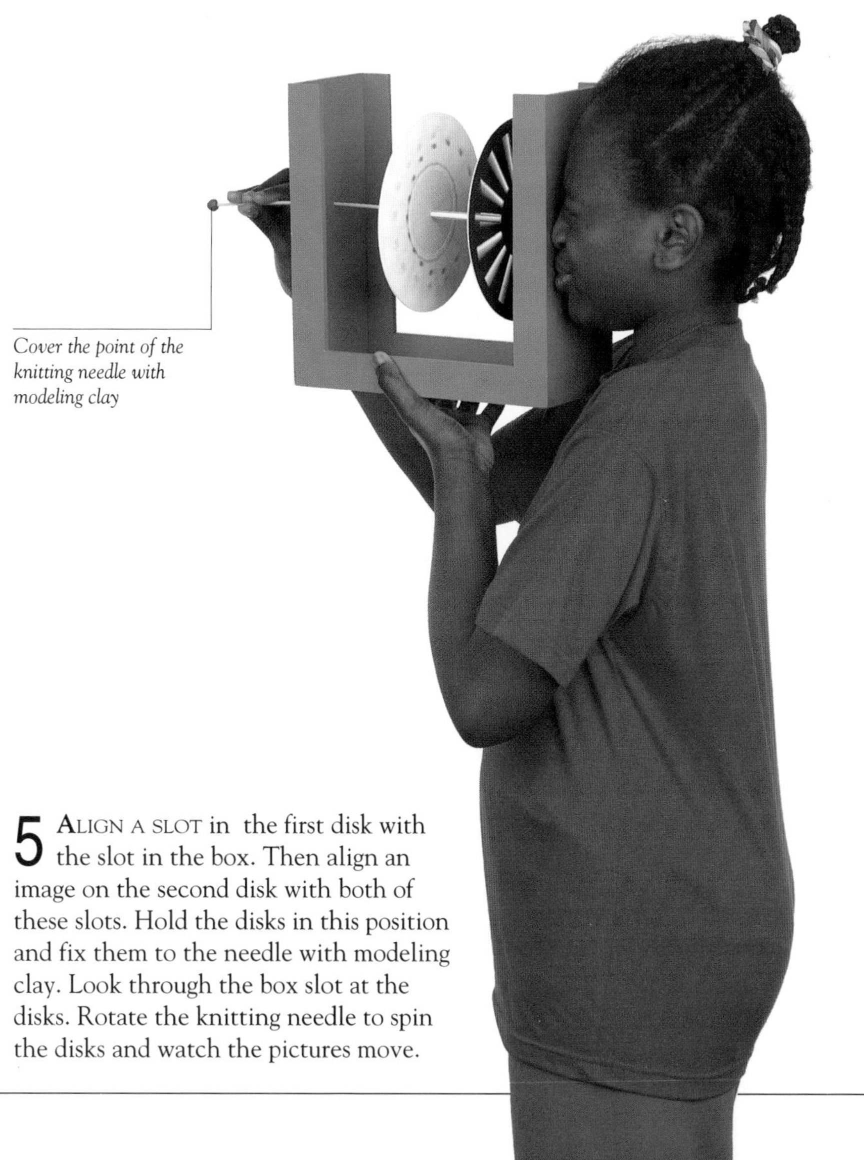

Cover the point of the knitting needle with modeling clay

5 Align a slot in the first disk with the slot in the box. Then align an image on the second disk with both of these slots. Hold the disks in this position and fix them to the needle with modeling clay. Look through the box slot at the disks. Rotate the knitting needle to spin the disks and watch the pictures move.

Cinema 2

A MOVIE CAMERA contains a reel of film that moves past a lens and a shutter similar to those in a still camera (p.134). The shutter opens and closes 24 times a second, and a sequence of still images appears on the film. At the movie theater, the film moves through a projector that flashes the still images on the screen one after the other 24 times a second. This is so fast that you see a moving picture (p.136). Slow-motion sequences are filmed at high speed and projected at normal speed.

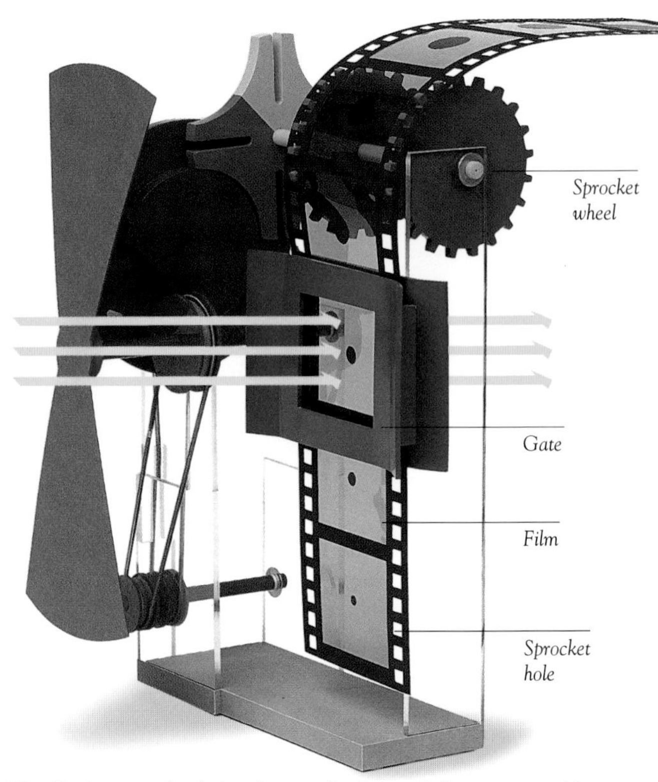

1. The film has sprocket holes along each edge, and these engage with a toothed sprocket wheel that moves the film. The film goes through a gate containing a window the same size as each image.

Film transport

The film does not move smoothly through the projector. It stops while each still image is projected onto the screen, and then moves on again to show the next image. To ensure that the picture is sharp, each image must remain stationary on the screen and you must not be able to see the film as it moves from one image to the next. To achieve this, a shutter in the projector cuts off light to the film as the film moves from one image to the next.

Sound and vision

A projector has a very bright lamp and a high-quality lens to project a bright, sharp picture. One system of recording sound on film is the optical system, in which a sound signal in the form of a wiggly line (the soundtrack) runs alongside the images. A bulb shines light through the soundtrack to a sound head. The wiggly track changes the intensity of light reaching the head, which converts the light into a varying electric signal. The electricity is converted to sound in loudspeakers. Many films have soundtracks made of magnetic material like that on audio tapes. The sound head is located away from the projector gate so that the film moves smoothly past the head. Feature films may consist of several reels, which are shown on two projectors, with one starting as the other stops. But in many theaters, all the films in a program are connected in one long roll.

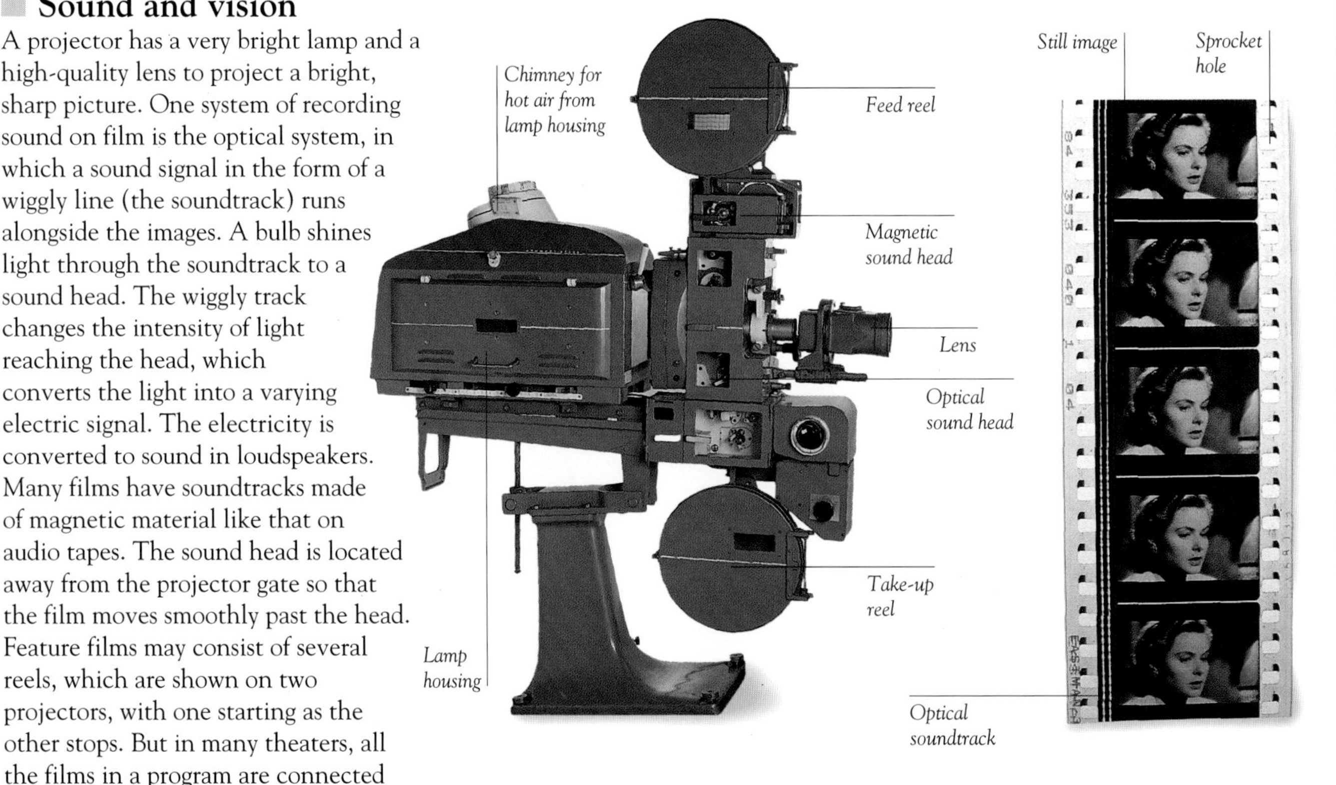

Cinema projector

Section of film

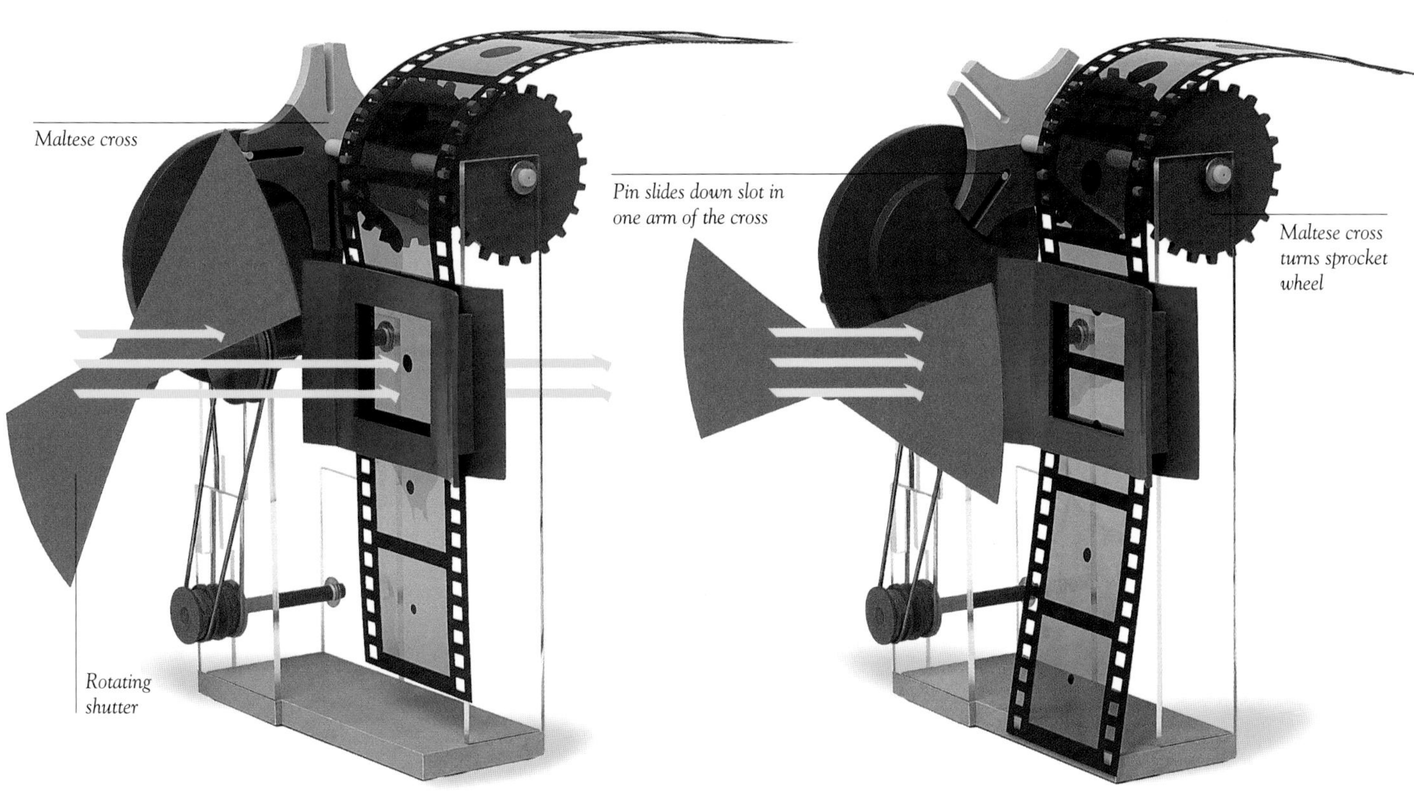

2. The film is held still in the gate as the lens projects the image onto the screen. In this model, a pulley wheel rotates continuously, driving a twin-blade rotating shutter that is linked to the pulley wheel by a belt drive.

3. A pin on the pulley wheel engages the Maltese cross mechanism and turns it. As the cross turns, it moves the film on to bring the next image into the gate. While the film moves, the shutter blocks off light to the film.

4. The pin leaves the slot in the cross after the cross has made a quarter turn. The cross comes to a stop, and the film remains stationary in the gate. The blade of the shutter moves past the gate.

5. The transport sequence starts again as the shutter blade passes the gate and the next image in the film is projected onto the screen. The complete sequence takes only $1/24$ of a second.

Microphones and loudspeakers

TAPE AND DISC players, movies, television sets, and radios all reproduce the sounds of voices, music, and the world about us. Sound reproduction usually begins in a studio, where a microphone turns sound into an electric signal. The reproduction process ends with loudspeakers, for example those in a stereo system. A loudspeaker turns the signal back into sound waves that finally reach our ears.

EXPERIMENT

Moving-magnet microphone

Adult help is advised for this experiment

Make a microphone that can record your voice. It has a balloon diaphragm that moves a magnet inside a wire coil. The magnet's field moves with the magnet, producing an electric signal in the coil. The signal goes to a tape recorder. The microphone can also receive a signal from a recorder to work as a loudspeaker.

YOU WILL NEED

• large balloon • insulated bell wire • tenon saw • cassette • thin insulated copper wire, about 32 SWG (the insulation may be transparent) • strong bar magnet, with poles at its smallest faces • vise • tape recorder with microphone and headphone jacks • cardboard 6 in (15 cm) square • cutting surface • pliers • wire strippers • tape • rubber band • mono plug that fits microphone and headphone jacks • scissors • craft knife • plastic funnel about 4 in (10 in) in diameter • sandpaper

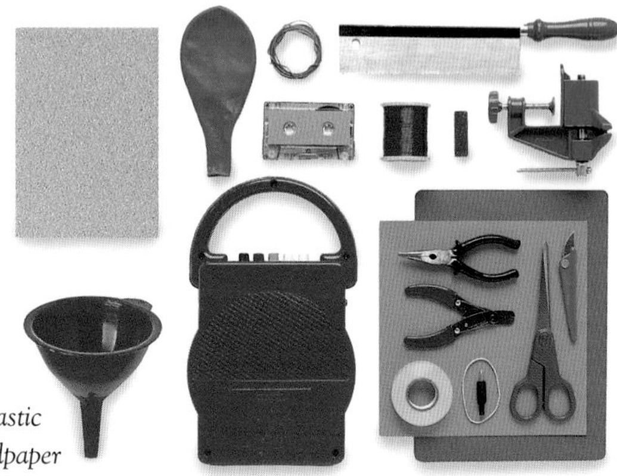

1 SECURE THE funnel in a vise. Saw off the spout, and as much of the body as necessary to leave a hole through which a pole face of the bar magnet can pass easily.

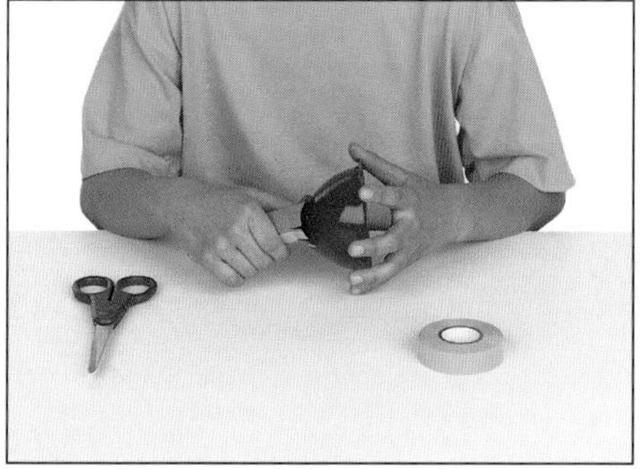

2 ROLL THE cardboard into a 6-in (15-cm) tube that fits in the funnel hole. Cut three 1-in (2.5-cm) slits in one end of the tube. Insert this end into the funnel. Tape the slits inside to secure.

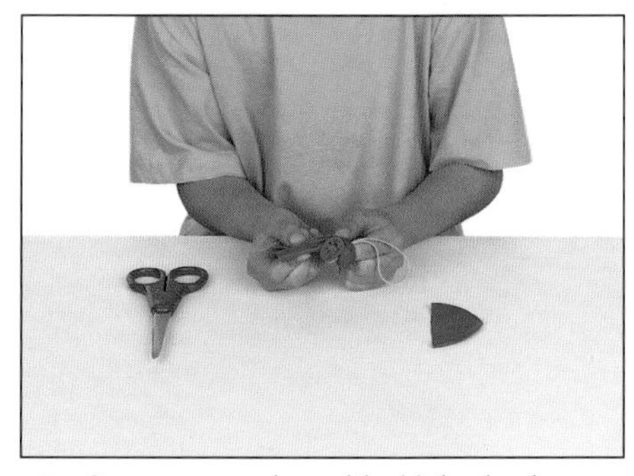

3 CUT THE neck and half the body from a large balloon and discard the remainder. Knot the neck of the balloon around a rubber band, so that the band hangs from the balloon neck.

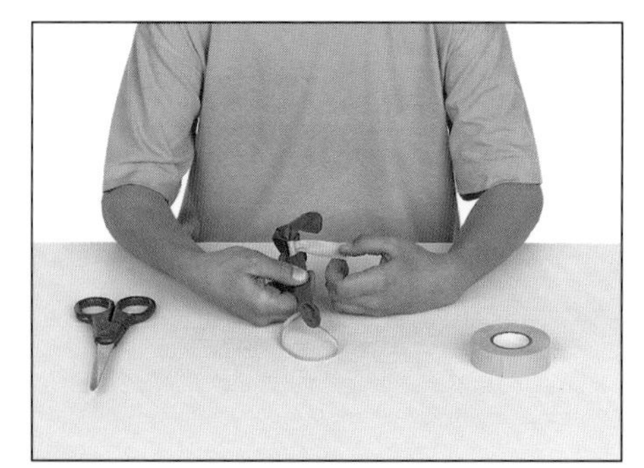

4 PUSH THE magnet into the body of the balloon up to the knot. Pull the balloon up around the magnet, draw it tight, and tape it closed over the magnet, leaving at least 2 in (5 cm) of balloon membrane projecting above the magnet.

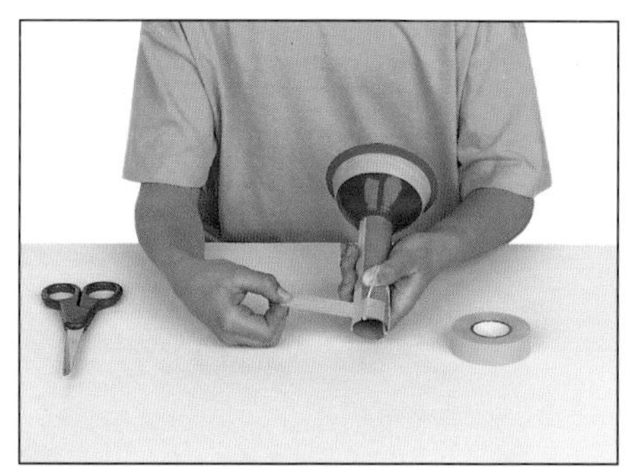

5 PASS THE knotted end of the balloon down the funnel into the tube. Tape the open end of the balloon over the funnel mouth. Stretch the rubber band out of the tube and tape it so that the magnet is freely suspended in the tube.

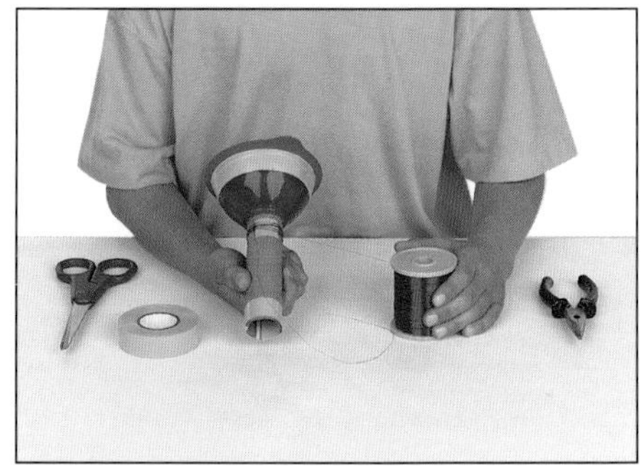

6 WIND ABOUT 250 turns of 32 SWG wire around the tube in the area of the magnet, leaving about 8 in (20 cm) free at each end. Secure the first few and the last few turns of the coil to the tube with tape. Cut off any excess wire.

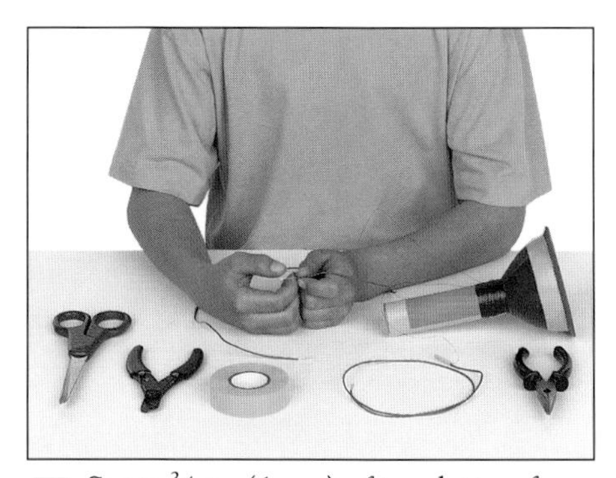

7 STRIP ³⁄₈ in (1 cm) of insulation from each free end of the coil. Strip the ends of two 24 in (60 cm) lengths of bell wire. Attach a wire to each end of the coil by twisting the wire ends with the coil ends and securing with tape.

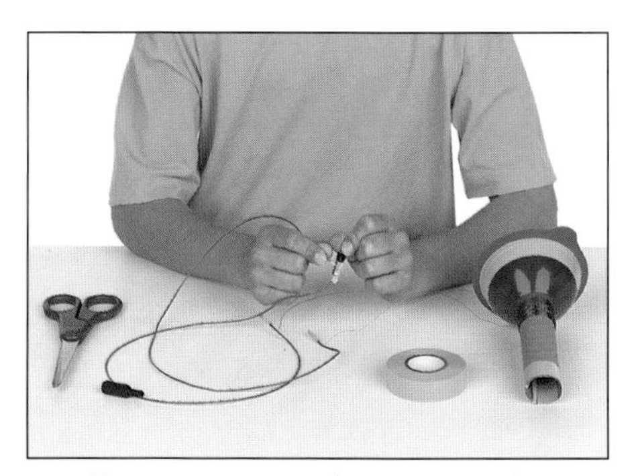

8 REMOVE THE insulating cover from the jack plug to expose the two contacts. Twist the free end of each wire into a contact. Use tape to keep the wires from touching. Replace the cover on the plug.

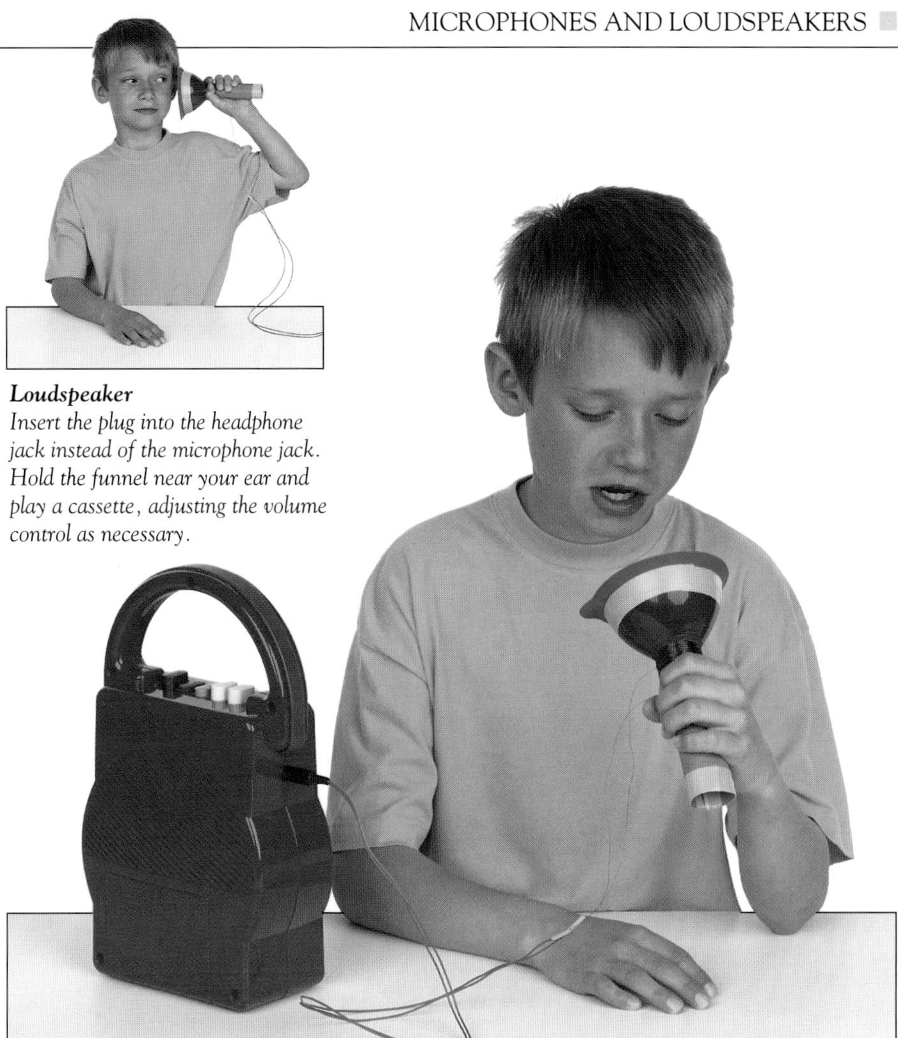

Loudspeaker
Insert the plug into the headphone jack instead of the microphone jack. Hold the funnel near your ear and play a cassette, adjusting the volume control as necessary.

9 CONNECT THE plug to the microphone input jack on the tape recorder. Then load a blank cassette into the tape recorder and press the "record" button (or buttons). Hold the microphone about 4 in (10 cm) in front of your mouth and speak or sing loudly into it. Stop the tape, rewind it, and play back your voice. Because this home-made microphone is not very sensitive, the recording may not be crystal clear, but your words will be audible.

Moving coils

Many microphones and most loudspeakers contain a stationary magnet and a moving coil. Sound waves striking the microphone cause the coil to vibrate in the magnet's field. This produces an electric signal in the coil, in which the current varies as the vibrations of the sound waves vary. When this signal goes to a loudspeaker coil, it produces a varying magnetic field in that coil. This field interacts with a magnet's field to make the coil vibrate. The coil moves the loudspeaker cone, which makes the air vibrate to produce the same sound waves. A moving-magnet microphone or loudspeaker works on the same principle.

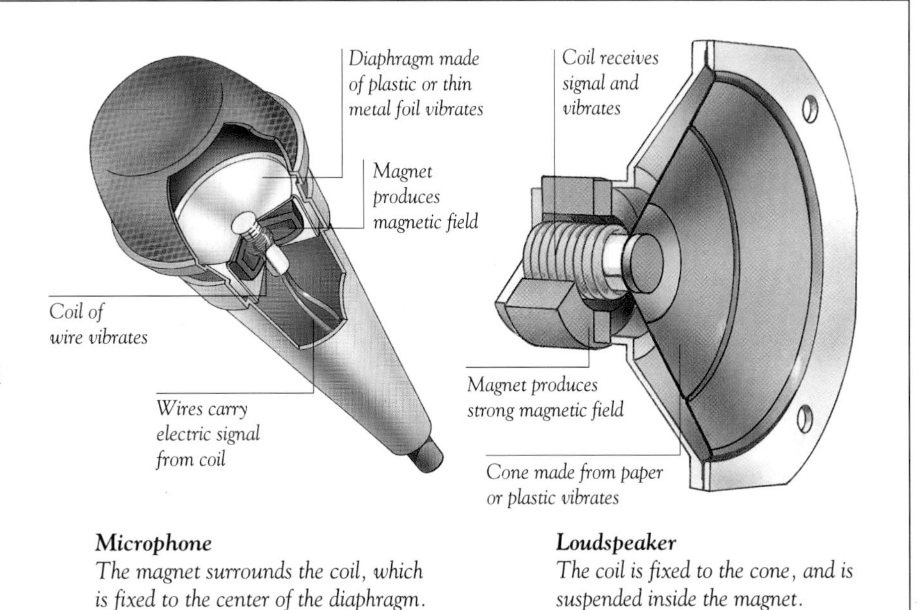

Microphone
The magnet surrounds the coil, which is fixed to the center of the diaphragm.

Loudspeaker
The coil is fixed to the cone, and is suspended inside the magnet.

Amplifiers

TAPE AND DISC PLAYERS, radios, and television sets all work with weak signals (pp.144–145). They contain amplifiers, which increase the strength of the signals enough to drive the loudspeakers or earphones that produce the sounds you hear. Amplifiers contain transistors, which boost weak signals using a strong electric current from a source such as a battery or an electric outlet. The weak signal and the strong current both go to the transistor. The weak signal, which varies in level depending on the sounds being reproduced, causes the strong current from the power supply to vary in level too. The strong current, now changing in proportion to the weak signal, becomes an amplified copy of the weak signal, and is strong enough to drive the loudspeakers or earphones. A volume control governs the amount of amplification by raising or lowering the overall strength of the amplified signal.

Current from power supply

Collector

Base

Emitter

Transistor casing

Strong emitter current

− + +

Weak signal

Amplifying a weak signal

A weak signal allows a strong current to pass.

How a transistor works

A transistor contains a sandwich of three pieces of a semiconductor, usually silicon. The outer pieces (the collector and emitter) are connected to a strong power supply. The central piece (the base) resists the flow of current through the transistor. When a weak electric signal is input to the base, however, the resistance of the base is lowered, allowing the strong current to flow into the collector, through the base, and out of the emitter. Variations in the level of the weak signal are transferred to the strong current, which emerges as an amplified copy of the weak signal.

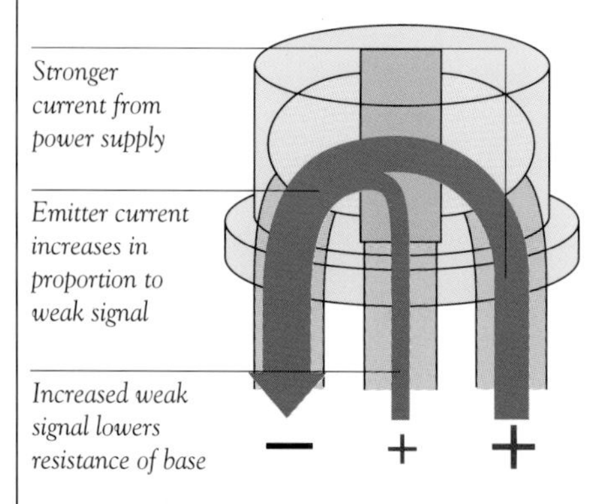

Amplifying an increased signal

As the weak signal increases, so does the current.

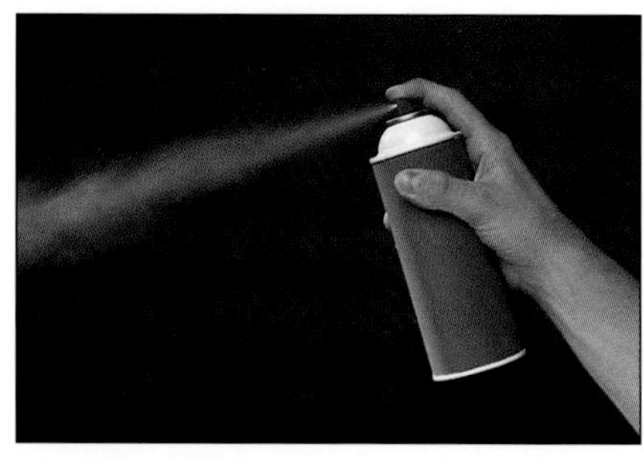

Controlling power

Amplifiers use a small force to control a large force. Aerosol sprays do this: a weak force from a finger controls the high-pressure spray.

EXPERIMENT

Transistor amplifier

Adult help is advised for this experiment

Show how a transistor works to amplify a weak current. First connect the circuit following the instructions opposite. This circuit uses a battery for a power supply. Note that the transistor's emitter, which connects to the negative terminal of the battery, is marked by a dot or tag. Then send a weak current around the circuit, bypassing the transistor, to make a light bulb glow dimly (the current is weakened by a salt water solution). Next connect the weak current from the water to the base of the transistor, causing the transistor to draw more current from the battery so that the bulb now glows brightly. A resistor keeps the weak current at very low levels so that the transistor is not damaged.

YOU WILL NEED

• 4.5V battery • scissors • tape • insulated copper wire • wire strippers • NPN transistor, BC441 or equivalent • foamcore base about 12 x 8 in (30 x 20 cm) • 4.5V bulb with holder • plastic dish • cardboard • salt • 5 alligator clips • 2 small screws • screwdriver • 220R resistor • pliers

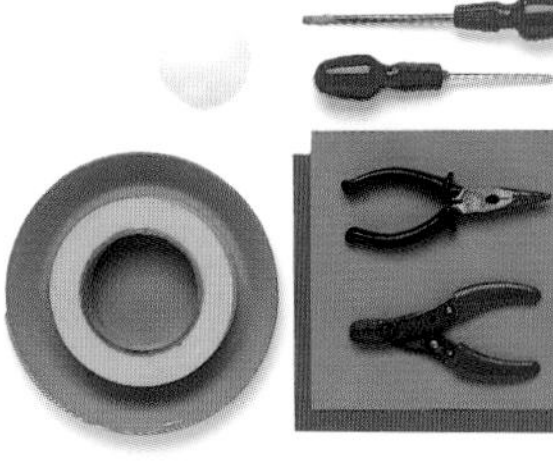

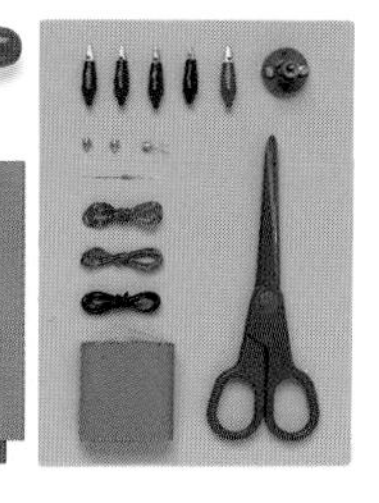

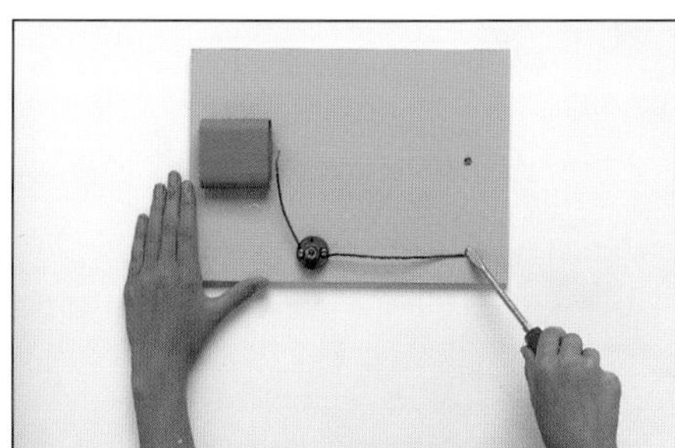

1 POSITION TWO screws on the board as shown above. Strip the ends of two 5 in (12 cm) lengths of wire. Use the wires to connect the negative battery terminal to one bulb terminal, and the other bulb terminal to the corner screw.

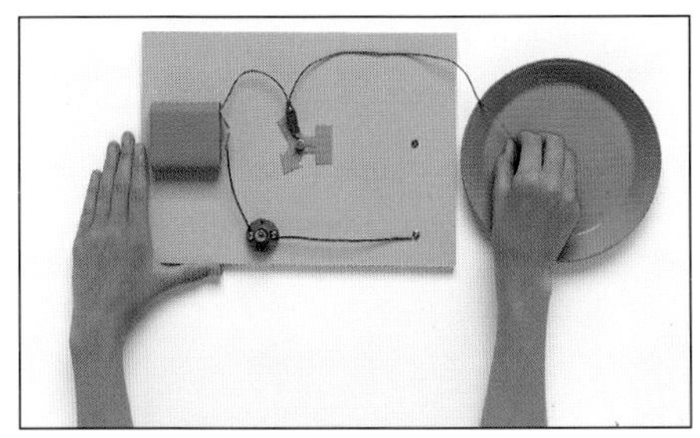

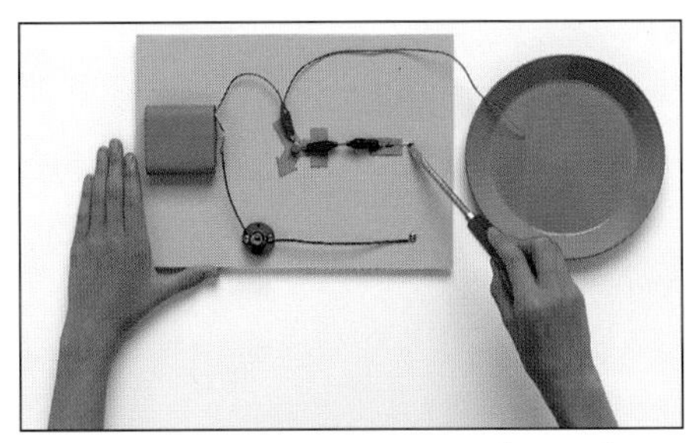

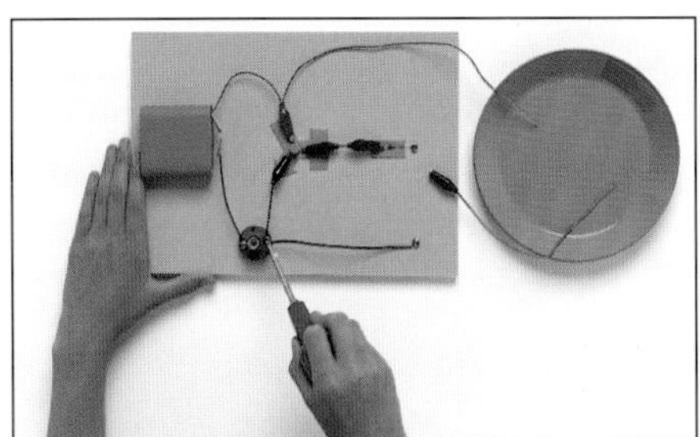

2 CONNECT TWO 8-in (20-cm) lengths of wire with stripped ends to a single alligator clip. Connect one wire to the positive battery terminal. Leave a stripped end of the other wire in the dish. Attach the alligator clip to the collector leg of the transistor.

3 ATTACH AN ALLIGATOR clip to the base leg of the transistor. Wind one resistor leg tightly around the free screw. Connect the other resistor leg to an alligator clip. Connect these alligator clips together using a short length of wire with stripped ends.

4 STRIP THE ENDS of two lengths of wire. Put a stripped end of one wire in the dish, and attach an alligator clip to the other end. Connect the second wire to the transistor's emitter leg with a clip, and attach the other endof this wire to the free bulb terminal.

5 CLIP THE free wire from the dish to the corner screw, which is connected to the bulb. Fill the dish with warm water, ensuring that the two stripped lengths of wire are submerged but not touching. Add salt to the dish slowly; stop adding salt when the bulb begins to glow dimly. At this point, no current is passing through the transistor, even though the transistor is connected to the battery.

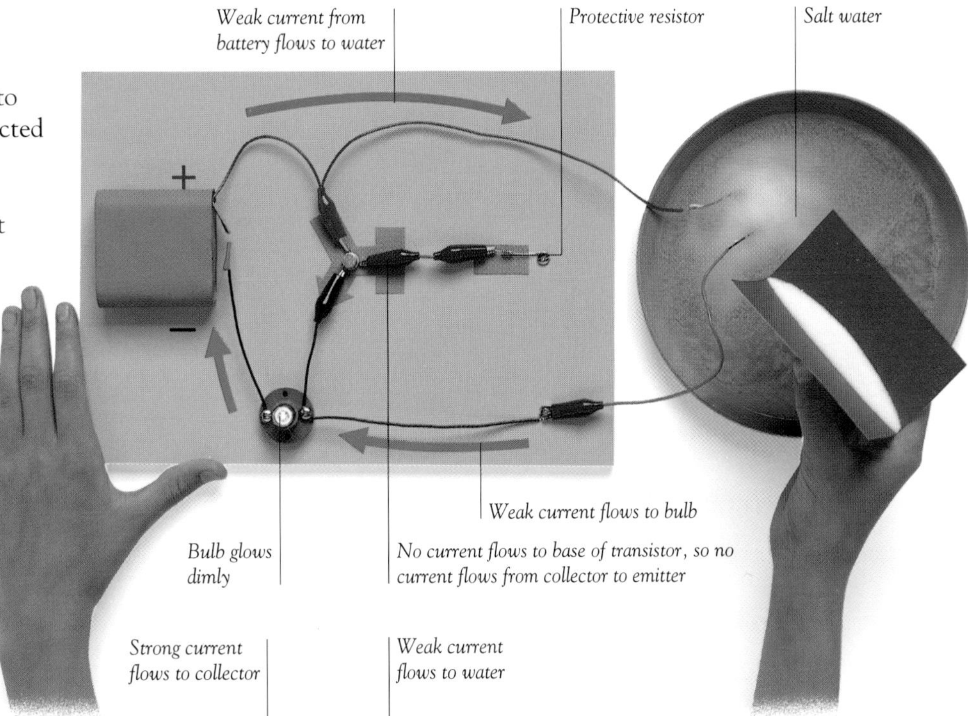

6 NOW DISCONNECT the wire between the dish and the corner screw, and clip it to the second screw where the resistor is connected. Ensure that the two stripped lengths of wire in the dish are still submerged but not touching. The new connection feeds the weak current into the base of the transistor. This causes the collector to draw a strong current from the battery, causing the bulb to glow more brightly.

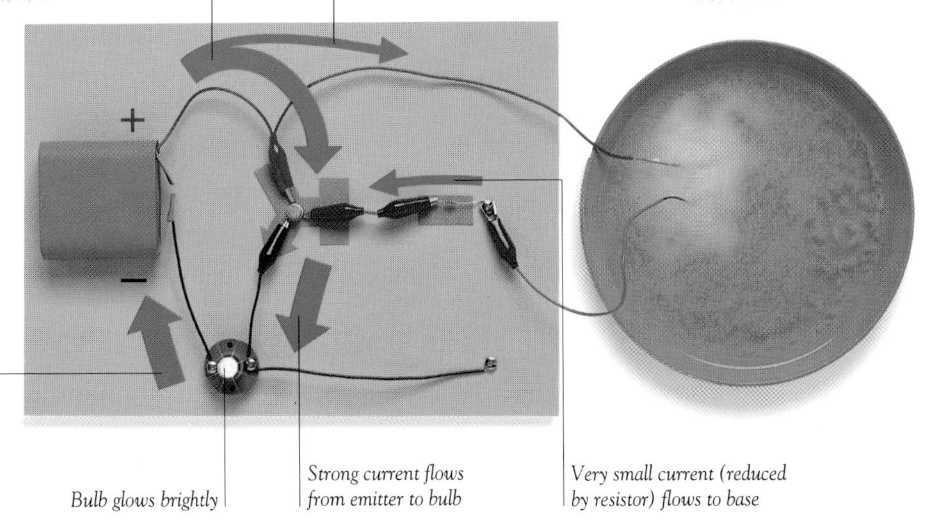

Sound recording

LISTENING PLAYS A large part in many people's leisure time. Portable stereos enable us to enjoy music anywhere, and audio systems in the home bring us sounds so lifelike that the singers and musicians could be performing in the room. All these sounds are recorded, as are the sound tracks of cinema films, videos, and much of the music and speech on radio and television. There are two main methods of recording—analog and digital.

Recording

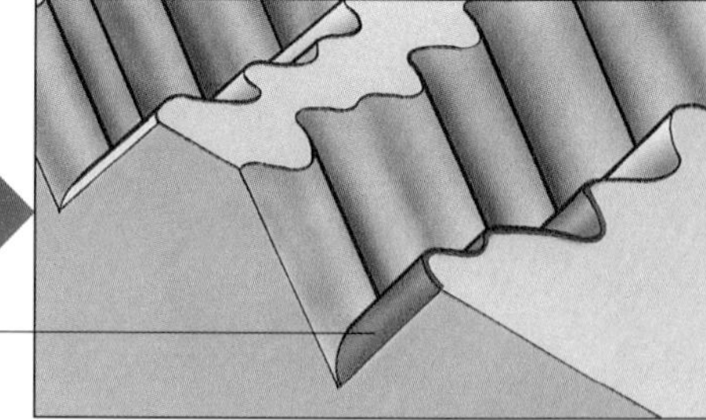

Phonograph record
On a phonograph record, the analog electric signals are stored in another analog form as one long wiggly groove on the record. A cutter records the signals from the microphones as a spiral groove in the surface of the record. The contours of the groove rise and fall continuously as they follow the variation in strength of the signals. The two signals are recorded on opposite walls of the groove.

Analog electric signal with varying strength

Original sound waves

Microphone

Magnetic field from recording head

Magnetic track on tape

Magnetic tape
On a tape, electric signals are stored in magnetic form. In analog recording, a recording head produces two magnetic fields that vary with the signals. The fields transfer changing levels of magnetism to the tape in two tracks. In digital recording, the signals are measured thousands of times a second. The measurements are turned into binary numbers that are stored in multiple tracks on the tape as pulses of magnetism.

Storing sound

Sound recording begins with microphones, which turn sound into electric signals to be stored by a recording system. In stereo recording, a pair of signals is produced. The signals are in an analog form: continuous and varying according to the frequency (pitch) and loudness of the sound. The signals are stored using either an analog system (records or tapes) or a digital system (CDs, mini-discs, or tapes).

Pit

On-off electric pulse signal

Binary measurements and on-off electric pulse signal

4 7 5 6 4 1 4

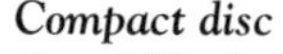

100 111 101 110 100 001 100

Compact disc
On a CD, the analog signals are converted for storage on a spiral track of pits and flat areas. The signal levels are "sampled"—measured thousands of times a second. The measurements are turned into binary numbers in the form of on-off electric pulses (binary 1's or 0's). Controlled by the pulses, a laser burns pits on the disc that represent one or more 0's and leaves unburned flat areas that represent one or more 1's. One track contains both signals.

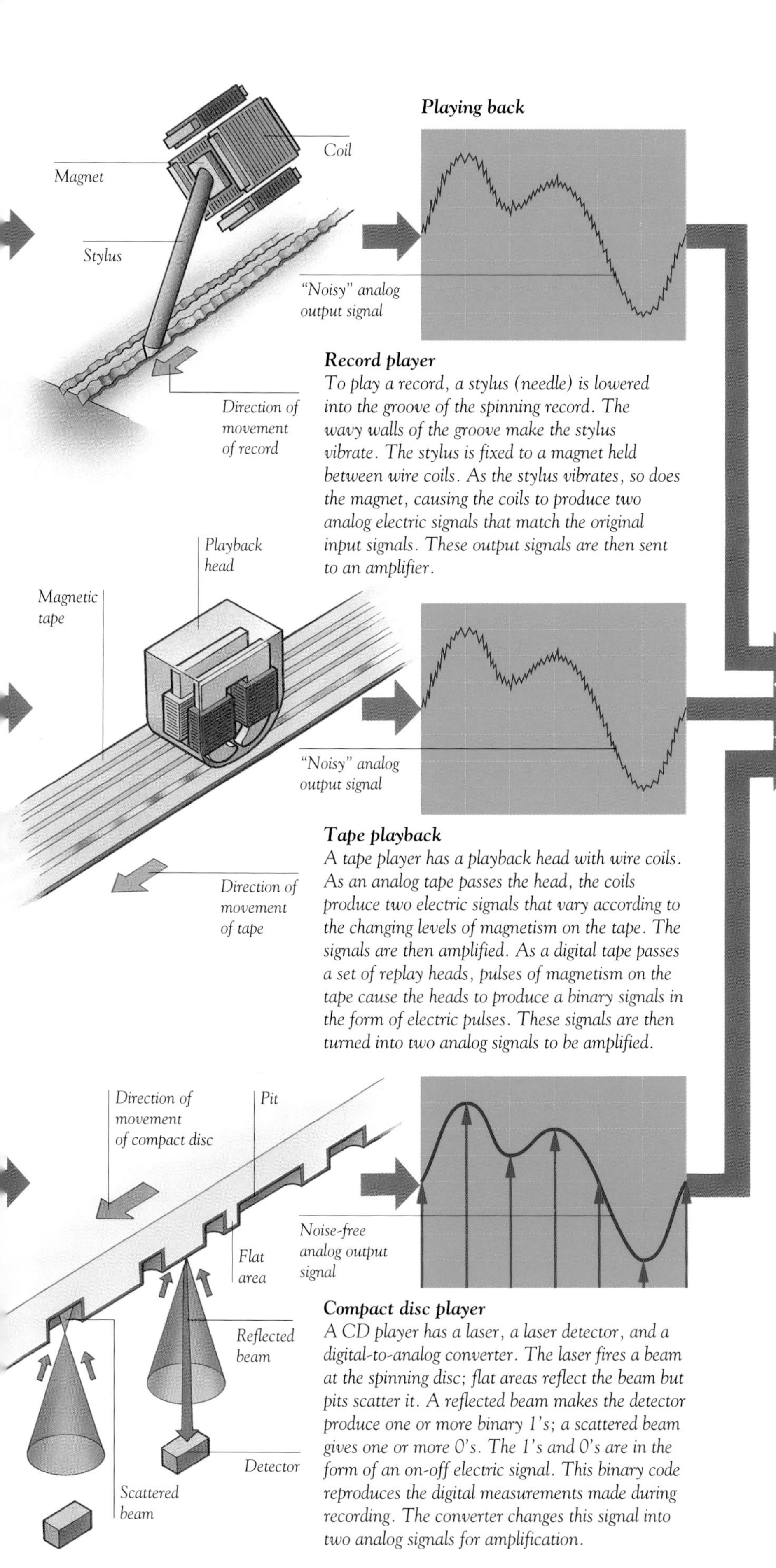

Record player
To play a record, a stylus (needle) is lowered into the groove of the spinning record. The wavy walls of the groove make the stylus vibrate. The stylus is fixed to a magnet held between wire coils. As the stylus vibrates, so does the magnet, causing the coils to produce two analog electric signals that match the original input signals. These output signals are then sent to an amplifier.

Tape playback
A tape player has a playback head with wire coils. As an analog tape passes the head, the coils produce two electric signals that vary according to the changing levels of magnetism on the tape. The signals are then amplified. As a digital tape passes a set of replay heads, pulses of magnetism on the tape cause the heads to produce a binary signals in the form of electric pulses. These signals are then turned into two analog signals to be amplified.

Compact disc player
A CD player has a laser, a laser detector, and a digital-to-analog converter. The laser fires a beam at the spinning disc; flat areas reflect the beam but pits scatter it. A reflected beam makes the detector produce one or more binary 1's; a scattered beam gives one or more 0's. The 1's and 0's are in the form of an on-off electric signal. This binary code reproduces the digital measurements made during recording. The converter changes this signal into two analog signals for amplification.

Reproducing sound

Both analog and digital machines output analog signals that should be identical to the original signals from the microphones. However, signals from analog systems may be distorted by noise caused by background magnetism on a tape, or by scratches in a record groove. Digital systems, however, read only binary 0's and 1's from the electric pulses in their signals. Even if the pulse levels are changed by background noise, the sequence of 0's and 1's remains unchanged; the measurements of the original sound waves are reproduced exactly in a noise-free analog output signal. The weak output signals must be amplified before being turned into sound by loudspeakers.

Analog and digital

Analog and digital systems use different types of signal. Analog signals are electrical replicas of the original sound waves. Digital systems convert sound into an on-off electric signal, which contains measurements of the original sound waves in binary code. This on-off signal is turned back into a smoothly varying analog signal by a digital-to-analog converter.

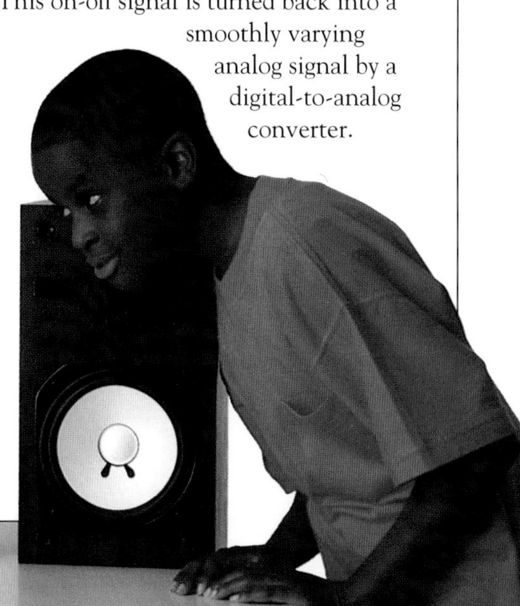

Radio

TUNE A RADIO to a station and sounds emerge from the set. The sounds originate from a microphone or sound-playing equipment at the station (pp.144–145). These devices produce an electric signal, which the station transmits as radio waves. The waves spread out at high speed and reach your radio, which turns them back into an electric signal. This signal powers the radio's loudspeaker, which converts the electricity back into sound.

Broadcasting

A radio station combines the electric sound signal from a microphone, CD-player, or tape player with a high-frequency electric carrier signal. The sound signal is used either to vary the amplitude (strength) of the carrier signal (amplitude modulation, or AM) or to vary its frequency (frequency modulation, or FM). The combined, or modulated, signal is converted into a radio wave that is broadcast from a transmitter. This varying radio wave produces a varying electric signal in the antenna of any radio within range. A circuit in the radio separates the high-frequency carrier signal from the low-frequency sound signal, then sends the sound signal to a loudspeaker.

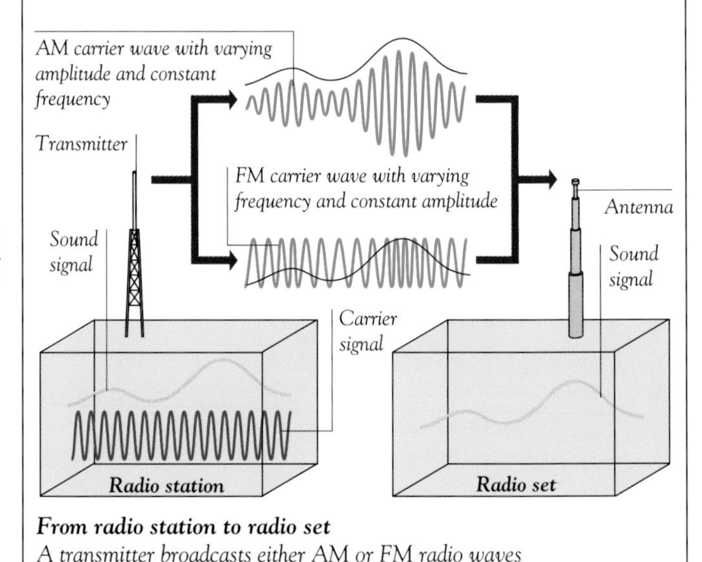

From radio station to radio set
A transmitter broadcasts either AM or FM radio waves

EXPERIMENT

Radio tuner

Every station sends out a radio wave within its own narrow frequency range. When the wave reaches an antenna, it creates an electric signal by causing electrons in the antenna to vibrate at the same frequency as the wave. An antenna receives hundreds of signals at any time, so the radio has a tuning circuit that filters out unwanted signals. Electrons in the circuit can vibrate within only one frequency range, which you set by tuning the radio. Only a signal within this range sets the electrons in the tuning circuit vibrating so that the signal passes the tuner. Show how this happens using pendulums to represent the electrons. One pendulum is set swinging only by another of the same frequency.

YOU WILL NEED

- *table*
- *string*
- *1 red and 4 green pieces of modeling clay*
- *string cut to lengths: 5 in (13 cm), 10 in (25 cm), 26 in (65 cm), two x 16 in (40 cm)*

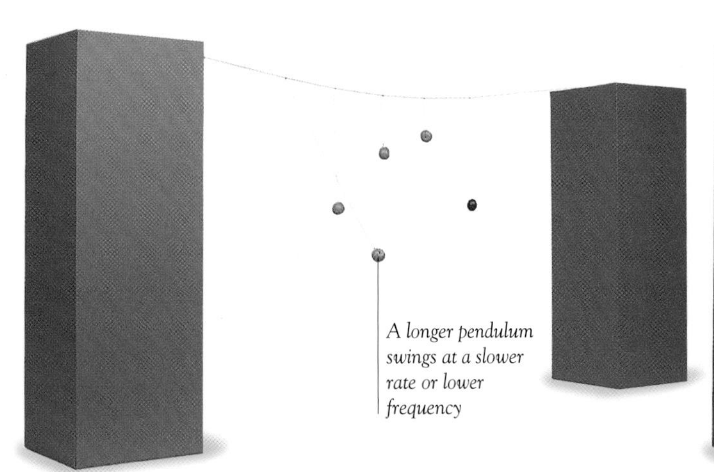

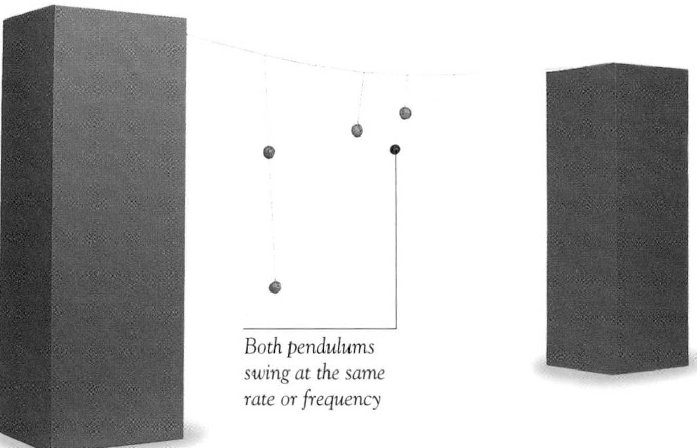

1 MAKE FIVE PENDULUMS by attaching red modelling clay to one end of a 40 cm (16 in) string, and green modeling clay to an end of each of the other four strings. Suspend them from a string stretched tautly between two table legs. Swing each green pendulum in turn. Because they are of different lengths, they swing at different rates or frequencies.

2 THE GREEN PENDULUMS represent electrons in the antenna made to vibrate by different signals; the red one represents electrons in the tuning circuit. Only a green pendulum of the same length (and the same frequency) as the red pendulum is able to transfer its movement to the red pendulum and set it swinging at this frequency.

EXPERIMENT

Sparks and crackles

Make and transmit some radio waves. First produce sparks by connecting two wires to a battery, connecting one of the wires to a metal file, and running the other wire over the file. The electric current goes on and off as the sparks fly, generating radio waves. These represent the radio waves transmitted from a radio station. A radio antenna picks up the waves from the file and produces crackling noises. You get noise instead of voices or music because the radio waves are not modulated like those coming from radio stations.

1 STRIP THE ENDS of two 12 in (30 cm) lengths of wire using wire strippers. Tape a stripped end of one wire to each of the battery terminals.

2 TAPE THE FREE end of one wire (from either terminal) just below the handle of the metal file, so that it makes an electrical contact.

YOU WILL NEED

- *4.5V battery* • *electrical wire* • *wire strippers* • *tape* • *scissors* • *portable radio* • *file*

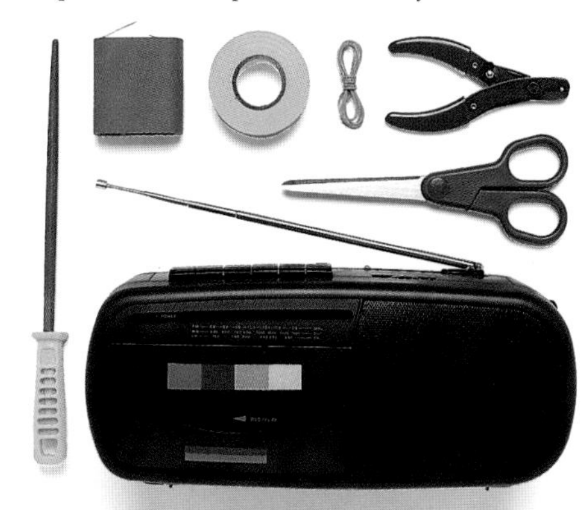

3 TUNE THE RADIO to an AM or FM station, and place it near the file. Fray the remaining free end of wire, then brush the frayed end over the file.

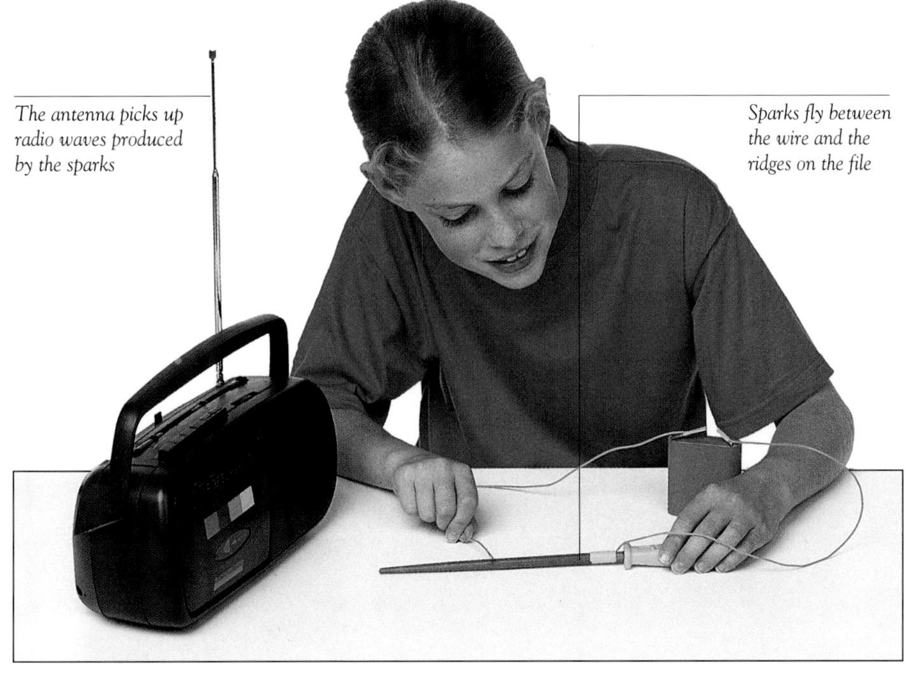

The antenna picks up radio waves produced by the sparks

Sparks fly between the wire and the ridges on the file

■ DISCOVERY ■

Lodge and Marconi

Using sparks, the German scientist Heinrich Hertz (1857–94) discovered radio waves in 1888. Six years later, the British scientist Oliver Lodge (1851–1940) used Hertz's discovery to send the first radio message, which was sent in Morse code between two buildings. The Italian inventor Guglielmo Marconi (1874–1937) resolved to develop radio as a means of instant worldwide communication. He began his work in 1895, and by 1901 Marconi was able to send a radio message across the Atlantic Ocean. Modulation, which enables sound to be broadcast by radio, was invented by the Canadian engineer Reginald Fessenden (1866–1932) in 1906.

Marconi with his radio equipment in 1896

Television 1

THE PICTURE YOU SEE on a television screen is generated by a television camera or a camcorder. The camera converts the light and color in the scene in front of it into an electric signal. This signal is carried to your home, usually by radio waves (p.146) broadcast from a transmitter. The radio signal may be relayed to a communications satellite in orbit before being picked up by a television antenna or dish, in which the radio waves recreate the original electric signal. This signal goes to the television set, which uses it to create the picture that you see. In cable television, the electric signal travels directly from the television station along cables to television sets. Video images are recorded on magnetic tape in much the same way as sound signals (pp.144–45).

Inside television
The interior of this early television set, dating from the 1930's, shows tubes and other components that process the electric signals coming from the antenna. These go to the picture tube (top) and loudspeaker (bottom).

EXPERIMENT

Scanning

Adult help is advised for this experiment

Although television appears to show a whole picture moving normally, what you are looking at is actually something very different. Television breaks the picture into pieces, flashing 30 still images onto the screen in sequence every second. The still images appear so quickly that the picture on screen appears to move. Furthermore, each still picture is made up of 525 horizontal lines. The camera scans the scene in front of the lens 30 times a second, splitting the image into lines and detecting changes of color and light along each line. For viewing, a dot of light moves across the screen, tracing each line in sequence from top to bottom 30 times a second. The dot's color and brightness vary and it moves so fast that a complete, moving picture is seen. This experiment shows how scanning breaks a picture into lines that appear so quickly in sequence that you see a complete picture.

YOU WILL NEED
- *slides*
- *large, white paper sheet*
- *slide projector*

1 LOAD A SLIDE into the projector and darken the room. Hold the paper in the projector's beam, a few paces away, and move it as necessary to project a complete, sharp image of the slide onto the paper. On the floor, mark the distance of the paper from the lens. Now roll the paper tightly to make a baton.

■ DISCOVERY ■

The beginnings of television

Several people were involved in the invention of television. The first person to send a picture using radio waves was the British inventor John Logie Baird (1888–1946) in 1924. He then developed a television system that used mechanical parts. However, this mechanical system had poor picture quality and was soon replaced by the electronic system that is still used today. Electronic television was invented in the 1920's in the United States by Vladimir Zworykin (1889–1982) and Philo T. Farnsworth (1906–71). The system was developed during the 1930's, and the first public television service opened in Britain in 1936. Color television was also developed during the 1930's, and the first public color broadcasts were made in the United States in 1941.

Vladimir Zworykin
Zworykin is holding the picture tube of an early television camera.

2 HAVE A FRIEND stand at the mark and rapidly move the baton down through the beam. A sequence of horizontal sections of the image forms on the baton. The sections appear to merge together so that you see a complete image.

How a television camera works

A television camera has a lens that forms an image of a scene on a light-sensitive tube. Camcorders (portable cameras) and some studio cameras form the image on an array of tiny light-sensitive elements called CCD's (charge-coupled devices). The tube or array scans the image 30 times a second to convert the light in the image into an electric signal. A color television camera contains three tubes or arrays that are sensitive to the amounts of red, green, and blue light in the image, and a system of semitransparent mirrors that separate these colors. Electronic components then combine the three color signals into a single electric picture signal, which is broadcast with a sound signal by radio waves from a transmitter. The signals modulate the radio waves so that they carry the picture- and sound-signals, using the same method used to broadcast sound signals by radio (p.146). The signals may be stored on videotape before transmission. In a camcorder, the picture- and sound-signals are recorded directly on videotape in the camera.

Broadcasting the signal

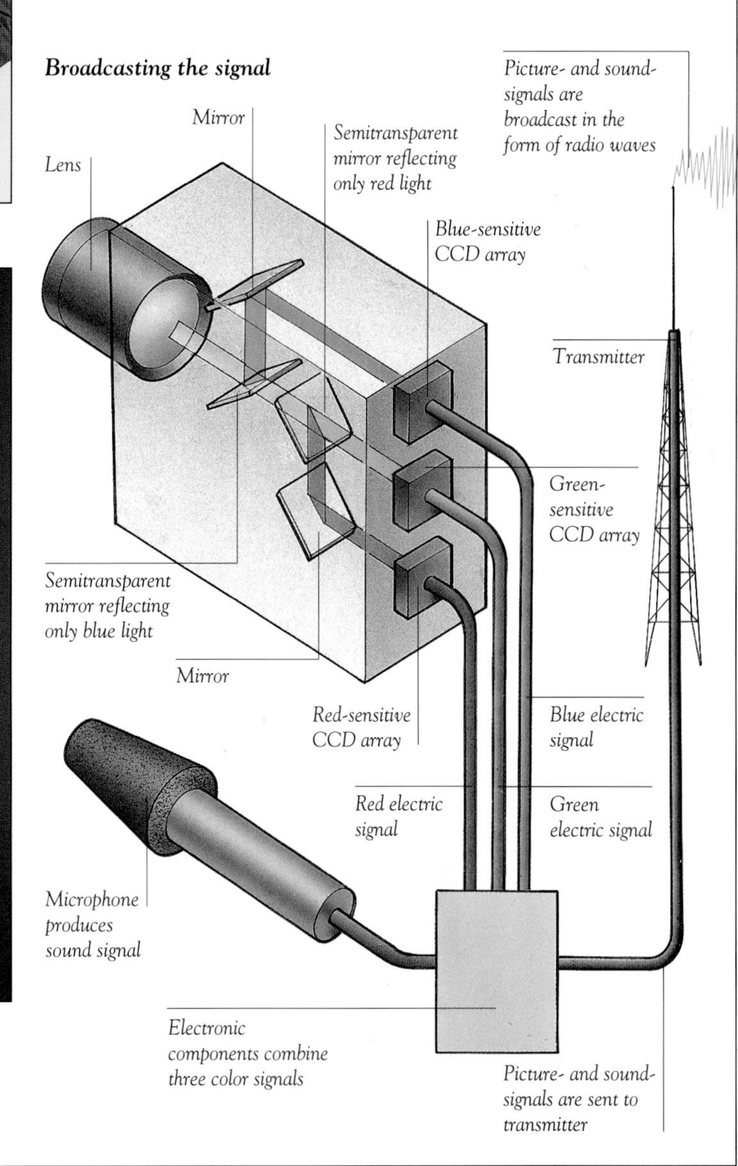

Television 2

PICTURE SIGNALS and sound signals come from the stations to which you can tune your television set. They may be broadcast using radio waves or may be sent along an electric cable. The signals for each channel have their own preset frequencies (p.146), and when you choose a channel, the set selects the pair of signals at those frequencies. A video recorder can also select a channel and record the pair of signals on a videotape. On replay, it sends the signals to the television set.

How a TV receiver works

Broadcast radio waves produce electric picture and sound signals in an antenna. The signals go to a receiver, where a circuit separates them. The sound signal is sent to a loudspeaker, while the picture signal is split into three color signals that represent the red, green, and blue parts of the picture. These color signals go to a picture tube, where they control the intensity and direction of electron beams fired at the screen by three electron guns, causing the beams to trace the scan lines of the picture (p.149). Each beam passes through slits in a shadow mask to strike tiny phosphor stripes on the screen, making them glow red, green, or blue. Three images in red, green, and blue appear simultaneously on the screen, and your eyes merge these images together so that you see a single full-color picture.

Radio waves containing picture- and sound-signals strike antenna

Antenna cable conducts signals to television set

Shadow mask
As each electron beam scans across the shadow mask, the slits let it through to reach phosphor stripes of only one color.

Electron gun

Vertical control signal

Vertical electromagnetic deflection coil

Slit in shadow mask

Electron beam

Phosphor stripe on screen glows when struck by beam

Horizontal electromagnetic deflection coil

Shadow mask

Color signals

Horizontal control signal

Electron beam

Loudspeaker

Sound signal

Picture tube

Screen

Glass cover

EXPERIMENT

Television screen

Adult help is advised for this experiment

Make a model picture tube using a flashlight to represent the electron guns and cellophane strips to represent the phosphor stripes. Show how the slits in a shadow mask allow each beam to pass so that it causes only red, green, or blue stripes on the screen to glow.

YOU WILL NEED

● *two sheets of poster board 14 x 4 in (35.5 x 10 cm), one black, one gray* ● *cardboard pieces as follows: 2 sides 11¾ x 4 in (30 x 10 cm); base 14 x 11¾ in (35.5 x 30 cm); lid 14 x 6 in (35.5 x 15 cm)* ● *cutting mat* ● *ruler* ● *steel rule* ● *4 each of red, green, and blue cellophane strips, 3 x ¾ in (7.5 x 2 cm)* ● *two 4 x 1½ in (10 x 4 cm) foamcore spacers* ● *red, green, and blue paper dots* ● *glue* ● *pencil* ● *flashlight* ● *craft knife*

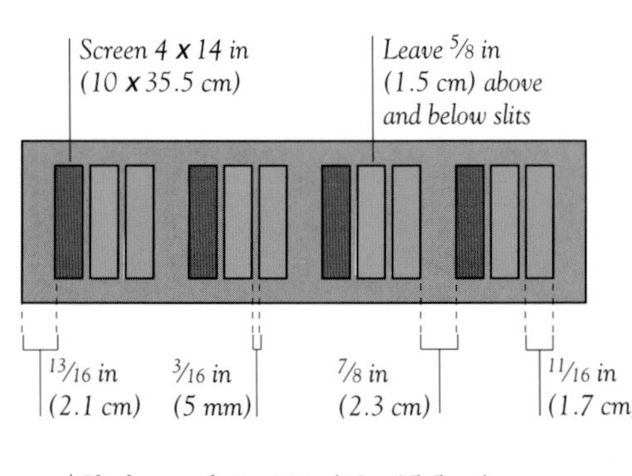

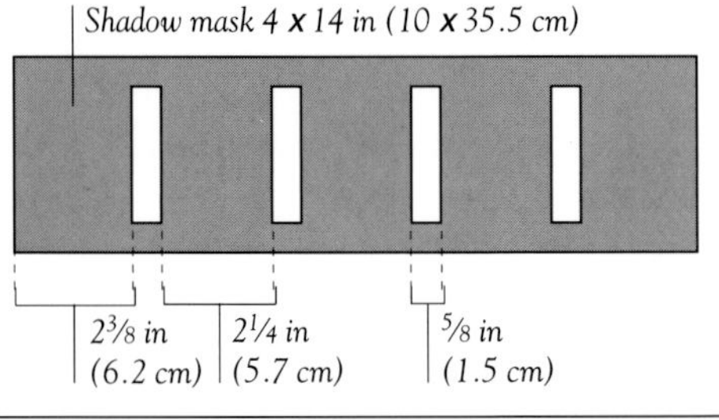

1 USE A STEEL ruler and craft knife to cut four parallel rectangular slits in the shut of black poster board, using the shadow mask template shown on the opposite page.

2 CUT TWELVE rectangular slits in the gray poster board, using the screen template on the opposite page. Glue different colors of cellophane over the slits, as illustrated.

3 GLUE A long edge of a spacer to each short edge of the shadow mask at right angles. Glue this to the base so that the spacers are flush with the corners of the front short edge of the base and the mask is set back inside. Glue the sides to the base. Glue the screen to the front of the base, parallel to the mask.

4 GLUE THE green paper dot halfway along the rear edge of the base, $\frac{3}{8}$ in (1 cm) from it. Glue the blue dot in line with the green dot, $\frac{3}{4}$ in (2 cm) from the side nearest the first red cellophane strip. Add the red dot $\frac{3}{4}$ in (2 cm) away from the opposite side. Glue the lid over the screen and mask.

Flicker-free pictures

Although a television set shows 30 complete pictures every second, each picture is made up of two successive images composed of alternate lines. The camera and set first scan the picture in 263 odd-numbered lines, then again in 262 even-numbered lines to form a complete picture. Although you see 30 complete pictures a second, these are made up of 60 separate images. This technique, called interlacing, eliminates flicker so that the sequence of still pictures appears to form a smoothly moving picture.

***The television** picture in this photograph is incomplete because the camera shutter closed before the full number of lines formed.*

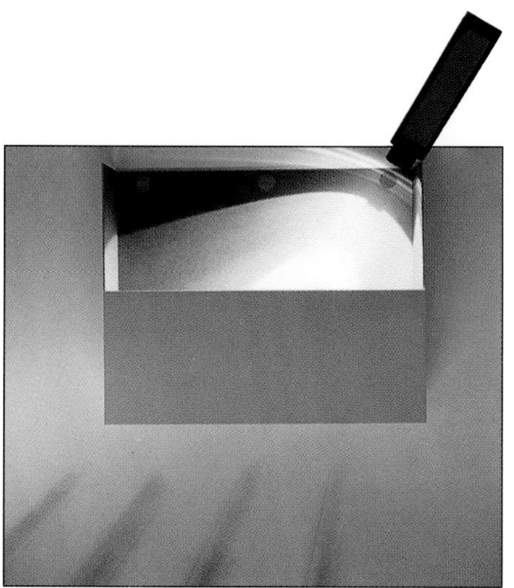

5 TURN ON the flashlight behind the red dot. Dim the lights. Angle the light to illuminate the whole shadow mask. Look at the screen from above.

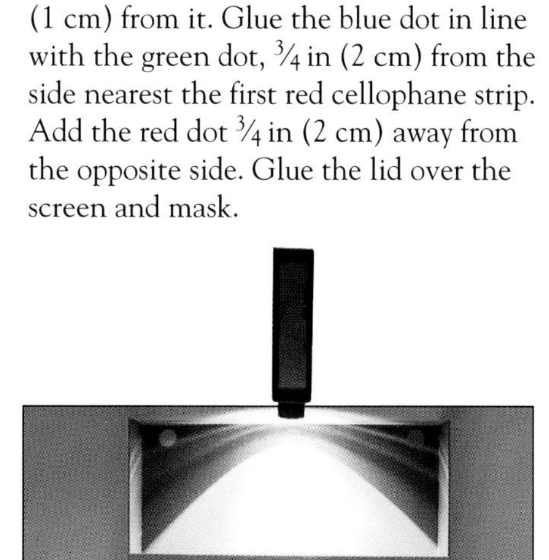

6 PLACE THE flashlight just behind the green dot, pointing at the center of the shadow mask. Observe the screen from above once again.

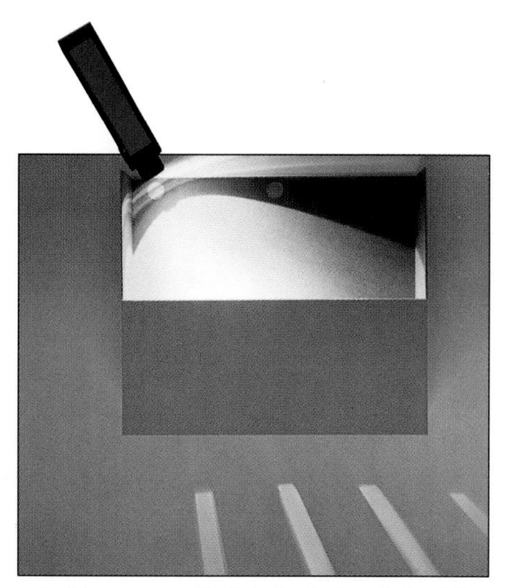

7 PLACE THE flashlight behind the blue dot. Angle the light to illuminate the whole shadow mask. Look at the screen from above once more.

INFORMATION TECHNOLOGY

Hyperactive machinery
A computer (above) handles information at lightning speed as electrical signals flash through its electronic components. Optical fibers (left) enable computers, telephones, and other electronic information machines to communicate by using light signals, which carry huge amounts of information through the tiny glass threads.

MACHINES SUCH AS telephones, fax machines, and computers are transforming people's lives. We can now send information—speech, words, pictures, numbers—anywhere almost instantly. We can also process this information in any way we want. Information technology helps reduce the workloads of many people, enabling them to live fuller and easier lives. We are starting to achieve massive and instant exchanges of information on a global scale, and this facility may one day help to bring the world together.

THE ELECTRONIC AGE

PERIODS OF HISTORY ARE OFTEN CALLED AGES. This century has seen the atomic age, marked by the discovery and development of nuclear power, and the space age, when space flight began. We are now living in the electronic age, in which machines have developed unparalleled and complex abilities and yet may be small enough to fit in a pocket.

Modern information technology began in the 1830's with the invention of the electric telegraph. This system used electric signals to send messages along a wire from a sending device to a receiving device. The first electric telegraph, made by the British inventor W.F. Cooke, used an electric current to move needles on the receiver; the needles pointed to letters of the alphabet, spelling out the message. It was quickly superseded by machines using Morse code. The telegraph was the first device to embody the basic concept behind information technology: that machines change information into a set of signals, send them somewhere else, then, at the receiving end, change them back into a form that reproduces the original information.

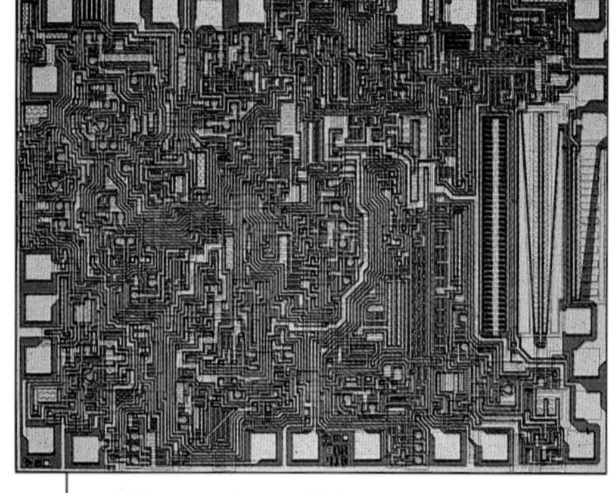

Electronic machines contain *miniature components, seen here packed into a microchip, that carry and direct signals made up of bunches of electrons. These electrical particles are tiny and travel extremely fast, so machines can be highly complex yet very small, and give nearly instantaneous results.*

Hardware

Since the end of the nineteenth century, information technology has increasingly affected our lives. This information revolution began in the 1870's with the invention of the telephone and record player, and continued with the radio, telephone, and tape recorder, all of which are now found in many homes. Telephones also became vital in business, and have been followed by photocopiers, computers, and fax machines.

All of these machines make up the hardware of information technology. They help us with communication, assist learning, bring us entertainment, allow us to exchange vital information when working, and enable us to work more effectively. The machines perform four principal tasks. The first is to send and receive information in the form of speech, written words, numbers, or images. The other main functions of these machines are storing, copying, and processing this information in ways that are useful to us.

Highways

The worldwide network of telephone lines and radio links is the main channel for the transmission of information.

These dishes relay signals *to satellites in orbit around the Earth, enabling almost instantaneous communication worldwide.*

It does more than just carry the sounds of voices: videophones send and receive images of their users, while fax machines transmit copies of pictures and documents over long distances. Computers can be interconnected through telephone lines, either

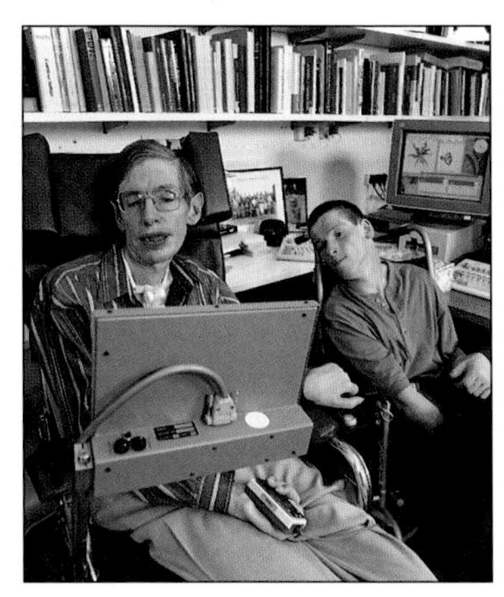

Information technology *can benefit the disabled. Unable to speak, the scientist Stephen Hawking uses a computer to voice his thoughts.*

one-to-one or in large networks. These networks carry messages (electronic mail) and business information, and provide access to computer databases that contain enormous amounts of information.

The telephone network carries some information in the form of electric signals travelling along metal wires. It also uses radio links and optical fibers. In radio networks, the signals are changed into radio waves or microwaves. In fiber-optic networks, lasers change information into light signals that travel along optical fibers (long glass threads). Optical fibers are becoming the most important element in the telephone network, because they can carry much more information than electric wires or radio links, and transmission is of very high quality. Some homes now have fiber-optic links that give direct access to the information superhighway. As a result, people can now do such things as buy goods or consult their doctors, all without leaving the house.

Copying and storing

Copying data is an important aspect of information technology. Photocopiers, for example, use static electricity to make black-and-white or color copies of pictures or documents. Storage of information is a major function of computers, which store data in three main forms: magnetic, optical, and electronic. The hard disks and floppy disks used by computers are both coated with magnetic material that stores data. CD-ROMs (compact discs that can hold still or moving images, sound, or text) use optical storage in the form of patterns of pits. The data is reproduced in its original form when a laser beam "reads" the disc. A single CD-ROM can store an entire encyclopedia—yet the computer can find any part of the data almost instantly. Finally, microchips contain tiny electronic components that store data as patterns of electric charges.

This "binary box" changes *decimal numbers into the binary numbers used by computers (p.164).*

Computers

A computer works by converting data into electric signals and back again. The signals are made up of binary numbers, which consist of a series of 0's and 1's ("bits")—the electric signals turn off and on to represent the 0's and 1's. Because the computer is basically handling numbers, it is known as a digital machine. Digital systems, which operate at high speed and produce high-quality results, are increasingly used in information processing. As well as processing information, some computers control other machines. Computers can also follow various sets of instructions, called programs, which tell them how to perform a wide range of tasks, such as word processing, playing games, calculating business figures, and playing music. These programs, and others that tell the computer how to operate itself, are known as software.

Business institutions such as this *stock exchange depend on information technology to bring them up-to-the-minute financial information.*

These young people *are taking part in a game that shows how a computer program works (pp.174–175).*

Photocopiers

HAVE YOU NOTICED HOW the screen of a television set quickly gets dusty? A charge of static electricity on the screen attracts particles of dust in the air, and they stick to the screen. A photocopier makes copies of documents in basically the same way. The machine uses static electricity to make particles of a black powder called toner stick to a sheet of paper and form a black-and-white copy of the document. The toner is then sealed on the paper so that, unlike the dust on a television screen, it cannot be wiped off. A color photocopier makes color copies in the same way, but uses three extra toners colored magenta (red-blue), cyan (blue-green), and yellow. The three extra images formed in these colors overlap with the black image to give a full-color copy.

Color photocopies

Full-color images are printed using four colors of toner: magenta, cyan, yellow, and black. In a color copier, a drum turns four times, once for each color. On each turn, a differently colored copy of the document is transferred from the drum to the same sheet of paper. Magenta, cyan, and yellow copies are made after passing white light reflected from the document through a green, red, or blue filter. The final copy is made with unfiltered white light and uses black toner. The four colors merge together in different combinations on the paper to give a single, full-color copy of the document.

EXPERIMENT

Balloon copier

Use a balloon and talcum powder to show how a photocopier works. Create a circle of static electricity by rubbing an area on the balloon and then discharging the center. The balloon picks up powder to form a large letter O, which you then transfer to a sheet of paper. Inside a real photocopier, an electrified drum acts like the balloon to pick up black powder and form of an image of a document and then transfer it to paper to make a copy.

YOU WILL NEED

• dark-colored paper • graphite pencil • talcum powder • tray • 2 balloons • pen • wool glove

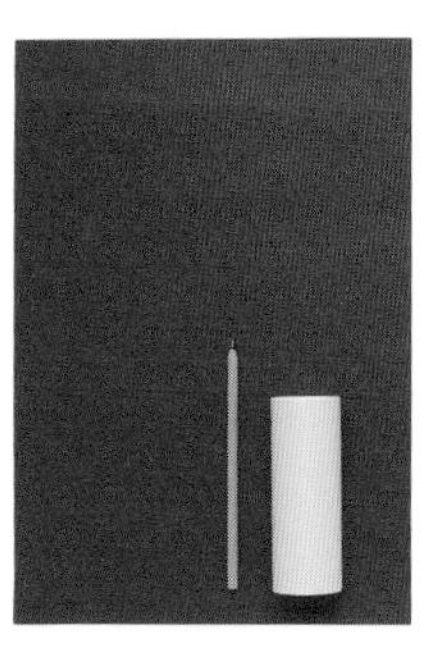

1 MARK A CROSS with a pen on an inflated balloon. Use a wool glove to rub the balloon around the cross, in a circular motion. Rub the surface of the balloon hard to create a circle of static electricity on the balloon.

2 PLACE SOME TALCUM powder on a tray. Ask a friend to agitate the tray vigorously to create a cloud of powder above the tray. Hold the balloon with the cross over the tray so that the charged part attracts powder.

3 ONCE ENOUGH powder has stuck to the balloon, carefully hold the balloon over a sheet of paper. Roll the balloon over the paper so that the powder is transferred to print a circle on the paper.

Colored images

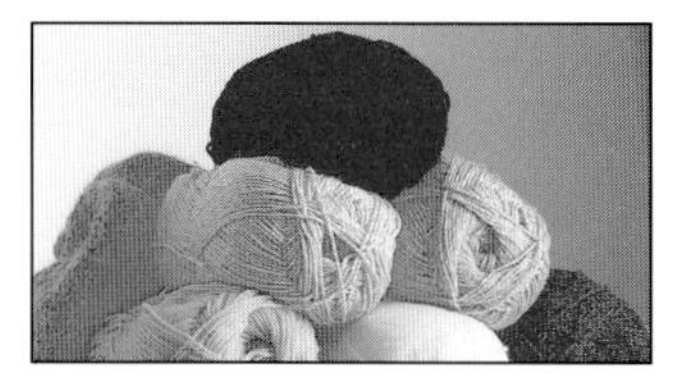

Magenta

Cyan

Yellow

Black

Printing sequence

Magenta

Magenta and cyan

Magenta, cyan, and yellow

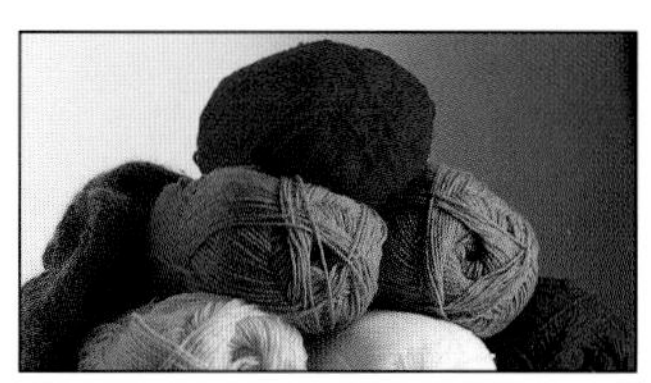

Magenta, cyan, yellow, and black

4 REPEAT STEP 1 with a new balloon. Once the balloon is charged, hold a graphite pencil near the cross marked on the balloon. You may hear a soft click as some electricity is discharged.

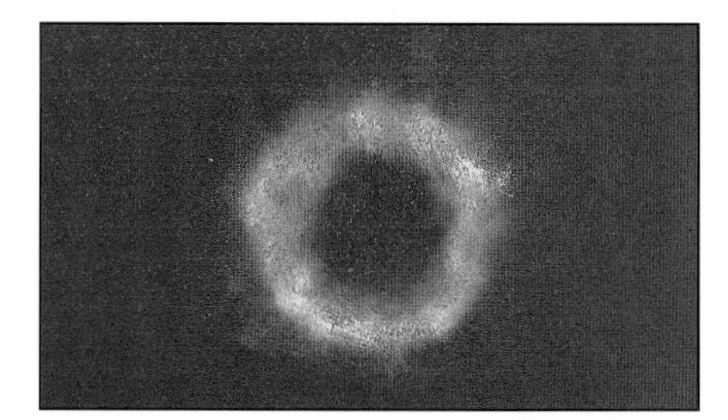

5 NOW REPEAT STEPS 2 and 3. The center of the charged circle has been discharged by the graphite. The remaining charged area around the edge of the circle attracts the powder and prints an "O" on the paper.

How a black-and-white photocopier works

When a black-and-white document is placed on a copier's glass window and the "copy" button is pressed, a lamp scans the whole document from beneath. The document's white areas reflect light from the lamp through a system of mirrors and a lens, while the black areas reflect no light. The lens projects an image of the document onto an electrically charged drum. The drum turns while the document is scanned, and light reflected from white areas of the document destroys the charge where it hits the drum. The drum is left with charged areas that correspond to the black parts of the document. These charged areas attract black toner powder from a roller to form a copy of the document on the drum. A strongly charged paper sheet attracts the toner from the drum. The paper is then heated to seal the toner and form a permanent copy. If the document is in color or has shades of grey, the colored or grey areas reflect light in proportion to how dark they are.

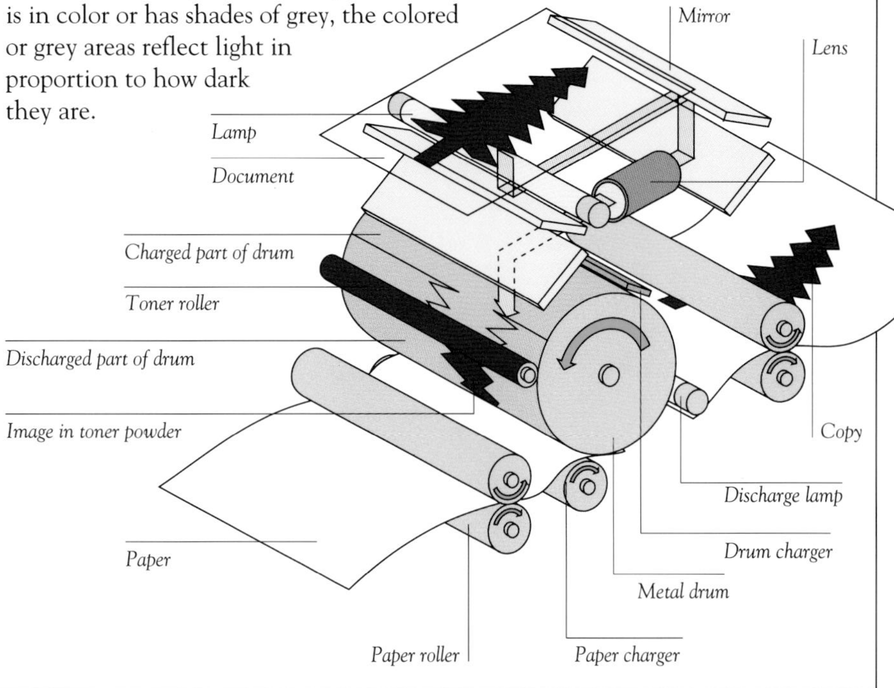

Telephone

PICK UP A telephone, dial a number, and within seconds you can be speaking to someone. The sound of your voice is turned into an electric signal at your end and back into sound at the other end. In most calls, the signal travels along wires to telephone exchanges that route the call to another telephone. Between exchanges, however, the signal may be turned into light pulses or radio waves. Light pulses are carried by fiber-optic cables. Radio waves are used for long-distance calls sent via satellites. The signal is turned into radio waves in a ground station; the waves are sent about 22,500 miles (36,000 kilometers) up to the satellite, then beamed to another ground station, where they are turned back to an electric signal for the receiver's telephone. Mobile phones send radio waves directly to a network of radio stations, which send them to the listener. Videophones work like television systems, and allow you to see your callers as well as talk to them.

How a telephone works

A microphone in the mouthpiece changes the sound of your voice into an electric signal. This electric signal travels through one or more telephone exchanges and finally to the other telephone. In the receiving telephone a vibrating membrane in a small loudspeaker in the earpiece converts the signal back into the sound of your voice.

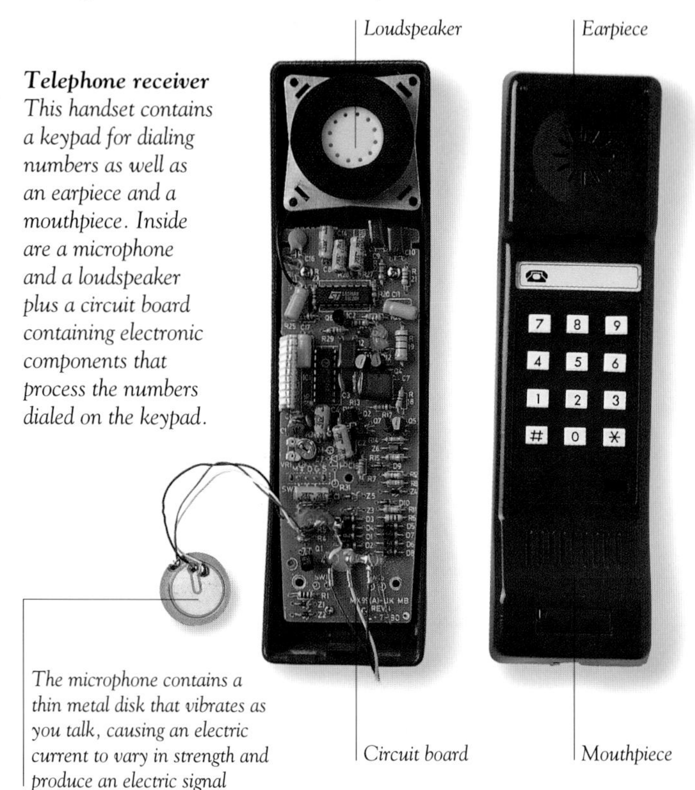

Telephone receiver
This handset contains a keypad for dialing numbers as well as an earpiece and a mouthpiece. Inside are a microphone and a loudspeaker plus a circuit board containing electronic components that process the numbers dialed on the keypad.

The microphone contains a thin metal disk that vibrates as you talk, causing an electric current to vary in strength and produce an electric signal

Fiber optics

Many telephone lines use fiber-optic cables. Instead of metal wires carrying electric signals, the cables hold very thin, clear glass fibers, each carrying a light signal. Electric signals from a telephone are converted to pulsed (digital) light signals by laser; digital signals are an efficient means of transmission, because they eliminate most background interference. The light signals are turned back into electric signals at the other end. Optical fibers transmit huge amounts of information, as each glass thread can carry several thousand calls at once. As information networks develop, such cables will bring video and other services directly to the home.

Reflection
See how a glass surface can reflect light. First stand facing a window with a friend. You see only a faint reflection because light rays from you strike the glass at 90° and most pass through the window. Then stand apart, near the glass. Now you can see each other clearly because the light rays strike the glass at a large angle and are reflected from the inner surface of the glass window.

Light cable
An optical fiber has a glass core inside a glass sleeve. Light entering the core is reflected at the wall between core and sleeve if it hits the wall at a critical angle (around 82° to the perpendicular or right angle). The reflected light continues to bounce along the length of the fiber.

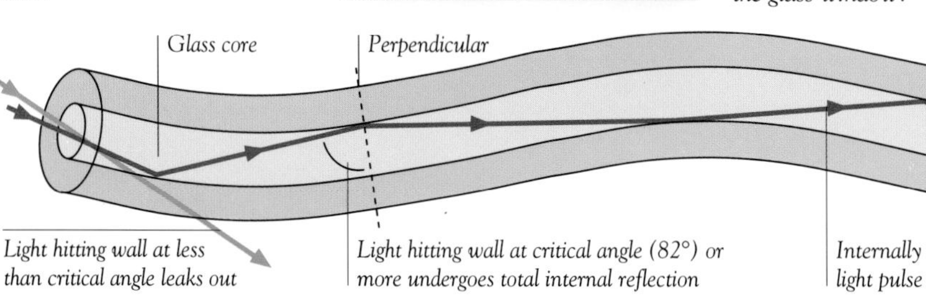

EXPERIMENT

Light carrier

Adult help is advised for this experiment

Using a plastic tube and water, show how an optical fiber works. The tube represents the outer glass sleeve of a fiber. It is filled with water, which represents the glass core. Some of the light entering the tube is internally reflected where the water and plastic meet. This internally reflected light travels along the tube.

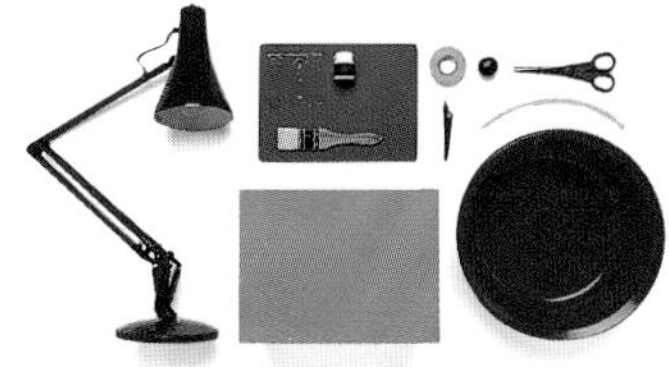

You Will Need

- *desk lamp*
- *cutting surface*
- *plastic wrap, about 6 in (15 cm) square*
- *pot of matte black paint*
- *paint brush*
- *tape*
- *small piece of modeling clay*
- *scissors*
- *craft knife*
- *1 ft (30 cm) long flexible plastic tube*
- *large cardboard box*
- *bowl of water*

1 Paint the inside of the box black. Use a craft knife to cut a hole in a narrow side of the box, near the base. The plastic tube should fit through this hole.

2 Immerse the tube in the bowl until it completely fills with water. Put plastic wrap over one end and secure this with tape for a watertight seal. Remove the tube, holding the uncovered end upright so that the water stays inside.

3 Push the tube through the hole so that the covered end is inside the box and the other end sticks outside by about 1 in (2.5 cm). Make a light-tight seal by pressing clay around the hole, inside and outside. Keep the uncovered end upright.

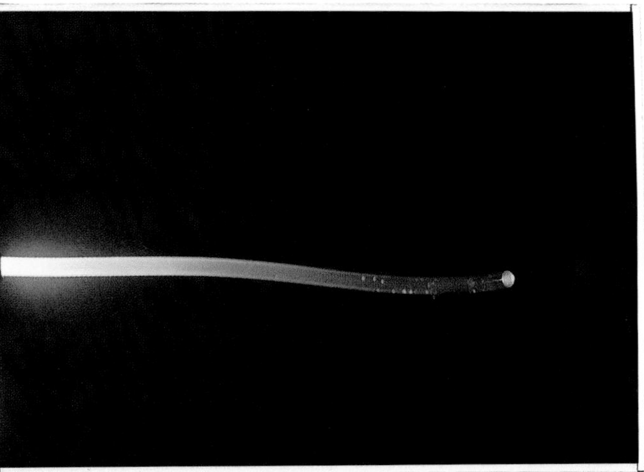

4 Position the tube inside the box so that it bends around and its end faces upward. Shine the desk lamp into the uncovered end. Do this in a dark room, or cover the box and peep inside without letting light into the box. The light travels through the water and is reflected by the tube's sides, in just the same way as light is conducted by optical fibers.

Fax machines

A FACSIMILE IS AN EXACT COPY of something. A fax machine (short for "facsimile machine") transmits a copy of a document along a telephone line to another fax machine. The document passes through the first machine, and the copy emerges from the second machine. Inside the first machine the document passes over a scanner that divides it into hundreds of rows of tiny squares and looks at each square in turn. As the document passes over the scanner, a light source shines on the document and the scanner measures the amount of light reflected from each square to see how dark the square is. The scanner sends out an electric signal if a square is dark or partially dark. The second machine receives this signal and uses it to control a printing head that marks dark squares on paper in time with the scanner. The copy is in black and white only, but future fax machines may be able to make color copies.

Circles and squares

The fax made in the experiment below is not very sharp. This is because the machine operates at a very low resolution and divides the document into only 64 squares. A real fax machine works at a high resolution and divides the page into about two million squares or more. This produces copies of written messages and drawings that are sharp enough to read easily.

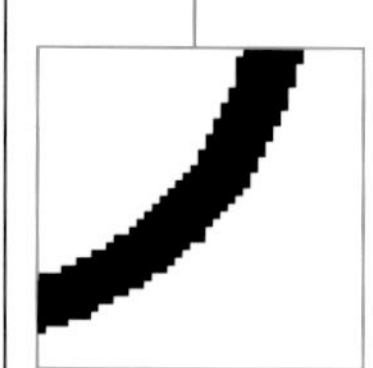

Image to be faxed

Fax copy of image

Magnified section of copy showing squares

EXPERIMENT

Send a fax

Two people can each make a simple fax machine and send a message to each other. The machine sends and receives only one letter or number at a time, but it works in the same basic way as a real fax machine.

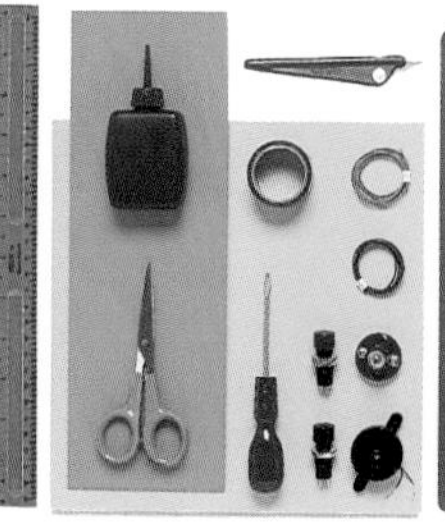

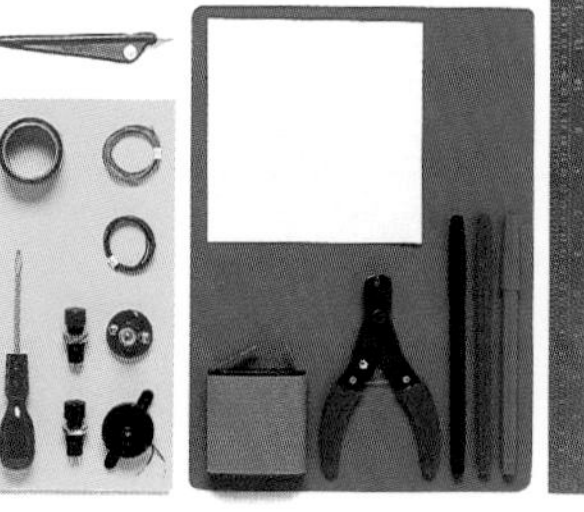

YOU WILL NEED

- ruler
- foamcore
- glue
- scissors
- craft knife
- tape
- insulated wire
- screwdriver
- 2 switches
- 3-9V DC buzzer
- bulb and bulb holder
- cutting surface
- paper
- 4.5V battery
- wire strippers
- colored pens
- steel ruler

Buzzer switch

Light switch

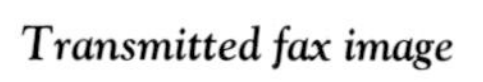

Transmitted fax image

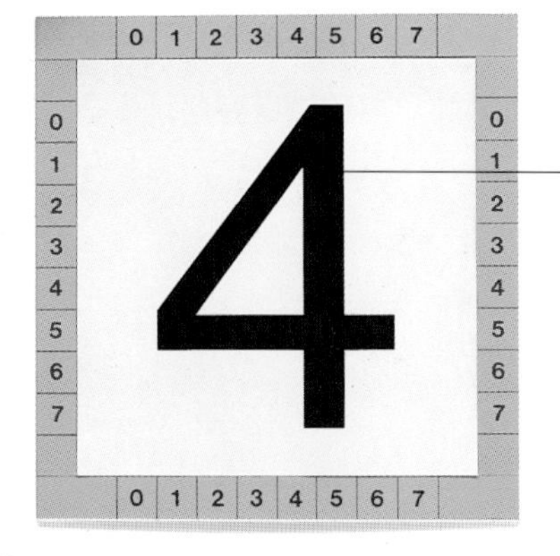

The image to be transmitted should have some curved or diagonal lines

Scanner

1 MAKE TWO fax machines as follows. Using a craft knife, cut two $9\frac{1}{2}$-in (24-cm) foamcore squares. Glue four $8\frac{3}{4} \times \frac{3}{4}$ in (22 x 2 cm) foamcore strips around the edges of each square.

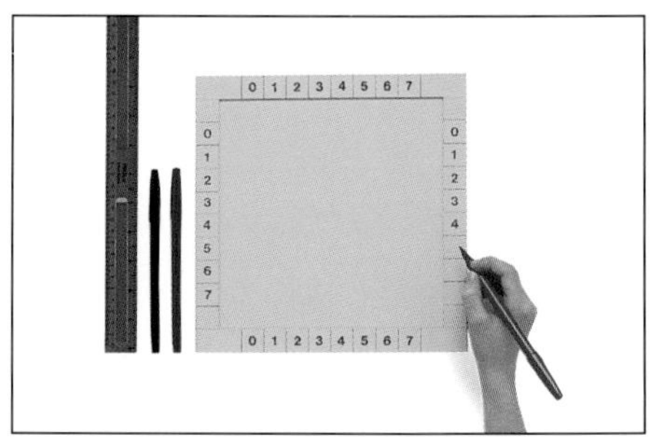

2 MARK EIGHT DIVISIONS along the four sides of each square as shown. Each division should be $\frac{3}{4}$ in (2 cm) wide. Number the divisions from zero to seven, moving from left to right on the horizontal strips and from top to bottom on the vertical strips.

Inside a fax machine

To send a document by fax, the pages are laid upside-down on the feeder tray of the transmitting fax machine, and the receiver's number is entered on the keypad. When the receiving machine answers, a roller in the first machine pulls in the pages of the document one by one and passes them over a scanner. Electric signals representing each page go to the second machine, which prints a copy on paper. The machine shown here contains a roll of thermal paper and a thermal printing head, in which a row of tiny electric heating elements heats areas of the paper as it passes from the roll under the head. Rows of dark dots form where the paper is heated, to produce a copy of the document. As the copy emerges from the rear, it is sliced into pages by a cutter.

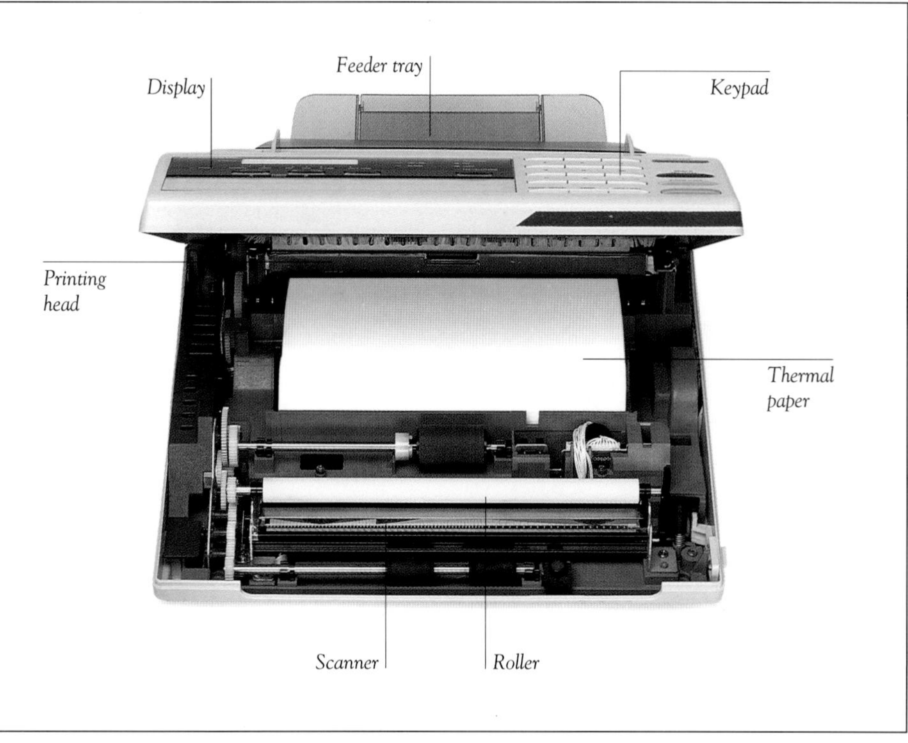

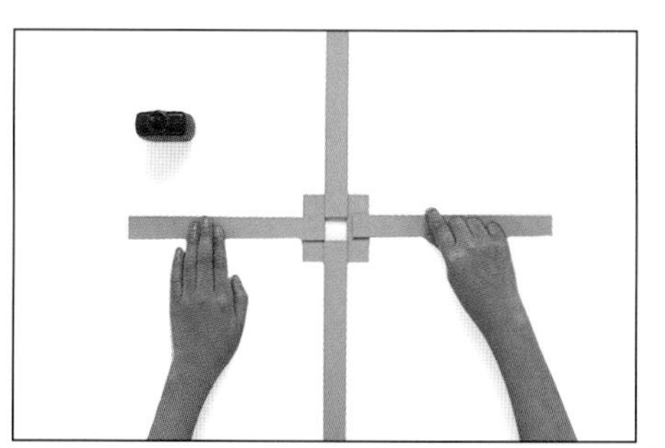

3 CUT TWO $2\frac{1}{4}$ in (6 cm) squares from foamcore. Cut a $\frac{3}{4}$ in (2 cm) central window in each. Glue four $7\frac{1}{4} \times \frac{3}{4}$ in (18 x 2 cm) foamcore arms to each square to make a scanner and a printer.

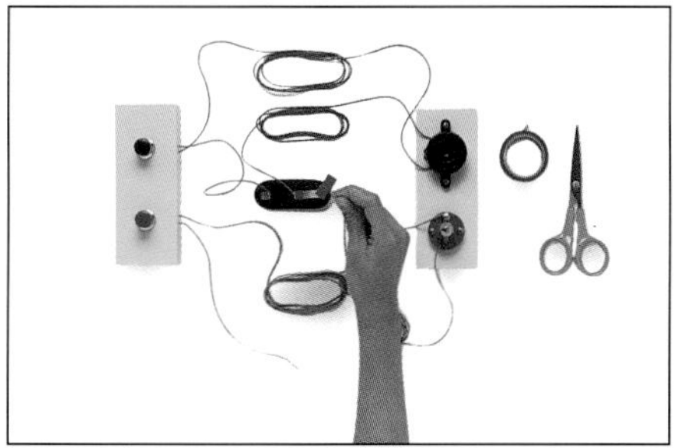

4 CUT FOUR LENGTHS of wire several yards (meters) long and strip the ends. Connect a bulb, buzzer, battery, and two switches to make a circuit as shown in the diagram at right.

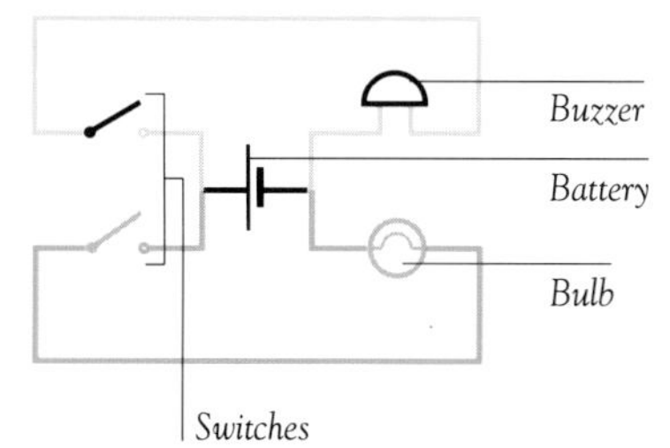

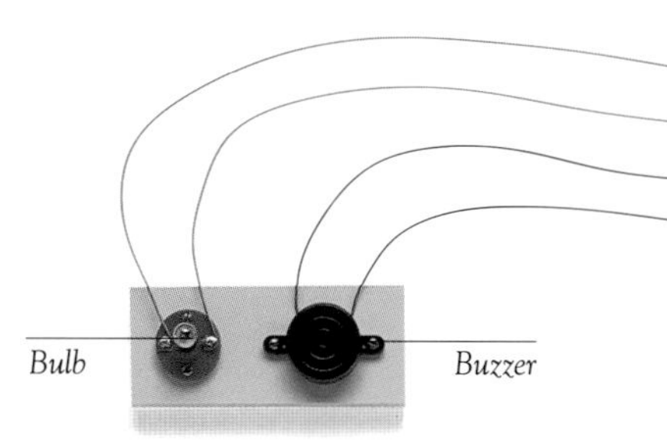

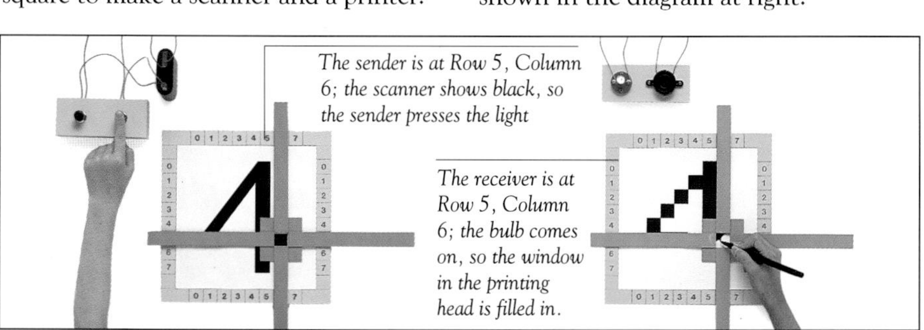

5 DRAW A THICK letter or number on paper and put this in the sending fax machine. Put blank paper in the receiving machine. Place the scanner and printing head at Row 0, Column 0 on both machines. The sender presses the buzzer switch to start. If the area in the scanner window is over 50% black, the sender presses the bulb switch. If the bulb comes on, the receiver fills in the square in the printing head window. The sender buzzes to signal a move to the next square. Continue square by square and row by row in the same way.

Received fax image

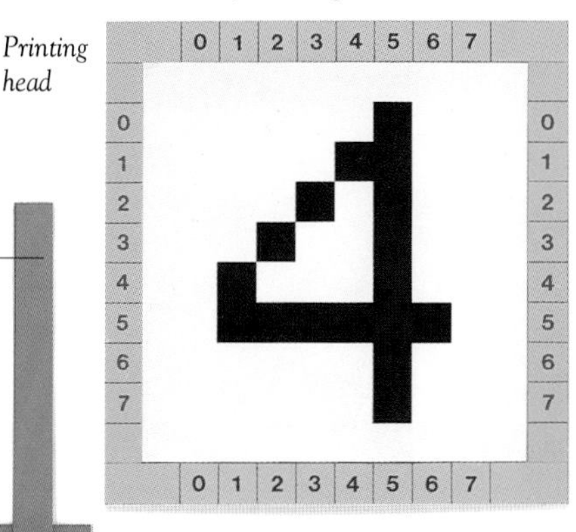

Computer systems

A COMPUTER IS NOT really a single machine, but a system of four main devices or units linked together. The input unit is used to operate the computer and feeds data (information) into it; a common input unit is the keyboard. The memory unit contains a hard disk and various chips, which together store data and programs (instructions that enable the computer to carry out particular tasks). More data and programs can be copied into the memory from portable floppy disks. The processing unit handles or processes the data in the way dictated by the program. Its results go to the output unit, which may display the results or use them to carry out useful actions, such as operating the moving parts in another machine. All the devices in a computer system are known as hardware. The programs that tell it how to carry out particular tasks are known as software.

Computer software

A computer system is useless without software. Most software comes on disks that you insert into the computer. The disks contain programs that tell the computer how to perform particular jobs. It may play a game, or do office tasks such as word processing or accounts, for example. Software may also contain useful information —a huge database in a library, for instance, or words, pictures, and sounds that you can learn from. Some software can be stored on a hard disk inside the computer. Software is also contained in microchips inside the computer.

Software disks
Floppy disks (top) contain a flexible plastic disk inside a plastic cover. A CD-ROM (bottom) is a type of compact disc that contains sounds and images as well as data.

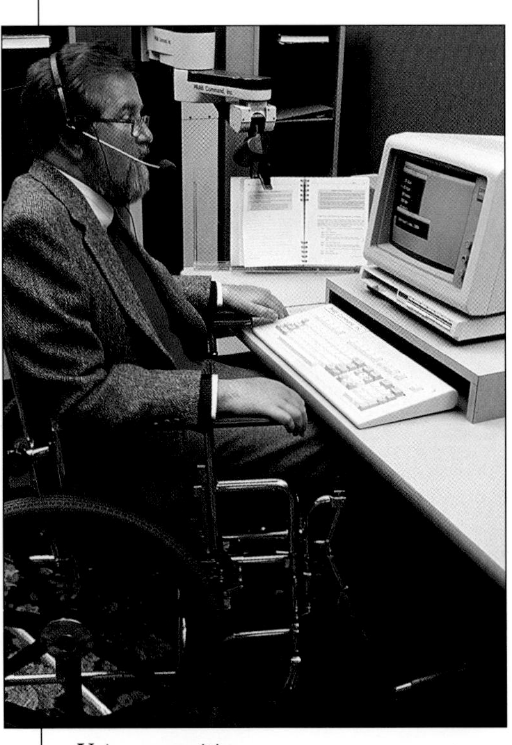

Voice recognition
A microphone feeds sounds into the computer, which then analyses them. The computer compares each sound with the sounds of words in its memory, and recognizes words that match.

Input devices

A keyboard and a mouse, which you operate using your hands, are standard input units supplied with many computer systems. It is also possible to give information or instructions to some computers by talking to them. Such a computer has a voice recognition device that can analyse speech and input the words into the computer. Some computers can also read: OCR (optical character recognition) devices scan printed matter or even handwriting and then input the letters or numbers. Some computers can even communicate with each other by using a modem—an input-output device that links computers together over telephone lines. It can input information and programs directly from other computers, and can also send such materials to other computers. A modem can connect a computer to world-wide computer networks.

Each key on a keyboard inputs a character (letter, number, or sign) to the computer

A mouse inputs movements and commands to the computer

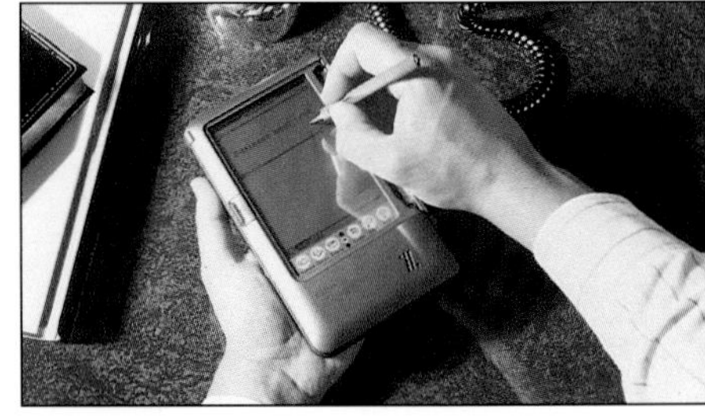

Character recognition
This portable notepad computer is operated by writing on the screen with a stylus. The computer changes handwriting into printed characters, and drawings into neat diagrams and geometric shapes.

Output devices

The most common output unit is a monitor, which may be a television screen or a liquid crystal display (LCD). Many computers also have a printer. These output units display and print words and pictures. Some computers can talk to us by turning their results into speech signals that go to attached loudspeakers. Output units may also perform physical movements, especially where computers control other machines.

Braille display
Blind people may use computers that have a Braille display connected to them. This is a special output unit containing a line of cells with studs that form Braille characters. The blind person feels the cells to read each line of words.

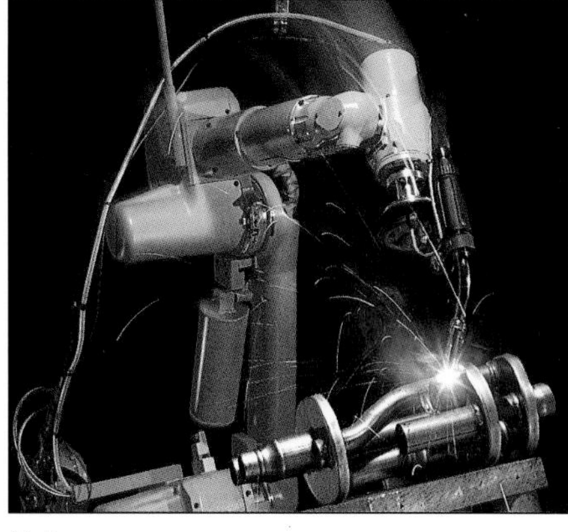

Robot arm
This robot is controlled by a computer, which moves the arm and enables it to perform sequences of precise actions.

The monitor can display moving pictures in full color, as well as words and diagrams

Floppy disks are placed in this slot; they can be used to store information and programs outside the computer, as a back-up to the internal hard disk

Memory and processors

A computer stores large amounts of data and programs on a hard disk in a sealed unit inside the computer. There, they remain ready for use. The computer's processing unit consists of one or more microchips, which form an integral part of its complex circuitry. Several processing units may be linked together to form a powerful supercomputer.

Supercomputer
The Cray supercomputer is a massive computer that can do complex calculations very quickly. One of its most important tasks is weather forecasting, which requires rapid processing of enormous amounts of data.

Complete systems

Many machines are controlled by their own computers, and are actually complete computer systems in their own right. For example, an automatic camera has a light detector as an input unit, a stored program and processor that calculate exposures, and output mechanisms that control the shutter and aperture. Other systems include electronic music systems that can store the notes you play, process them, and then play them back in various ways; and automatic navigation systems in aircraft. Machines that are highly independent of humans are often called smart machines.

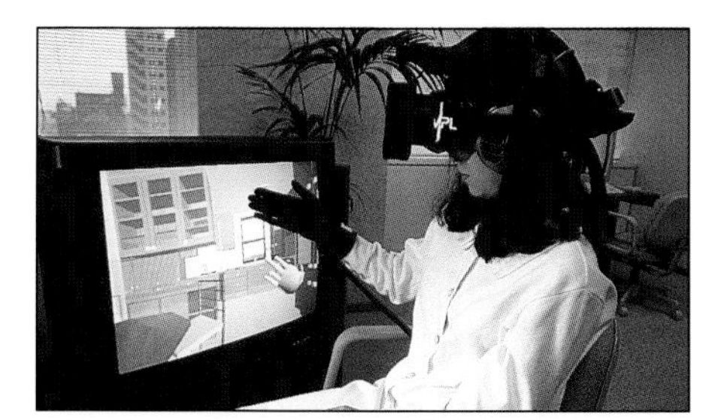

Virtual reality
Using body sensors as input units and a headset as an output unit, virtual reality lets users "enter" and appear to move about in a computer-generated world—here, one of kitchen designs.

Music system
This keyboard can store music and play it back with the sound of almost any instrument.

Computer keyboard

PEOPLE WHO USE A COMPUTER for such jobs as word processing mostly operate it with a keyboard. Tapping the keys feeds letters, numbers, punctuation marks, and other symbols into the computer. These may simply appear on the screen, but the keys can also make the computer carry out actions, especially in computer games. The keyboard produces an electric code signal for each key. The codes go to the computer's processor (pp.172–173), and may then be stored in the memory or sent to the screen or a printer to appear as letters or numbers. Each signal is encoded as a binary number. Computers use binary numbers because they are easier to turn into electric signals than decimal numbers.

EXPERIMENT

Counting binary box

Binary numbers use only two digits, 0 and 1, but decimal numbers use ten digits, from 0 to 9. In the decimal system, the digits in a number represent (from right to left) ones, tens, hundreds, thousands, and so on. So the decimal number 1,001 has 1 one, 0 tens, 0 hundreds, and 1 thousand. In the binary system, the digits in a number represent (from right to left) one, two, four, eight, and so on, doubling each time. So the binary number 1001 has 1 one, 0 two, 0 four, and 1 eight. It is equivalent to the decimal number 9. Each digit in a binary number is called a bit, so 1001 is a four-bit number. The binary box below can be used to change any decimal number from 0 to 15 into its binary equivalent. In a computer, electric current goes on and off to represent binary code numbers. On means 1 and off means 0, so the binary number 1001 (decimal equivalent 9) is an on-off-off-on electric signal.

YOU WILL NEED
• pen • 15 ping-pong balls • cardboard box and lid slightly wider than four balls and slightly longer than eight balls • craft knife • glue

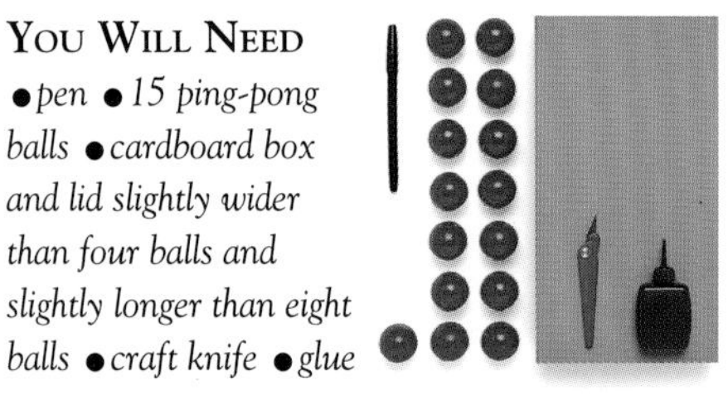

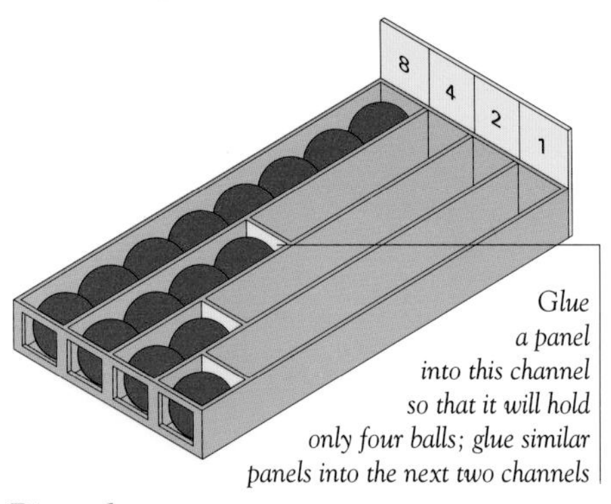

Binary box
Trim the box to the height of the balls, leaving a taller flap at one end. Cut three strips from the lid and glue them in the box to form four channels. The channels (from left to right) should hold eight, four, two, and one ball respectively. Cut and glue three panels to divide three of the channels so that they hold the right number of balls. Write the number of balls that each channel can hold on the flap above the channels. Cut a window at the other end of each channel.

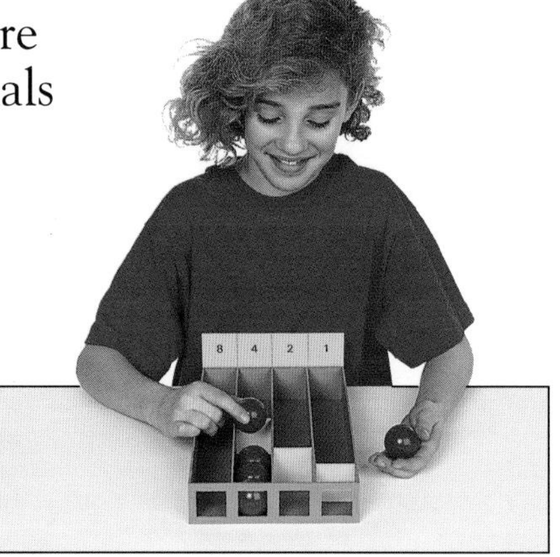

1 TAKE A NUMBER of balls (the girl in the picture has chosen five balls). See if you can fill the eight-ball channel. If this channel fills, try to fill the next channel with the remaining balls.

2 IF YOU CANNOT fill a channel, leave it empty and try the next channel. In the picture above, the last remaining ball will not fill the two-ball channel, so it goes into the one-ball channel.

3 THE WINDOWS of the box show the binary equivalent of the decimal number of balls. A ball in a window represents a 1 and an empty window represents a 0 in binary code. The binary equivalent of five, shown here, is 0101.

EXPERIMENT

Basic keyboard

Adult help is advised for this experiment

Build a keyboard that produces an electric code signal for each key, just like a real computer keyboard. It contains eight keys, and lights up bulbs with the binary code for each key.

You Will Need

- *steel ruler*
- *cutting surface*
- *three 2.5V bulbs in bulb holders and screws*
- *compass*
- *paper fasteners*
- *two cardboard rectangles 8 x 3 in (20 x 8 cm)*
- *screwdrivers*
- *insulated wire*
- *foamcore base for bulb holders*
- *aluminum foil*
- *pen*
- *double-sided adhesive tape*
- *adhesive tape*
- *craft knife*
- *3V battery*
- *wire strippers*
- *scissors*
- *ruler*

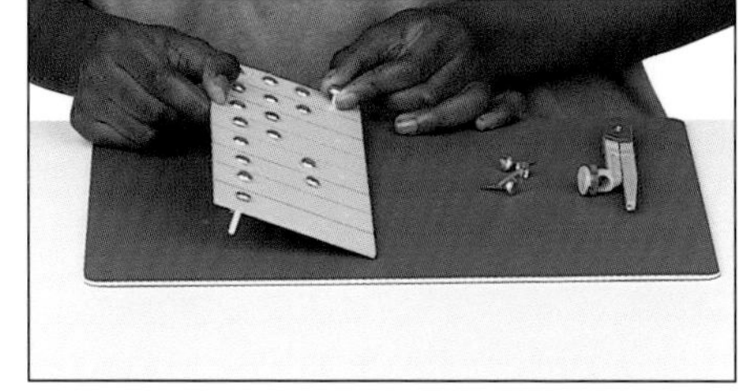

1 Mark one cardboard rectangle to outline eight keys. Pierce holes with the compass in the other rectangle where contacts are shown on the template below. Insert paper fasteners in the holes.

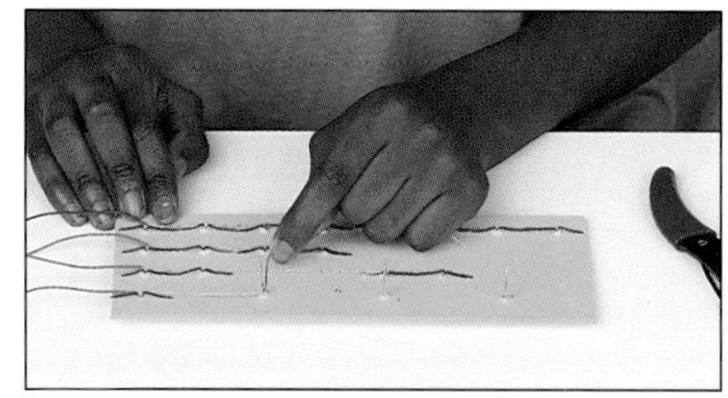

2 Connect the four lines of paper fasteners with short lengths of stripped wire on the underside of the card. Leave about 12 in (30 cm) of each wire trailing from the side.

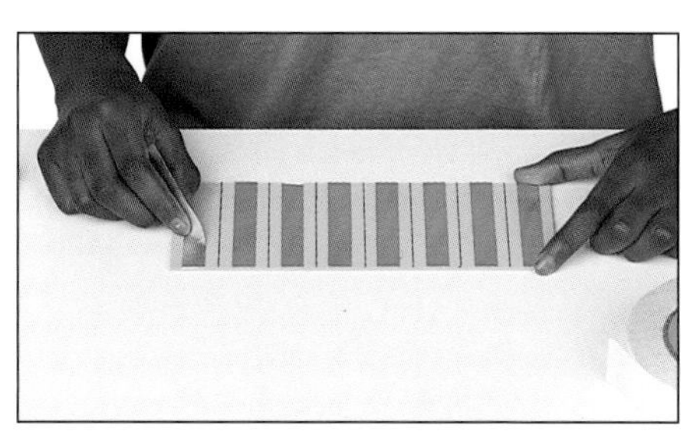

3 Use double-sided adhesive tape to stick a strip of foil, shiny side up, over each of the eight keys on the first piece of cardboard.

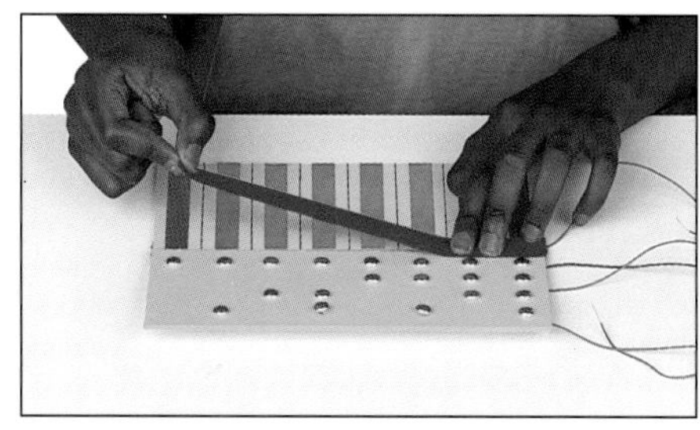

4 Join the two rectangles with a tape hinge so that the foil and paper fastener heads face upwards. Make sure that the two cards can be folded together.

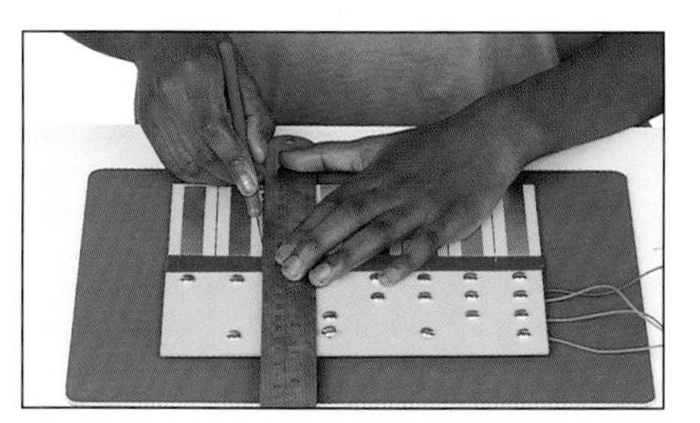

5 Use the craft knife and steel ruler to cut seven gaps between the foil strips to form eight keys. The cuts should start at the tape hinge.

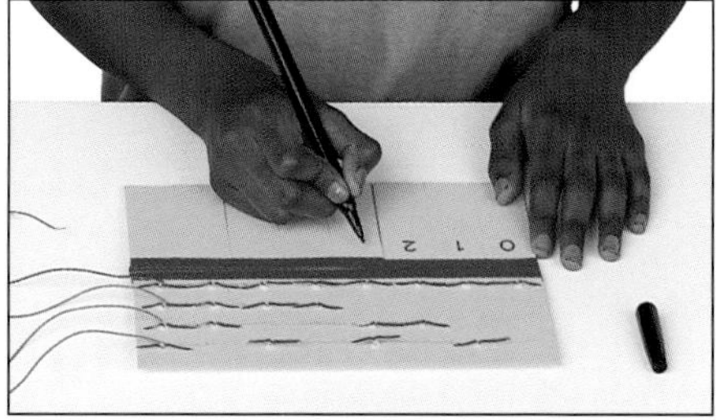

6 Mark the keys 0 to 7 from left to right. Fold the two halves of the keyboard together. The keys must touch only the first row of paper fasteners.

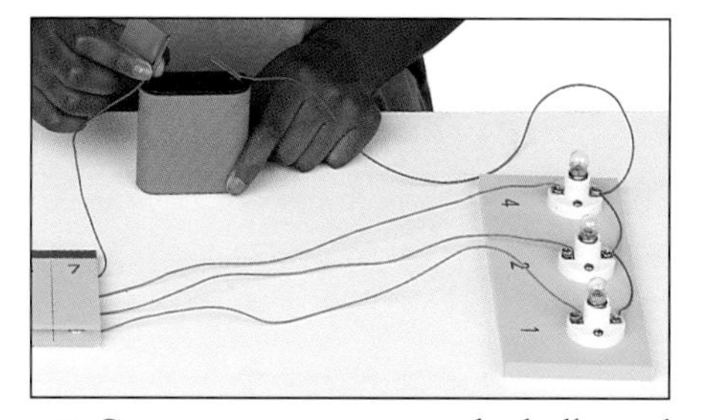

7 Connect the wires to the bulbs and battery as shown in the diagram below left. When no key is pressed, no bulb should be lit.

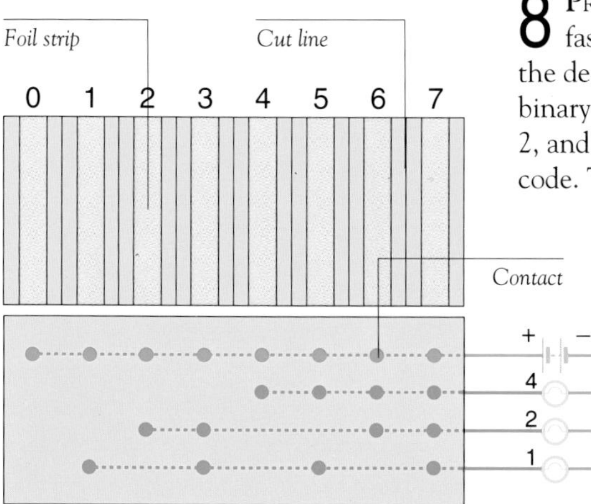

8 Press each key separately against the fasteners. Watch the circuit convert the decimal number on each key into its binary equivalent. The bulbs (marked 4, 2, and 1) go on or off to give the binary code. The foil under each key touches certain fasteners, completing a circuit between the right bulbs. In a real keyboard, each key connects wires to produce the on-off electric signal that forms a binary code for the key.

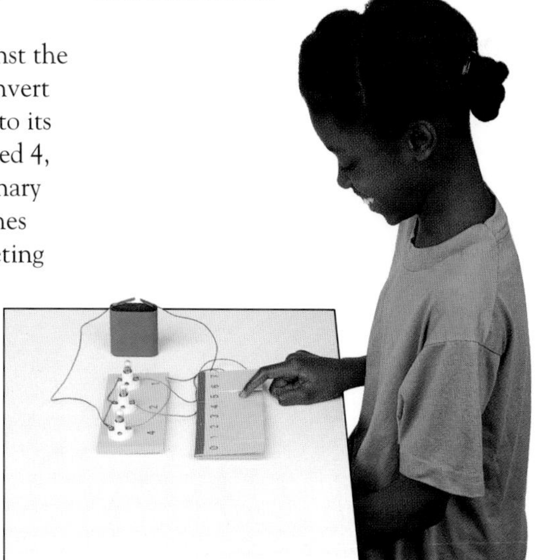

Bar-code reader and mouse

AT THE CHECKOUT in a store, as the clerk passes each of your purchases over a special light, its name and price immediately flash up on a display. The light is part of a machine that reads bar codes, which are the patterns of black-and-white bars on product labels. The bar-code reader is an input unit for the store's computer, which holds a list of all the products in the store and their current prices. Each product has a bar code with a unique pattern that represents a code number for that product. The reader converts the pattern into the code number. This goes to the computer, which finds the product in the list and sends its name and price to the checkout.

Handling bar codes
Store clerks may use hand-held readers, or the bar-code reader may be under a window in the checkout counter.

Computer mouse

Operating a computer is often easier with a mouse than with a keyboard. You move the mouse over the desk, and a cursor, often shaped like an arrow, duplicates its movement on the screen. You can position the cursor at an icon, a small picture representing an action. When you click the mouse by pressing its button, the computer carries out the action. Like a bar-code reader, the mouse may use on-off light flashes to send signals to the computer.

Flashing lights
As you move a mouse, a ball inside rolls in the same direction. The ball is connected by rollers to two slotted wheels that move with it. As the wheels turn, LEDs shine through the slots to produce beams of flashing light. The beams are converted into electric signals by photodiodes on the other side of each wheel. These signals move the cursor on the screen.

Pressure roller
Roller ball
Side-to-side roller
Side-to-side slotted wheel
Button
Switch
Connector cable
Printed circuit board
Microchip
Photodiode
Up-and-down slotted wheel
Light-emitting diode (LED)
Up-and-down roller

EXPERIMENT

Cracking the code

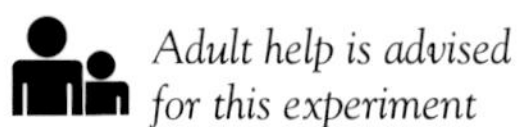

Adult help is advised for this experiment

Make a bar code and a bar-code reader. This bar code has four black or white data bars to represent a four-bit binary number (p.164), and two black start/stop bars. A black bar means a 1, and a white bar means a 0. The reader moves a light beam over the bar code. A light detector turns light reflected from the bars into an on-off electric signal. This goes to a buzzer, which sounds the binary number in the bar code. *Read pages 10–11 before starting this experiment.*

YOU WILL NEED

● *breadboard wire* ● *ruler* ● *pen* ● *pliers* ● *wire strippers* ● *colored cardboard* ● *black, and white, sheets of paper* ● *tape* ● *double-sided tape* ● *scissors* ● *pencil and pad* ● *9V buzzer* ● *4011B quad NAND chip* ● *NPN transistor, BC441 or equivalent* ● *5K variable resistor* ● *2 alligator clips* ● *breadboard and base* ● *9V battery and connector* ● *flashlight* ● *LDR*

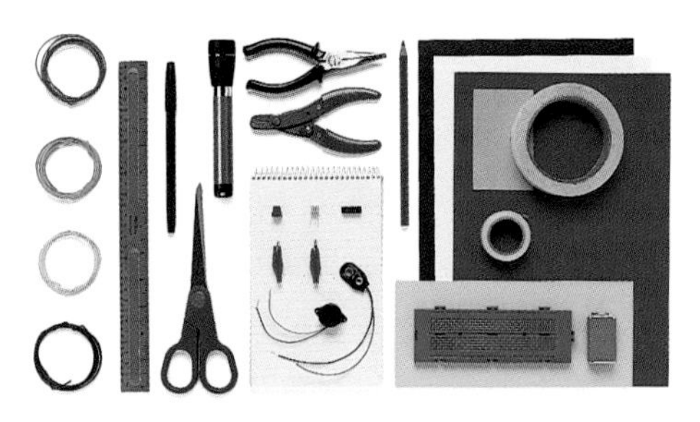

Bar-code template

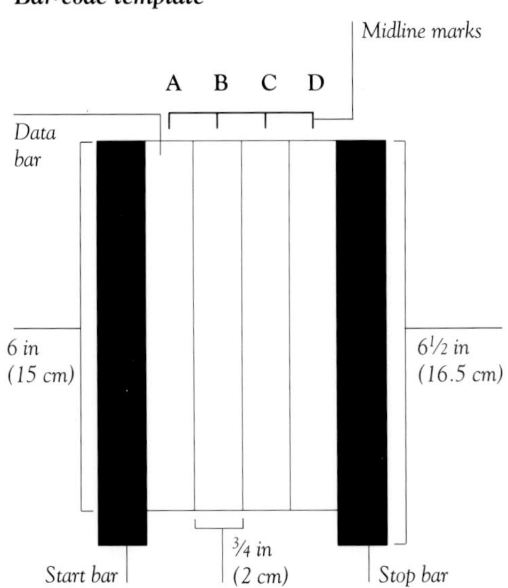

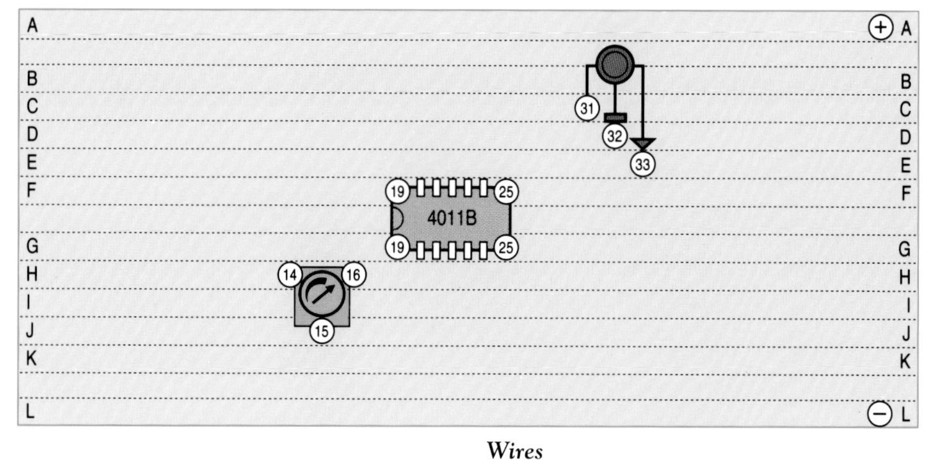

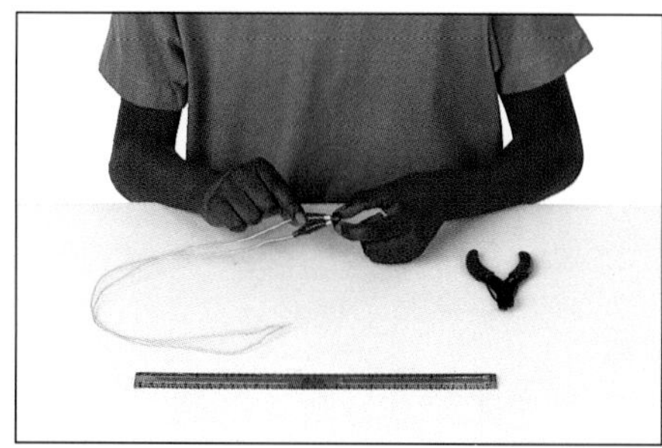

1 MEASURE AND CUT two lengths of breadboard wire 18 in (45 cm) long. Strip the insulation from the ends using wire strippers. Attach one end of each wire to an alligator clip and clamp one clip to each leg of the light-dependent resistor (LDR).

Wires

A19–B19 A31–B31 D20–D21 E14–E20
E22–E32 F14–G14 K15–L15 K25–L25

A14 B14

F33 L39

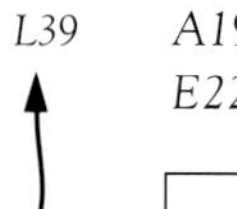

Flashlight

Buzzer

Light-dependent resistor (LDR) with hood

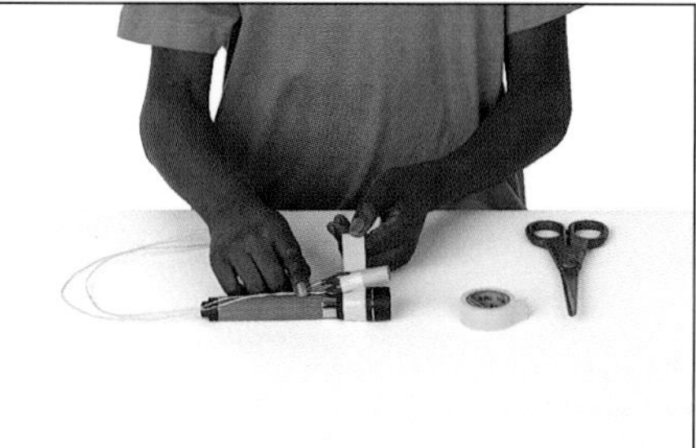

2 MAKE A TUBE of cardboard and tape it around the LDR to make a hood. Attach this to the flashlight so that the LDR and flashlight bulb are parallel to each other.

3 COPY THE TEMPLATE (opposite) onto white paper. Mark the four central data bars with 1s or 0s and stick black paper strips over those bars designated 1. Label the bars **A–D** as shown opposite.

4 CONNECT the wires from the reader to the breadboard. Switch the flashlight on. Point the reader vertically down, 1 in (2.5 cm) from the white paper. Adjust the 5K variable resistor until the buzzer just turns off. Check to be sure the buzzer sounds as the reader moves over a dark bar and is silent when over a white bar.

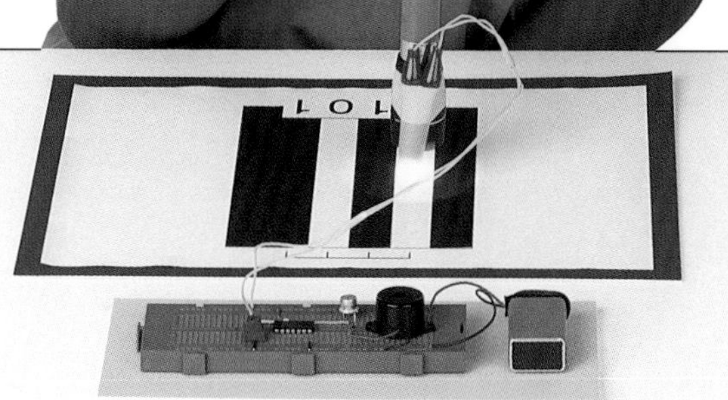

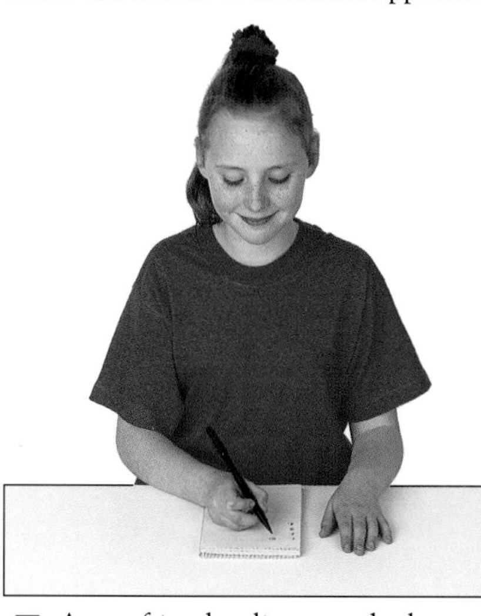

5 ASK A friend to listen to the buzzer. Move the reader left to right across the code. The first and last buzzes signify when the code starts and stops. As you cross each data bar, call its letter. As each letter is called, the friend writes a 1 if the buzzer is on, and a 0 if it is off.

Microchip

COMPUTERS AND MANY other electronic machines contain microchips, often called chips. Most chips look like centipedes, with rectangular black bodies and rows of metal legs along the sides. The legs, or pins, fit into a circuit board where they connect the chip to wires. The wires carry electric code signals that the chip may store or process. Parts of the memory and processing unit of a computer consist of microchips. The heart of the chip is a small piece of a semiconductor material (p.187), usually silicon, which contains thousands or even millions of tiny electronic components connected together to form a circuit. This is called an integrated circuit (IC) because the components are not separate but manufactured as one integrated unit.

Microchip

This magnified view of a chip's integrated circuit (IC) shows a maze of miniature electronic components and pathways. The huge number of components in chips allows them to store large amounts of information and make complex calculations. A chip in a computer is fed with a constant stream of electric pulses by the computer's clock. The pulses are altered inside the chip to become code signals that flash along the IC's pathways and in and out of the components. The chip stores and processes the signals to make calculations almost instantly. The clock's speed is measured in megahertz (MHz)—a 33-MHz computer works at 33 million pulses a second. A fast clock enables the computer to operate at high speed.

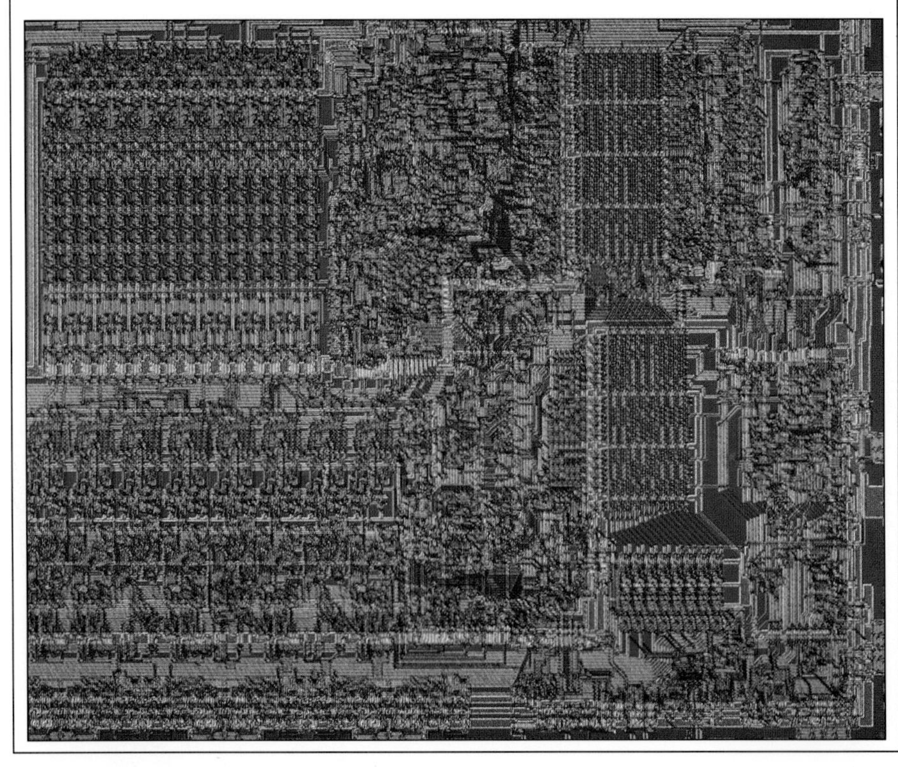

Logic gates

Microchips contain logic gates made of combinations of components such as transistors. They process a stream of electric pulses sent out by the computer's clock, opening to allow some pulses to pass or closing to block others. The gates are operated by pulses from other parts of the computer, often sent by other logic gates. They convert the clock pulses into output sequences of on-off pulses that form binary code signals. Combinations of logic gates perform tasks, such as adding up binary numbers. The truth tables below show the operation of three common logic gates, each with two input channels.

Input channel (light shading) — ***Output channel*** (dark shading) — 0 ***No pulse*** — 1 ***Pulse***

Input	Input	Output
0	0	0
0	1	0
1	0	0
1	1	1

AND gate
This gate opens to produce an output pulse if it receives two input pulses. The gate closes if it receives no input pulse or one input pulse.

Input	Input	Output
0	0	0
0	1	1
1	0	1
1	1	0

EOR gate
This gate opens to produce an output pulse if it receives one input pulse. The gate closes if it receives no input pulse or two input pulses. EOR stands for "exclusive or."

Input	Input	Output
0	0	1
0	1	1
1	0	1
1	1	0

NAND gate
This gate opens to produce an output pulse if it receives no input pulse or one input pulse. Two input pulses close the gate. It is the reverse of the AND gate. NAND stands for "not and."

How an AND gate works

This model represents the action of an AND gate inside a microchip. Colored balls are used to represent pulses of electricity—yellow balls for input pulses and a green ball for a pulse from the computer's clock. The clock pulse passes the gate to become an output pulse only when two input pulses arrive at the gate. The 0's and 1's in a truth table represent electric pulses turning off and on—0 means that no input pulse arrives or no clock pulse passes; 1 means that an input pulse arrives or a clock pulse passes through the gate.

Input channel

Output channel

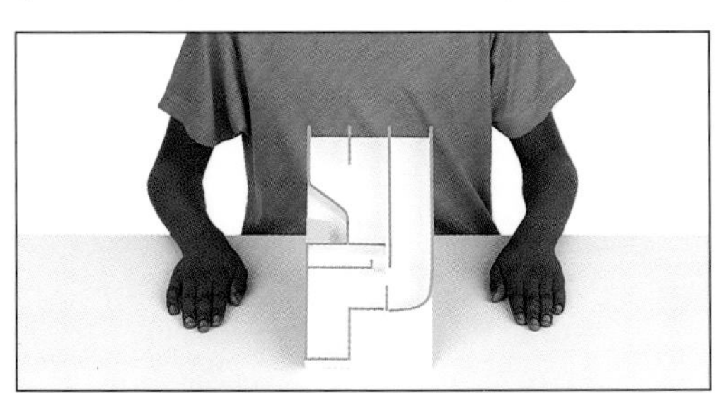

No input pulses
In the first line of the truth table, no electric pulse enters either of the two input channels in the logic gate.

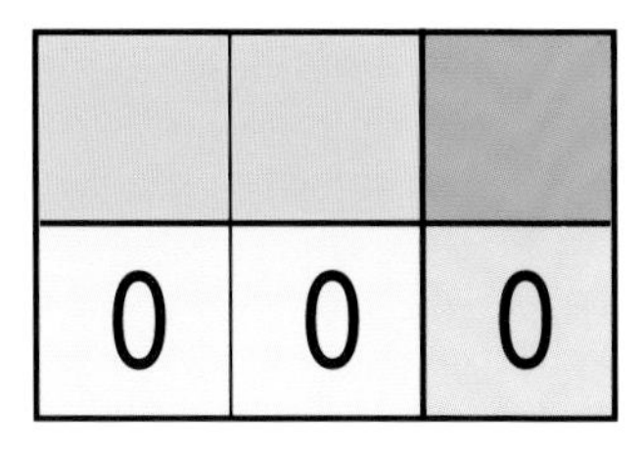

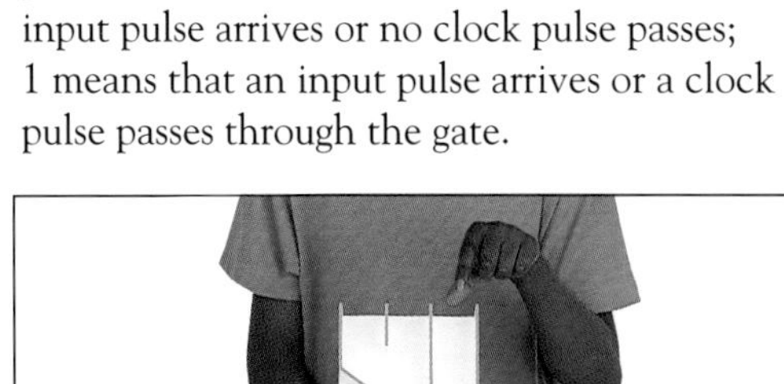

No output pulse
The gate remains closed, so the clock pulse cannot pass through the gate to become an output pulse.

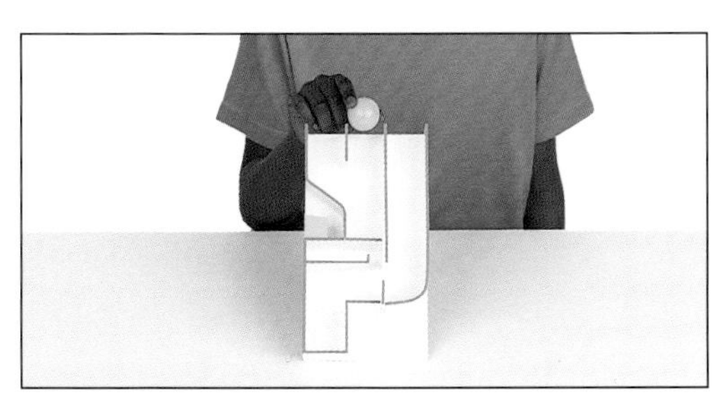

Right-channel input pulse
In the second line of the truth table, one electric pulse enters the logic gate through the right input channel.

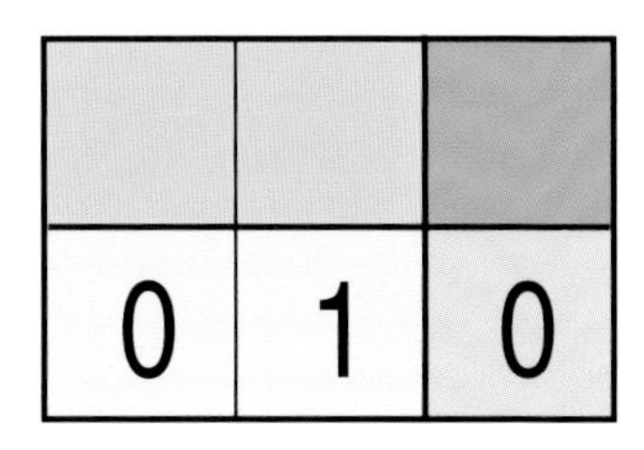

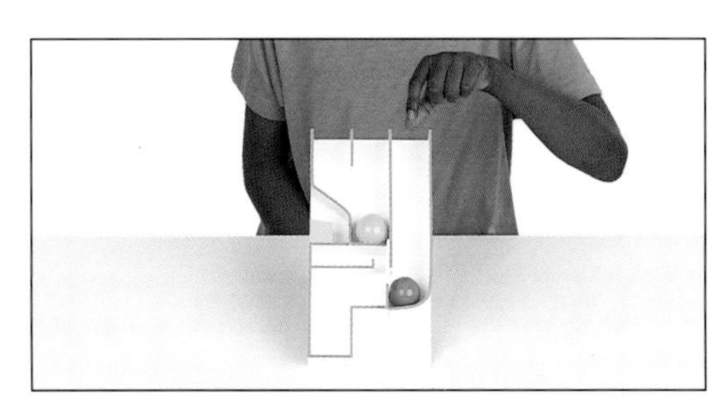

No output pulse
The gate remains closed, and once again the clock pulse cannot pass through the gate.

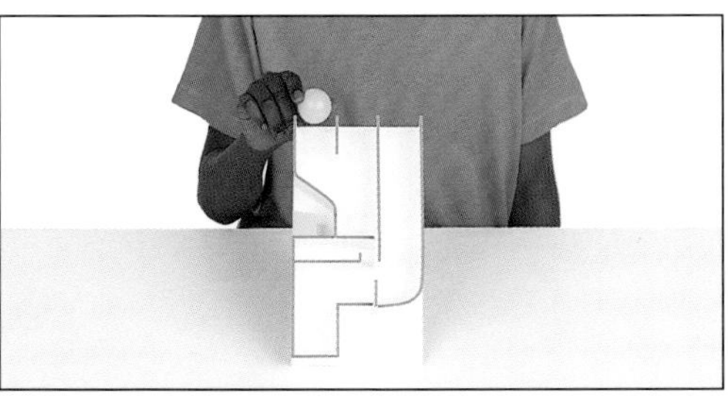

Left-channel input pulse
In the third line of the truth table, one electric pulse enters the logic gate through the left input channel.

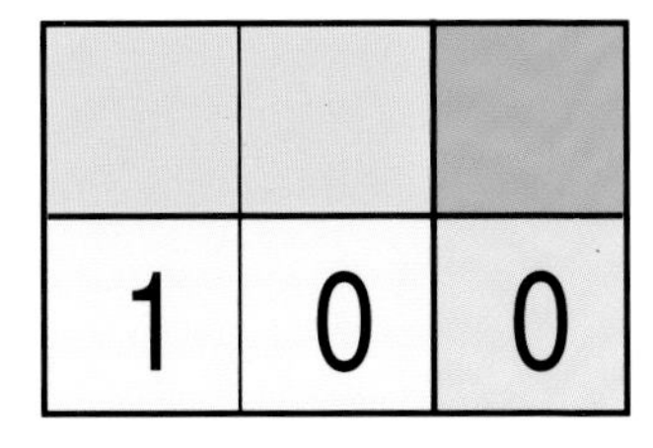

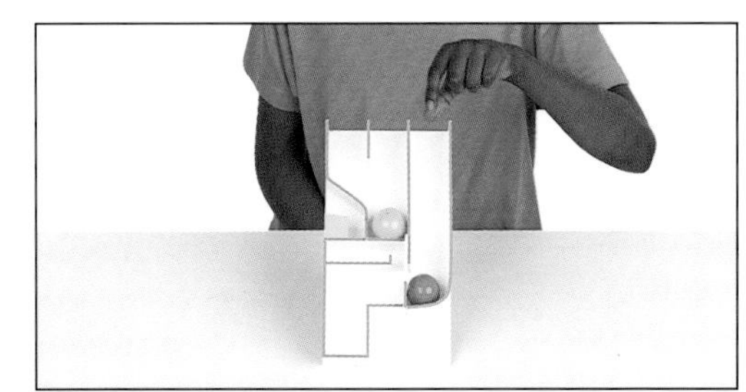

No output pulse
The gate remains closed, and once again the clock pulse cannot pass through the gate.

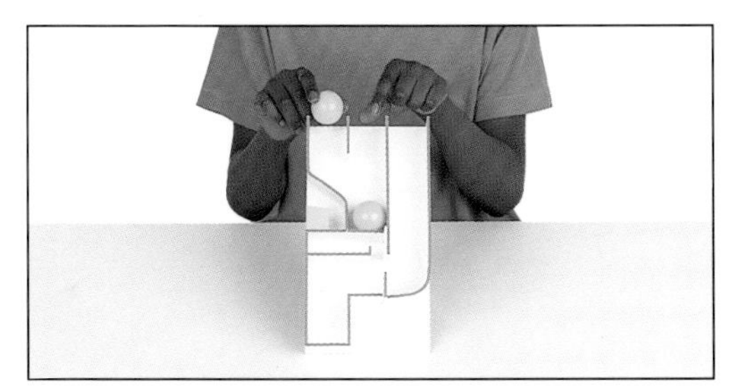

Two input pulses
In the fourth line of the truth table, an electric pulse enters each of the input channels.

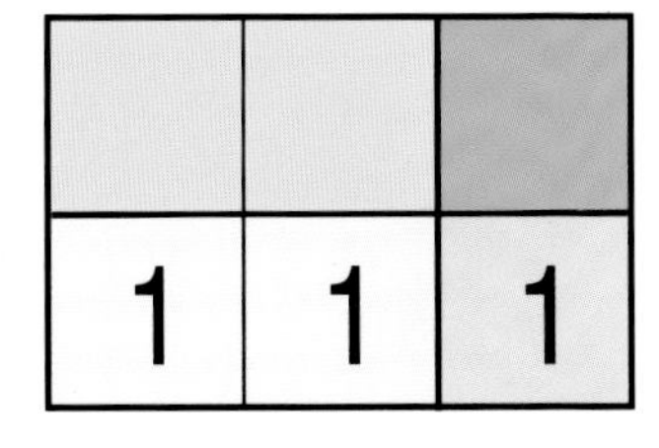

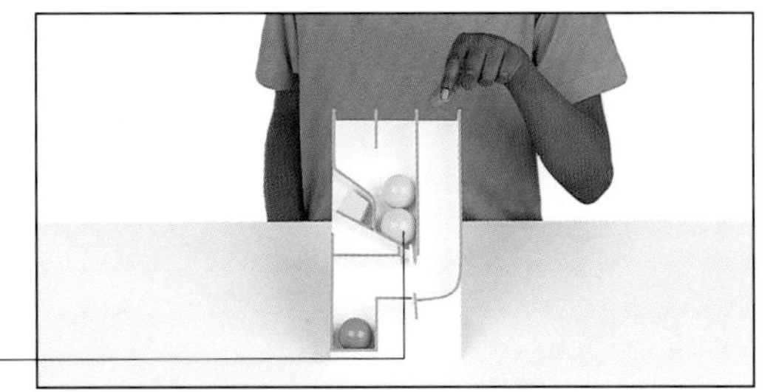

Two input balls are heavy enough to swing the seesaw and open the gate

One output pulse
Two input pulses open the gate, allowing the clock pulse to pass through the gate to become an output pulse.

Computer memory

A COMPUTER'S MEMORY CAN STORE vast amounts of data (information)—an entire encyclopedia, for example. But the memory, which includes several different units, does much more than just hold data. ROM (read-only-memory) chips permanently hold basic instructions that tell the computer how to work. RAM (random-access-memory) chips hold only the programs and data being used by the computer. The hard disk, the computer's main memory, stores programs and data as long as necessary; they can be deleted or more can be added. Portable compact discs (CD-ROM's) permanently hold huge amounts of data. Floppy disks are also portable but can be erased and can record new data; programs and data held on them can be copied onto the hard disk.

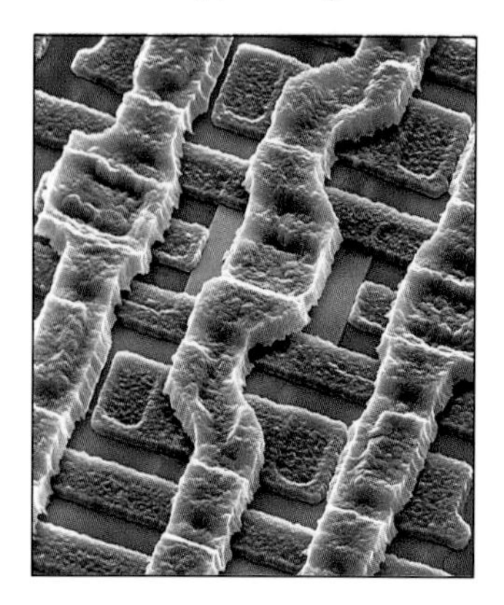

RAM chip closeup
This RAM chip shows a grid of memory cells and data lines.

Inside a disk drive

A computer has two disk drives, one for floppy disks that you insert, and a hard-disk drive containing sealed disks. To store data, a read-write head moves to a part of the disk's magnetic surface. While a motor spins the disk, the head receives electric signals in binary code, and records them on the disk's surface in the form of magnetic signals. To retrieve data, the head moves to a part of the disk and turns the magnetic signals there back into electric signals. The head can also erase magnetic signals on the disk.

Hard disk drive

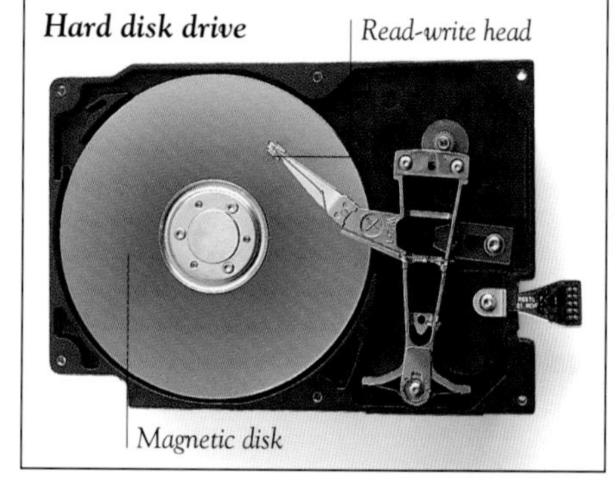

RAM chip

A computer uses information (data) in the form of on-off electric signals that represent binary numbers; each number is divided into groups of eight bits (0's or 1's). Eight bits make up a byte, and a RAM chip may store millions of bytes. This model shows how a chip has rows of eight memory cells, each row storing one byte (represented here by colored balls). Each row is identified by an address number. The signals flow along one set of data lines to be stored in the selected row, and leave along another set of lines.

1. The computer writes (sends) an eight-bit binary number 01000101 (decimal 69) to the memory in the form of on-off electric pulses. The memory already contains the numbers 10101101 in address 2 and 11110010 in address 4.

2. The computer now selects an available (empty) address at which the data is to be stored. Above, the binary number 01000101 is being written to (stored at) the row of eight empty memory cells at address 3.

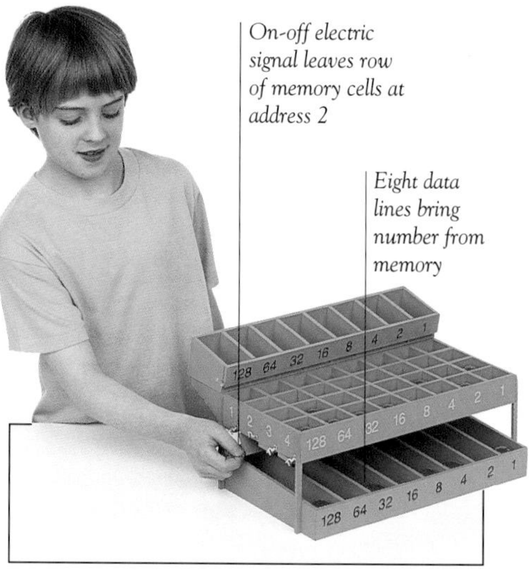

3. The computer is now reading (retrieving) data from its memory. The byte (eight-bit number) leaves the selected memory cells along another set of data lines. Here, 10101101 (decimal 173) is being read from address 2.

EXPERIMENT

Spinning disk drive

A computer splits data into small units (for example, eight-bit binary numbers, or bytes, p.170) and stores each unit separately on a magnetic disk. Build a model of a floppy disk drive and see how a computer stores data on a disk and then retrieves it. You can rotate the disk, and write data onto it or read data from it through a window that represents the read-write head of a real disk drive.

You Will Need

- *9 in (24 cm) diameter cardboard disk* • *scissors* • *craft knife* • *cutting mat* • *3 skewers* • *straw* • *compass* • *1½-in (4-cm) cardboard square* • *notepad* • *pencil* • *pen* • *paper cup* • *ruler* • *steel ruler*

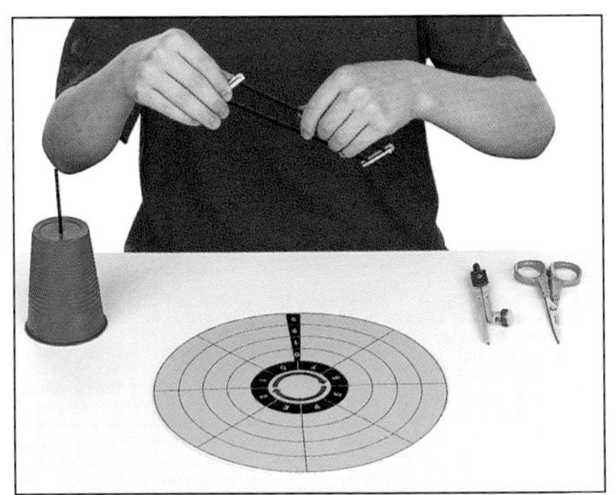

1 Copy the template below onto the card disk. Cut two straws 1½ in (4 cm) long. Make a tracker arm by pushing two skewers through the sides of both straws 1 in (2.5 cm) apart. Push a third skewer through the base of the paper cup.

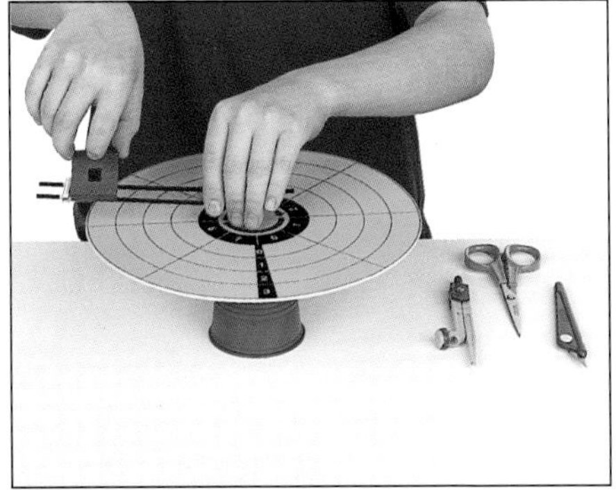

2 Upend the cup and push the center of the disk onto the skewer. Cut a ¾-in (2-cm) square window in the cardboard square. Fold the square around the skewers with the window centered. Stick one straw on the skewer in the cup.

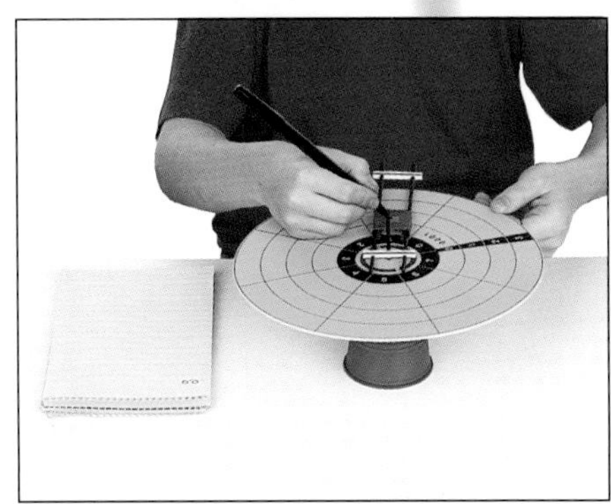

3 Choose a decimal number. Convert each of its numerals into a four-bit binary number (p.182). Write each four-bit number in a separate block on the disk. On a notepad, record the identity numbers of the track and sector of each block in sequence to make a directory.

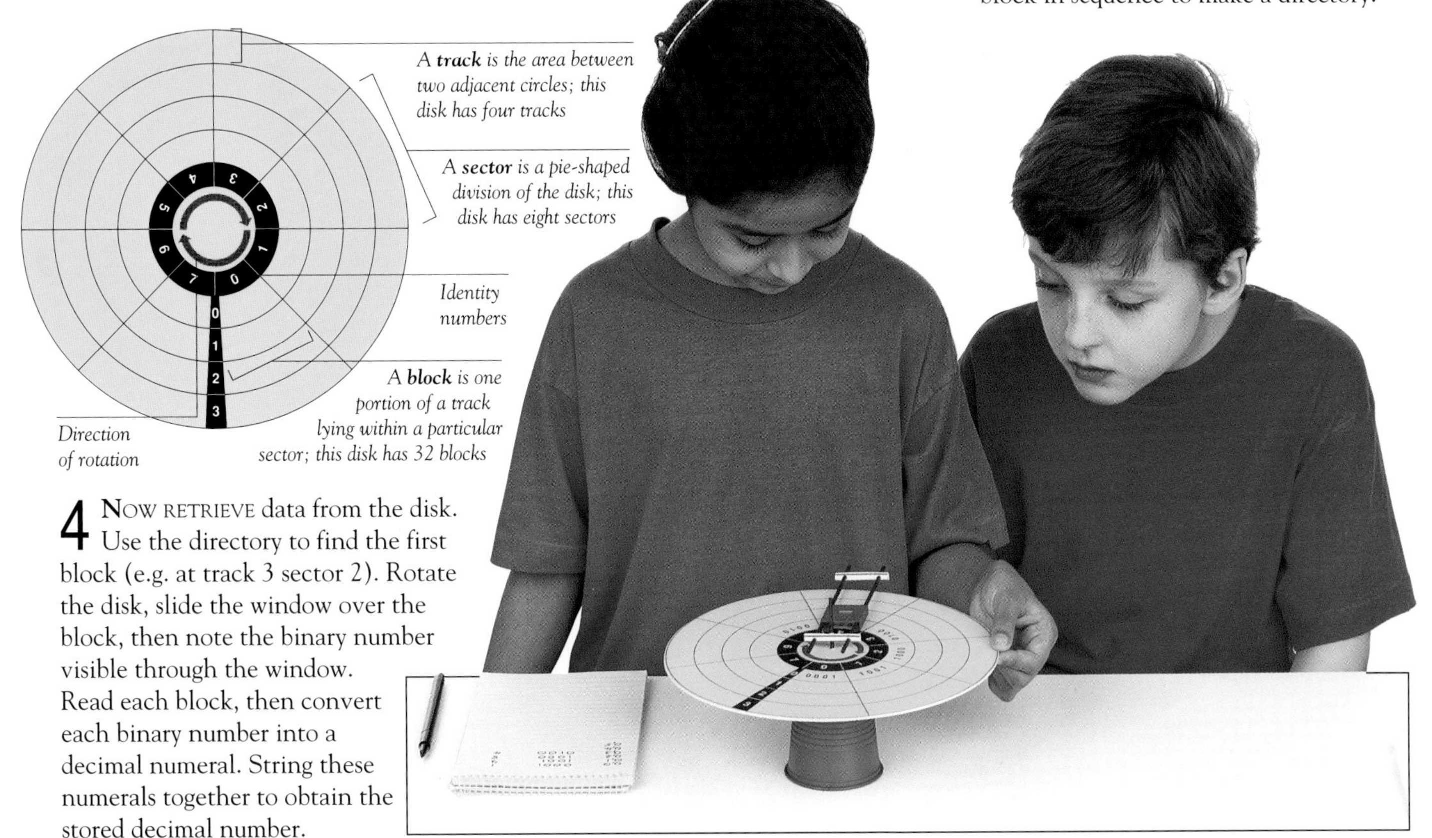

4 Now retrieve data from the disk. Use the directory to find the first block (e.g. at track 3 sector 2). Rotate the disk, slide the window over the block, then note the binary number visible through the window. Read each block, then convert each binary number into a decimal numeral. String these numerals together to obtain the stored decimal number.

Processing unit

WHATEVER THE TASK, a computer breaks it down into a series of binary-number calculations (p.164) carried out by a processing unit (or microprocessor). The numbers, in the form of electric signals, are fed to the processing unit by an input unit. The processing unit contains many tiny electronic components. Some of these are connected to form logic gates (p.168), which perform calculations at lightning speed by changing the electric signals as they pass the gates. The results go to an output unit.

EXPERIMENT

Binary abacus

Make an abacus that adds binary numbers in the same way as a processing unit. Add two four-bit numbers, each a combination of 0's and 1's, to a maximum total of 1111 (15 in the decimal system) by moving each slider so that a 0 (dark green), 1 (light green), or carry 1 (red) shows in each window. The abacus uses the rules of binary addition. In these, two input bits make one total bit and a carry bit.

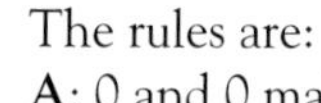

The rules are:
A: 0 and 0 make 0 and carry 0
B: 0 and 1 make 1 and carry 0
C: 1 and 0 make 1 and carry 0
D: 1 and 1 make 0 and carry 1

below, the binary number 0100 (decimal 4), shown in green, is input. Then 0101 (decimal 5), in yellow, is input and added to the first number. Start at the right side and add each pair of numerals separately.

YOU WILL NEED
- *craft knife*
- *pen*
- *cutting mat*
- *ruler*
- *steel ruler*
- *scissors*
- *glue*
- *colored cardboard*
- *colored paper*

Sliders and squares
Make four sliders that fit between the frame and base. Make four dark yellow 0 squares and four light yellow 1 squares.

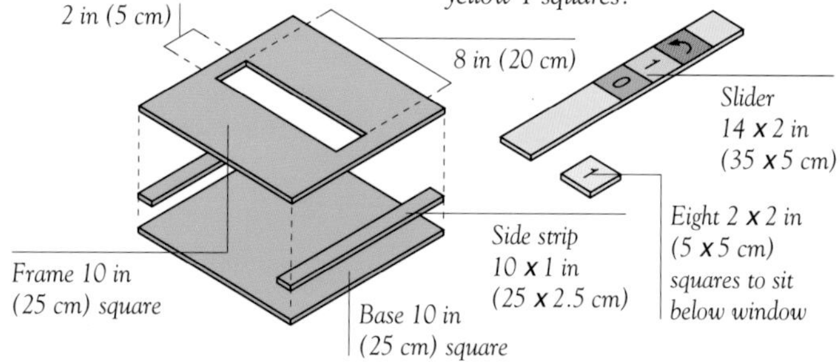

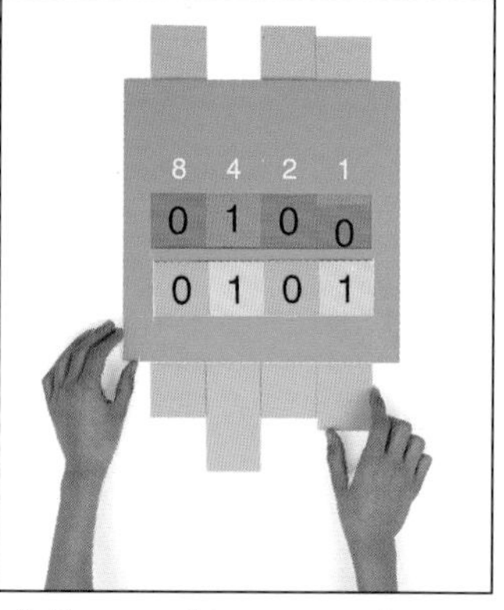

1 SET THE binary numbers to be added in the green and yellow rows. When adding a yellow 1, pull the slider down one square. When adding a 0, leave it in the same position.

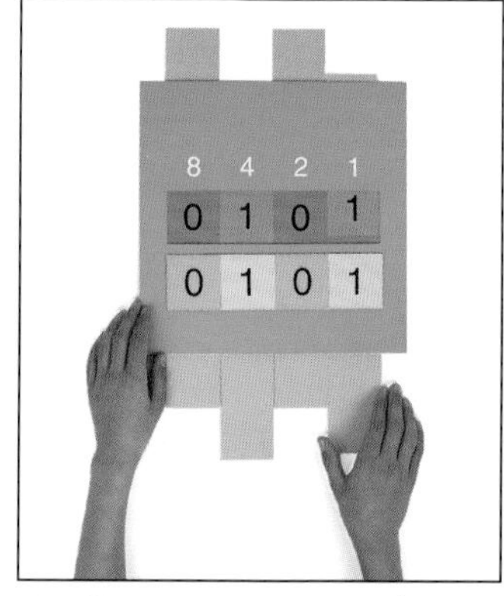

2 START ADDING on the right in the first window, which shows 0. Add a 1 by pulling down the slider. The 0 in the window changes to 1. This is Rule **B**.

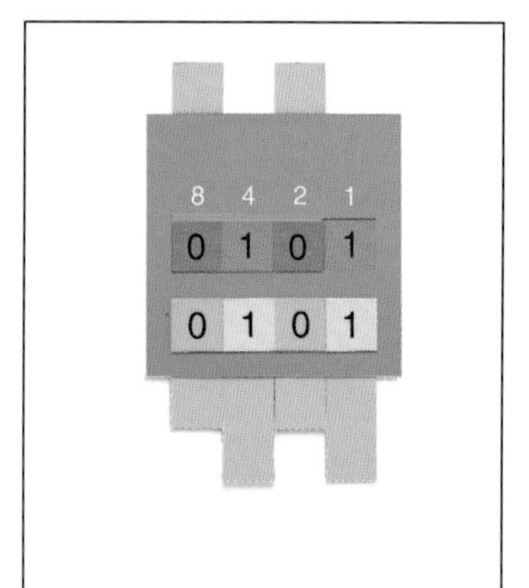

3 MOVE TO Window 2, which reads 0. Since you are adding a 0, do not move the slider. The 0 in the window does not change. This is Rule **A**.

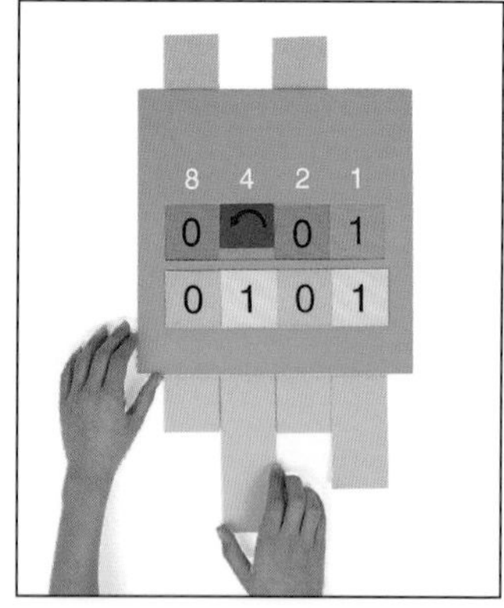

4 MOVE TO Window 4, which shows 1. Add a 1 by pulling down the slider. The red arrow shows, meaning you must carry 1 to the next window. This is Rule **D**.

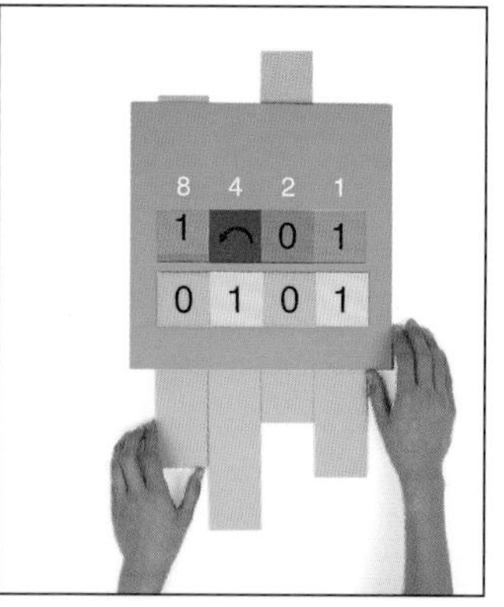

5 TO CARRY 1, pull down the slider in the next window to the left, then push up the first slider to show 0. The 0 in Window 8 changes to 1, and Window 4 reads 0.

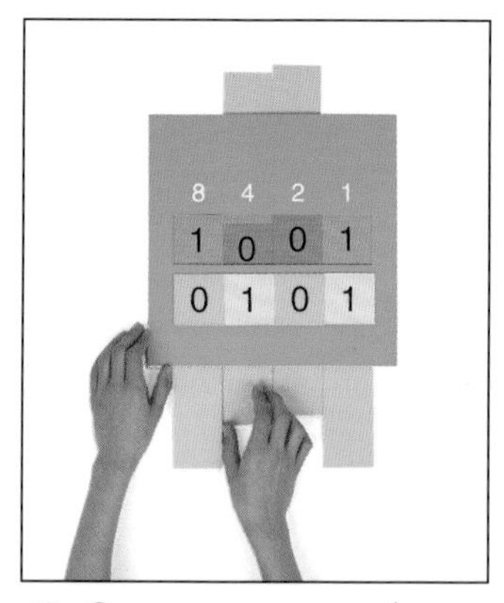

6 COMPLETE THE sum by adding a 0 to the 1 showing in Window 8. Do not move the slider; the 1 remains. This is Rule **C**. The final result is 1001 (decimal 9).

EXPERIMENT

Half adder

Inside a processing unit, half adders follow the four rules of binary addition (left) to add the bits (0's or 1's) in binary numbers. Build a half adder in which EOR and AND gates produce the total and carry bits respectively (see tables on p.168). Pressing an input switch (yellow) inputs a 1; leaving it unpressed inputs a 0. If an input-LED (yellow), the total-LED (green), or the carry-LED (red) is lit, it signifies a 1. An unlit LED signifies a 0. *Read pages 10–11 before starting this experiment.*

You Will Need

• breadboard and base • 9V battery and connector • breadboard wire • pliers • wire strippers • 2 normally-open momentary SPST switches • 2 transistors, BC108 or equivalent • CMOS 4070B quad EOR chip • CMOS 4081B quad AND chip • 1 red, 1 green, and 2 yellow LED's • two 10K resistors • four 220R resistors

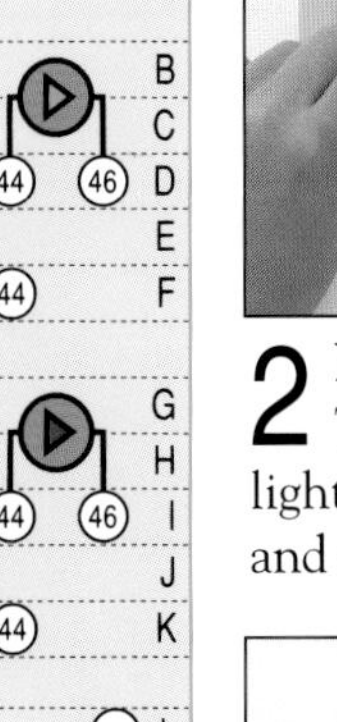

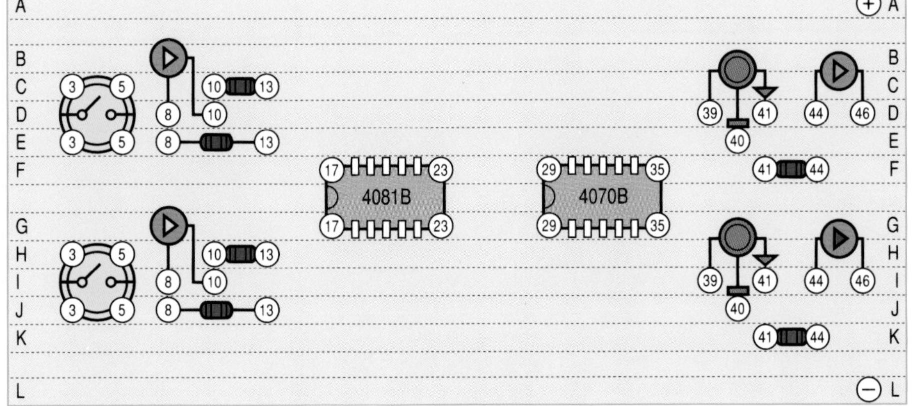

220R resistors

C10–C13 H10–H13 F41–F44 K41–K44

10K resistors

E8–E13 J8–J13

Wires

A3–B3 A17–B17 A29–B29 A39–B39 B5–B19 C19–C30 C32–C40 D18–D31 E18–G5 E20–H40 F3–G3 F5–F8 F13–G13 F39–G39 F46–G46 K5–K8 K13–L13 K23–L23 K35–L35 K46–L46

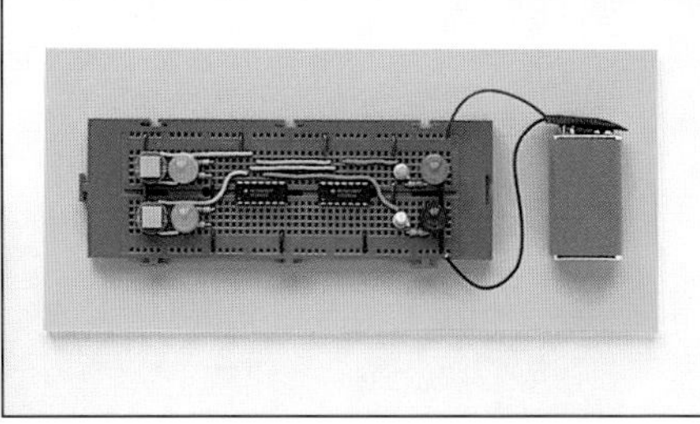

1 Do not press either of the yellow input switches on the left. All the LED's remain dark, indicating that 0 and 0 make 0 and carry 0. This is Rule **A**.

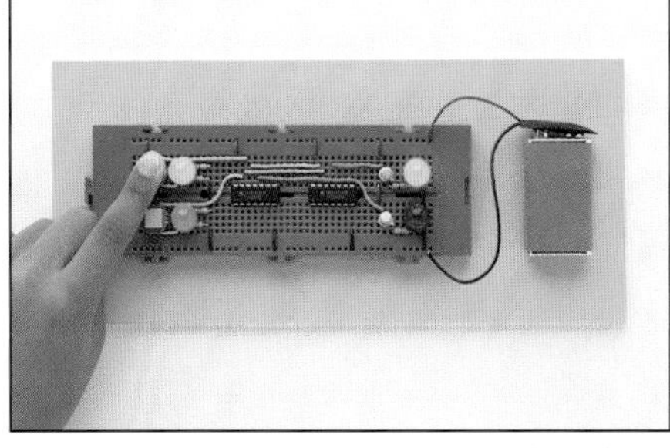

2 Press the upper input switch only. The upper input-LED and total-LED light up, indicating that 1 and 0 make 1 and carry 0. This is Rule **C**.

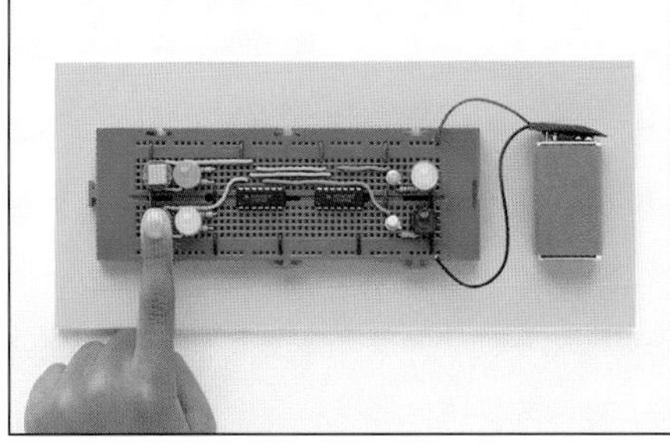

3 Press the lower input switch only. The lower input-LED and total-LED light up, indicating that 0 and 1 make 1 and carry 0. This is Rule **B**.

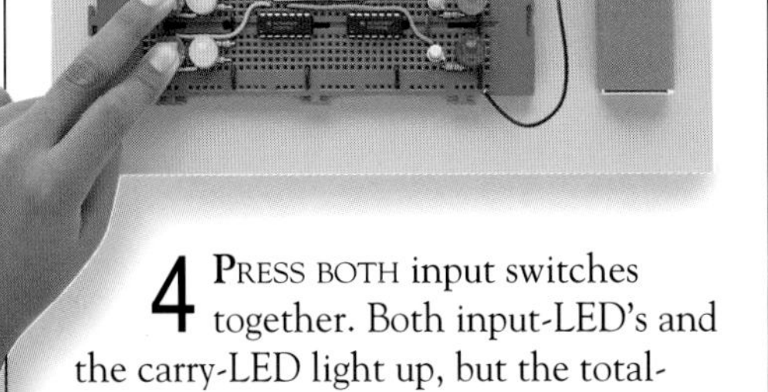

4 Press both input switches together. Both input-LED's and the carry-LED light up, but the total-LED remains unlit, indicating that 1 and 1 make 0 and carry 1. This is Rule **D**.

Full adder

Each half adder in a processing unit can add only two bits (binary 0's or 1's) at a time, one from each of the two binary numbers being added together. In order to add two complete binary numbers, a chain of half adders and OR gates are connected together to make a full adder. A carry bit from one half adder goes to the next stage in the chain. This full adder can add two four-bit numbers.

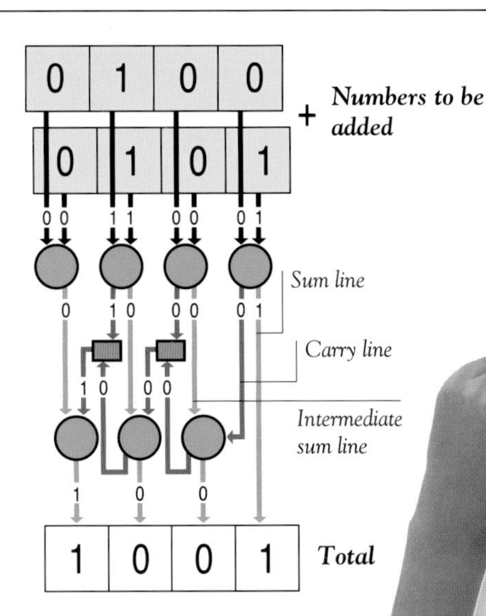

Half adder

OR gate (outputs 0 only if both inputs are 0)

Computer program

IN ORDER TO PERFORM any task, a computer needs a program. This is a list of instructions that tells the computer how to perform the task. The program consists of a set of code signals stored in the computer's memory, on a disk, or in a microchip. The signals direct the computer's processing unit to make the calculations involved in the task, and to control the workings of other units such as screens or printers. Program instructions are originally written into a computer by a programmer using a special computer language. The program can then be installed in other computers, which translate the instructions into electric code signals. A programmer must consider carefully the sequence of steps a program requires. Shown below are two search programs, which both find a particular item in a long list of items. One is slow, but the other searches the list quickly.

EXPERIMENT

Search game

Follow two search programs to see how they work. One is simple, but usually slow, and the other is a little more complex but usually fast. At least 15 people write their ages in months on a piece of paper, then stand in age order. Two searchers pick one age from the list, then use the two programs, one after the other, to find the person with that age. The instructions of both programs are given in the flow charts opposite. Each searcher decides if two ages are the same by subtracting one from the other. If the result is zero, then they are the same. Computer programs make decisions by comparing numbers in the same way.

YOU WILL NEED

- *cardboard* • *scissors* • *string* • *pen* • *notepad*

Slow search program
This program checks each member of a group in sequence to see if he or she is the specified age. It is very slow for large groups. A group of one million members would need, on average, 500,000 checks to find one member.

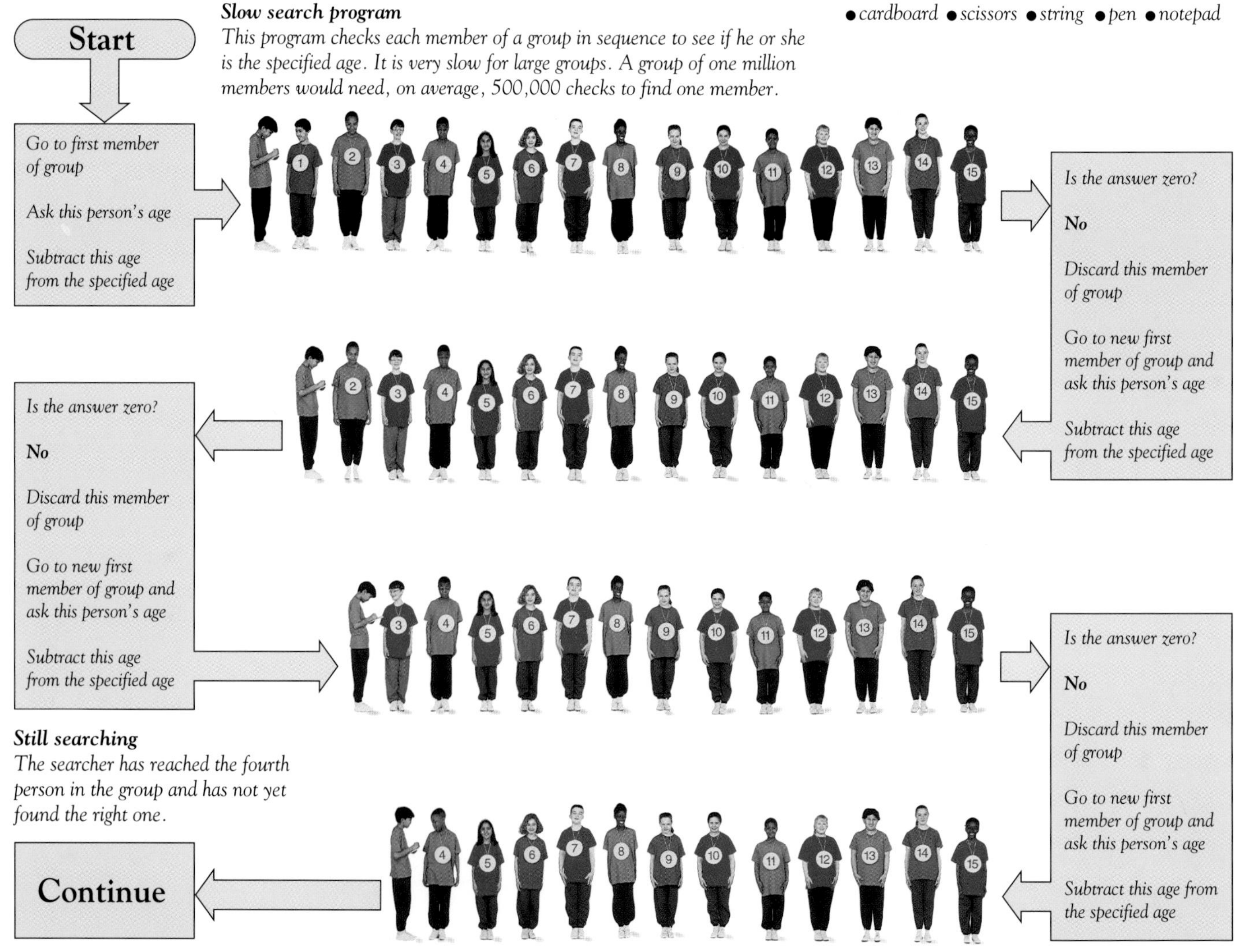

Still searching
The searcher has reached the fourth person in the group and has not yet found the right one.

Flow charts

A flow chart is a basic diagram of a computer program. These two flow charts give the basic instructions used in the two search programs below and the order in which you carry them out. Both charts contain a loop, in which a set of instructions is repeated. At the start of each loop, the program requires the answer "yes" or "no" to a question. If the answer is "no", the program follows the loop. If it is "yes", the program leaves the loop and you have found the answer.

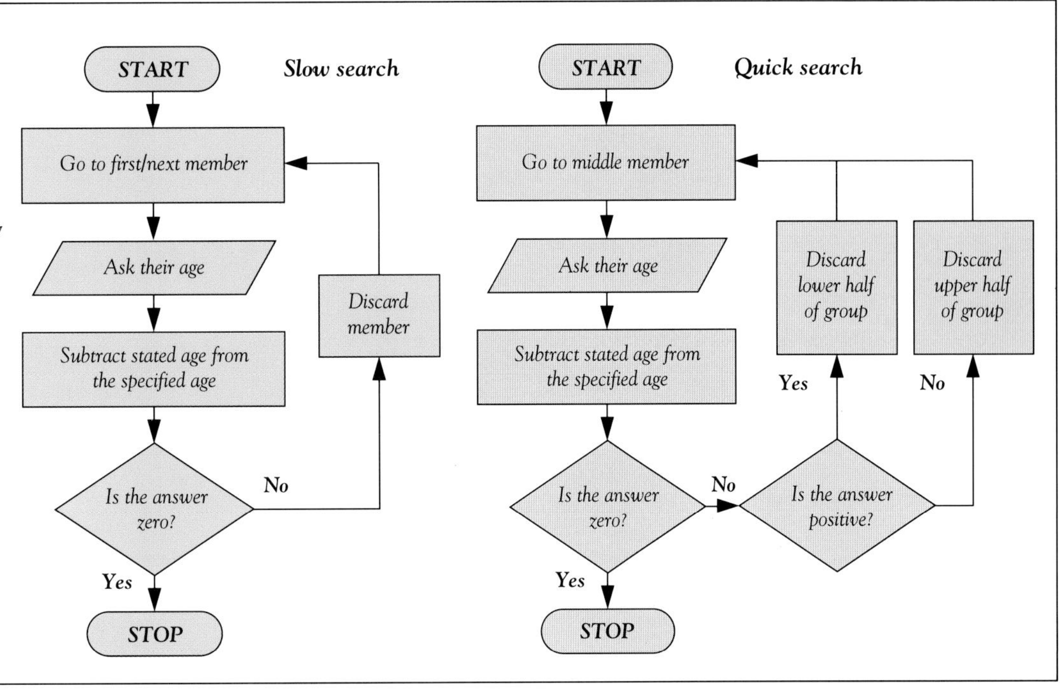

Quick search program

With this search method, the group to be searched is halved in number at each step. For groups of a substantial size, it requires fewer steps than the slow search. For example, a group of one million members would need, at most, 20 checks to find a particular member.

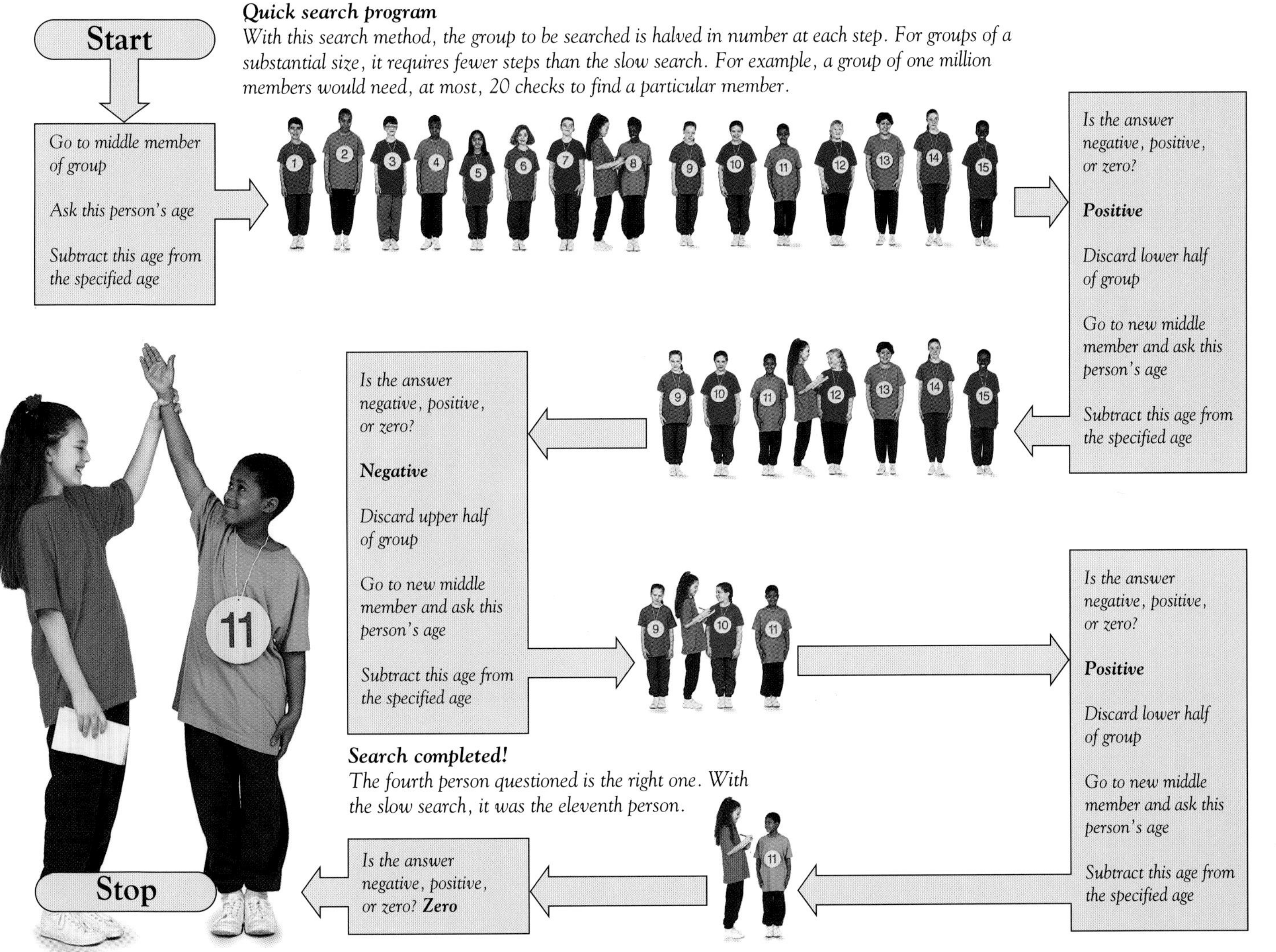

Search completed!
The fourth person questioned is the right one. With the slow search, it was the eleventh person.

Computer display and printer

MANY COMPUTERS display the results of their calculations as words or pictures on a screen, or by printing them on paper. The resulting images consist of tiny dots. The computer produces its results in binary numbers (1's and 0's) that go to the display or printer in the form of an on-off electric signal. The display or printer shows or prints a dark or colored dot for each binary 1 and leaves a dot blank for each 0. The dark or colored dots combine to form letters, numbers, or pictures. Most screens use electron guns to darken or color the dots in the same way as television sets (p.150). Portable computers and calculators have liquid-crystal displays (p.185), which use polarized light to make each dot light or dark.

YOU WILL NEED

• 24-in (60-cm) length of 3/4-in (2-cm) square wood strip • plywood strip 28 x 1 3/4 in (70 x 4.5 cm) • plywood sheet 10 x 7 1/2 in (25 x 19 cm) • two plywood spacers 2 x 2 1/4 in (4.8 x 5.5 cm) • green cardboard 4 x 6 in (10 x 15 cm) • pencil • saw • glue • ruler • vise • scissors

EXPERIMENT

Seven-bar display

Adult help is advised for this experiment

The liquid-crystal display (LCD) of a calculator shows numbers consisting of dark numerals on a light background. Each numeral from 0 to 9 is made up, at most, of seven dark segments, or bars. In the calculator's microchip each decimal numeral is represented by a seven-bit binary code number (a combination of seven 0's and 1's). This code is sent, as an on-off electric signal, to a group of seven segments. Those segments that receive a current darken to display the numeral.

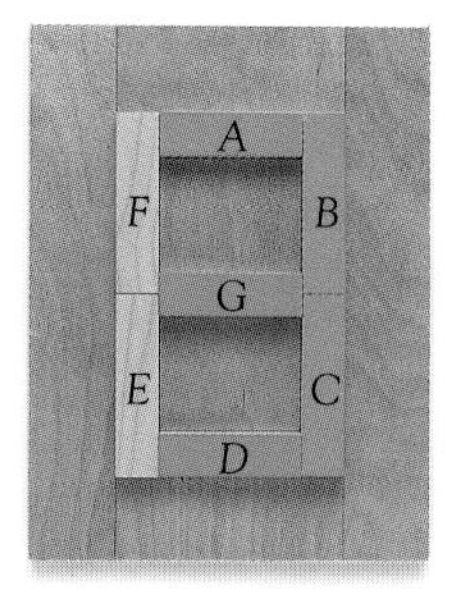

	A	B	C	D	E	F	G
0	1	1	1	1	1	1	0
1	0	1	1	0	0	0	0
2	1	1	0	1	1	0	1
3	1	1	1	1	0	0	1
4	0	1	1	0	0	1	1
5	1	0	1	1	0	1	1
6	0	0	1	1	1	1	1
7	1	1	1	0	0	1	0
8	1	1	1	1	1	1	1
9	1	1	1	1	0	1	1

1 CUT THE plywood strip into two 10-in (25-cm) and two 4-in (10-cm) lengths. Glue them to the plywood sheet, flush with its edges, to frame a rectangle 6 1/2 x 4 in (16 x 10 cm). Cut the wood strip into four 3 3/16-in (7.9-cm) blocks and three 2 1/4-in (5.5-cm) blocks.

2 GLUE GREEN cardboard to one face of each of the blocks. Arrange in the frame with the longer blocks at the sides to form a figure–8. Glue the spacers in the gaps between the blocks. Set each block with a plain face upwards. Choose a number between 0 and 9 from the

3 IN THE TABLE above, each of the seven blocks is given a letter from A to G. Follow the binary code in the row beside the chosen number, turning a block to green if its column shows 1 and leaving it blank if its column shows 0.

Dot-matrix printer

A print head containing a hammer and a column of pins moves across the paper, printing one line after the other. The image to be printed is encoded as binary numbers made up of 0's and 1's. Each 1 sends an electric pulse to an electromagnet, causing the hammer to strike a pin. The pin moves forward to hit an inked ribbon, making a dot on the paper. Columns with varying numbers of dots are printed beside each other to reproduce the image.

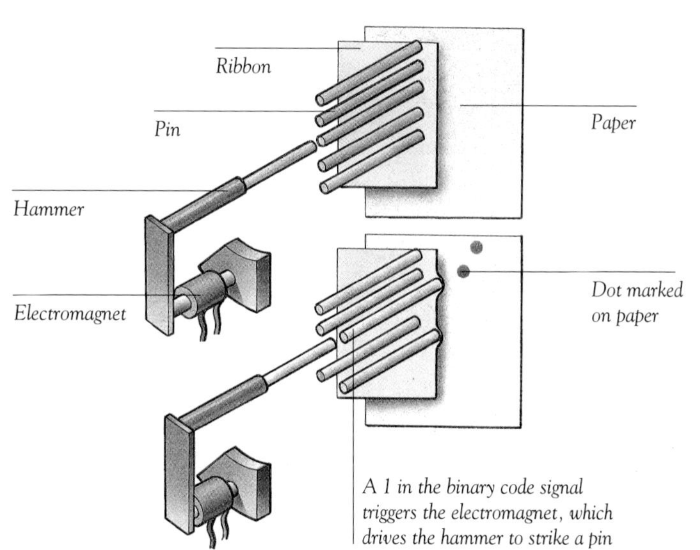

A 1 in the binary code signal triggers the electromagnet, which drives the hammer to strike a pin

Low to medium resolution
Printers with nine pins give low resolution (sharpness). Better printers have 24 pins.

Ink-jet printer

An ink-jet print head moves across paper in the same way as a dot-matrix print head to mark columns of dots on paper. The head contains a column of ink-filled tubes, each having a heating element. The image to be printed is encoded as binary numbers made up of 0's and 1's. Each 1 sends an electric pulse to an element in an ink tube. The element heats the ink, producing an expanding bubble, which sends a tiny jet of ink spurting from the tube.

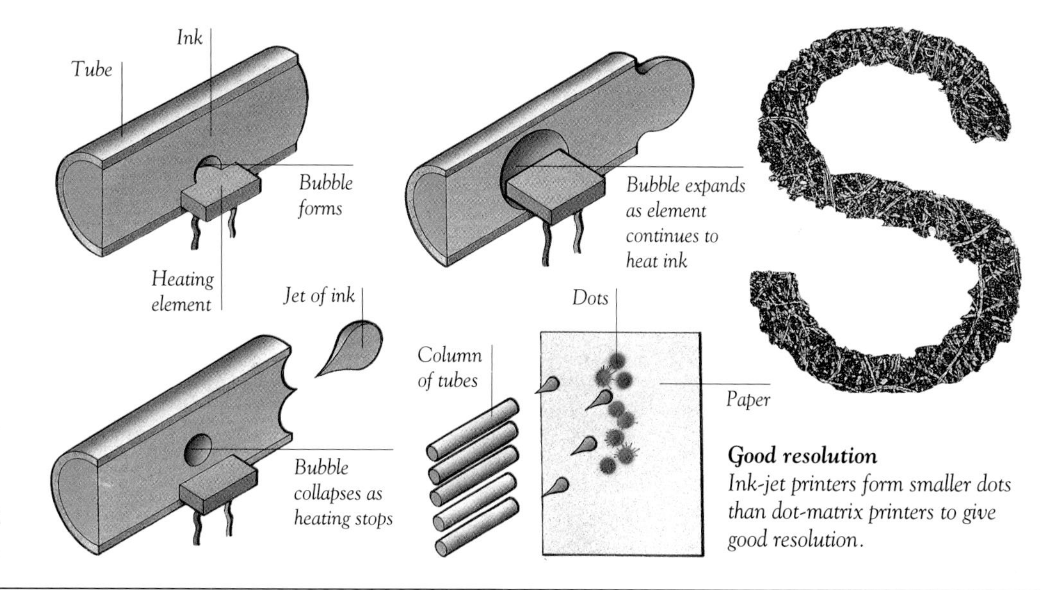

Good resolution
Ink-jet printers form smaller dots than dot-matrix printers to give good resolution.

Laser printer

A laser fires a beam of light, which is repeatedly turned on and off by a binary code signal that represents the image. A spinning mirror causes the on-off beam to scan across a rotating drum. The beam is focused on the drum by lenses. Where the beam hits the drum, it changes a negative charge of static electricity to a positive charge. Ink particles stick to the positively charged parts of the drum, building up a complete image, which is transferred to paper.

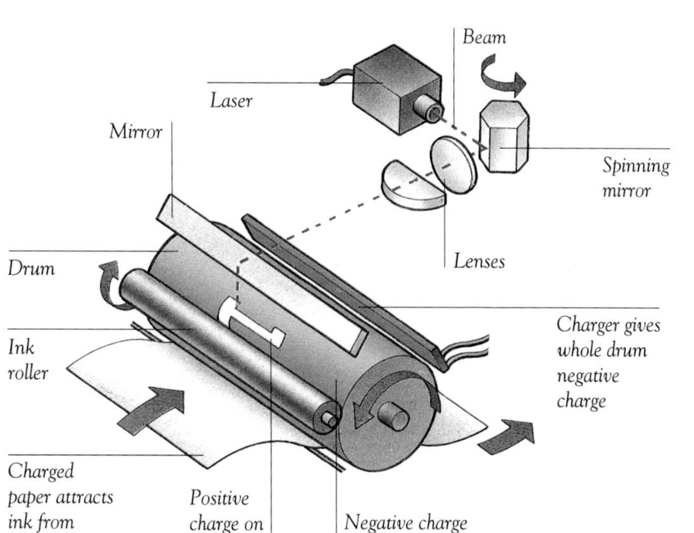

Very high resolution
Laser printers form very small dots; among common printer types they give the best resolution.

Homebuilt computer 1

BUILD A SIMPLE working computer that demonstrates how all the main units of a computer fit together and operate. The computer has an input unit that you use to feed in two numbers, which are then stored in the computer's memory. The numbers are then added by a processing unit, and the result appears on the computer's display unit. The program—to add two numbers—is built into the computer and cannot be changed.

YOU WILL NEED

• 2 cardboard pieces 4 x 2 in (10 x 5 cm) • steel ruler • scissors • 5 foamcore pieces 2¾ x 2 x 3/16 in (7 x 5 cm x 5 mm) • cutting mat • black marker pen • foamcore base 12 x 8 in (30 x 20 cm) • tracing paper 4 x 2¾ in (10 x 7 cm) • 2 normally open momentary SPST switches • 4508B dual four-bit latch • 7 green LEDs • four-pole changeover DIL switch • single-pole changeover DIL switch • 4011B quad NAND gate • 4511B BCD to seven-segment latch/driver • 4008B four-bit full adder • 9V battery with connector • 2 breadboards • seven 220R resistors • sixteen 10K resistors • tape • wire • double-sided tape • compass • craft knife • wire strippers • pliers

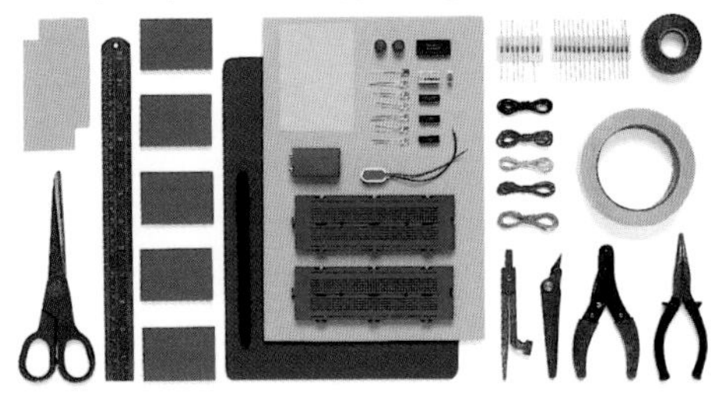

EXPERIMENT

Assembling your computer

First fasten the two breadboards to the base with double-sided tape. Then plug the microchips, switches, and resistors into the two boards, and install the wires neatly around the components.

Construct a seven-segment LED display unit as shown on the opposite page, then connect the wires from this display to the top breadboard. The computer will then be ready to use.

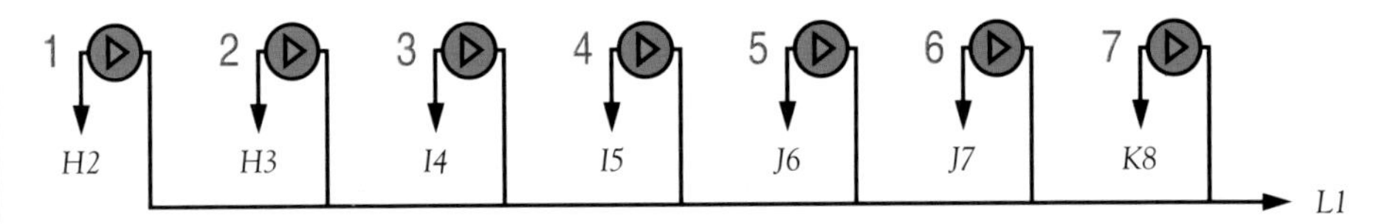

Upper board

Lower board

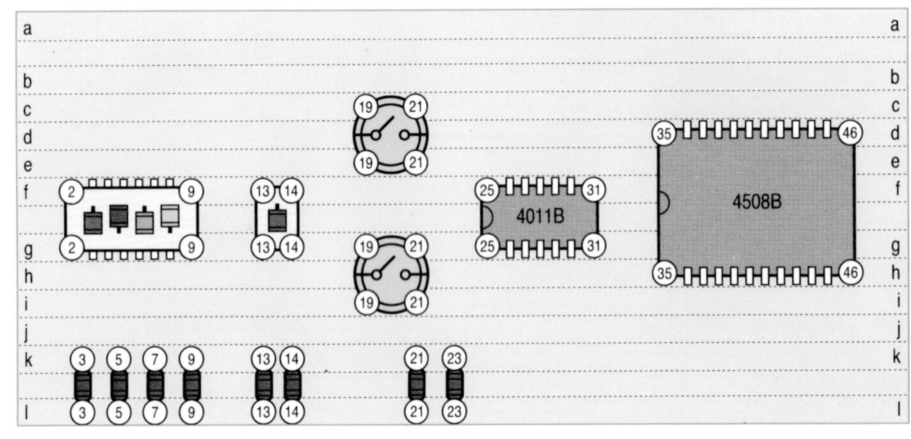

Upper board 220R resistors

F2–G2 F3–G3 F4–G4 F5–G5
F6–G6 F7–G7 F8–G8

Upper board 10K resistors

K32–L32 K34–L33 K35–L34 K36–L35
K37–L37 K38–L38 K39–L39 K40–L40

Lower board 10K resistors

k3–l3 k5–l5 k7–l7 k9–l9
k13–l13 k14–l14 k21–l21 k23–l23

Upper board wires

A16–B16 A34–B34 B2–B19 B3–B20
B37–I21 C4–C21 C5–C22 C38–J17
D6–D23 D39–K16 E7–E17 E8–E18
E16–H18 E35–G32 E40–K22 E41–H41
H20–H23 I18–I19 K23–L23 K41–L41

Lower board wires

a2–h2 a13–b13 a19–b19 a25–b25
a35–b35 b3–c37 b5–c39 b21–c26
b37–j44 b39–j42 b41–j40 b43–j38
b45–i13 c7–c41 c9–c43 c28–c44
c46–i46 d26–d27 e2–e3 e4–e5
e6–e7 e8–e9 e13–e14 f19–g19
f21–g23 g21–h25 h4–h6 i2–i4
i6–i8 i25–i26 i35–j46 j14–k36
j27–j37 k31–l31 k46–l46

Interboard wires

A1–a1 H32–k45 H34–b36 H36–b38
H38–b40 H40–b42 I35–k43 J37–k41
J39–k39 L2–l1

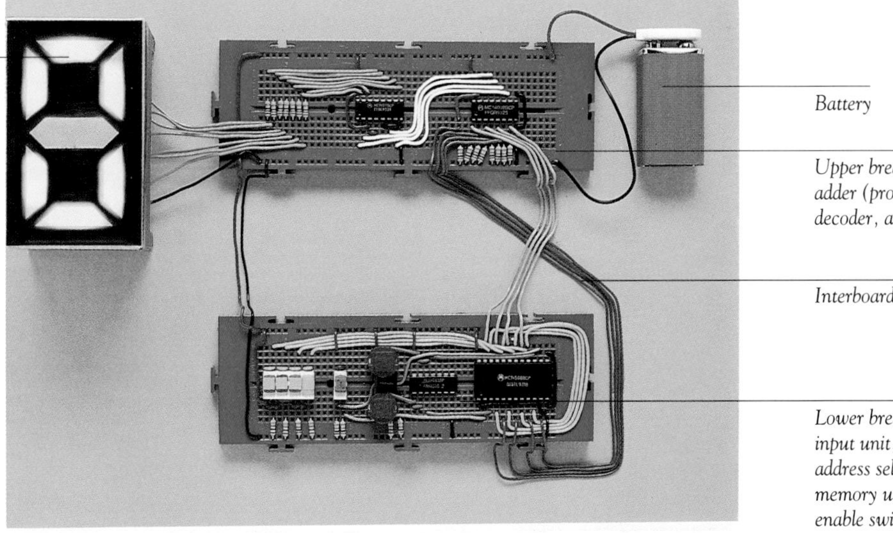

Seven-segment LED display shows numbers between 0 and 9

Battery

Upper breadboard holds adder (processing unit), decoder, and display unit

Interboard wires

Lower breadboard holds input unit, memory address selector, memory unit, and enable switches (p.181)

Complex circuits, simple sums
Your homebuilt computer has the equivalent of all the main units of a manufactured computer. All these units and wiring are needed for the homebuilt computer to perform even the simplest of mathematical calculations. A microchip in a real computer, however, has circuits so complex that it can perform millions of calculations a second (p.168).

Template for LED base (between red lines) and seven-segment display

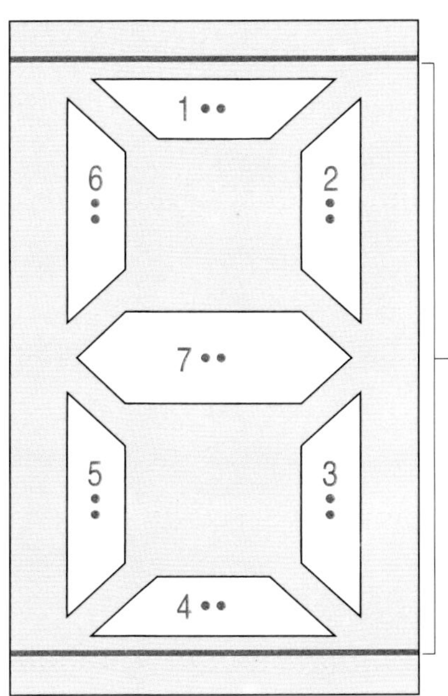

Diagram for LED display

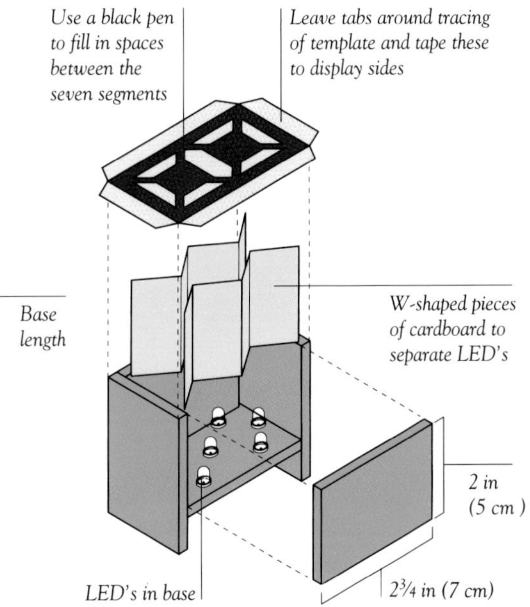

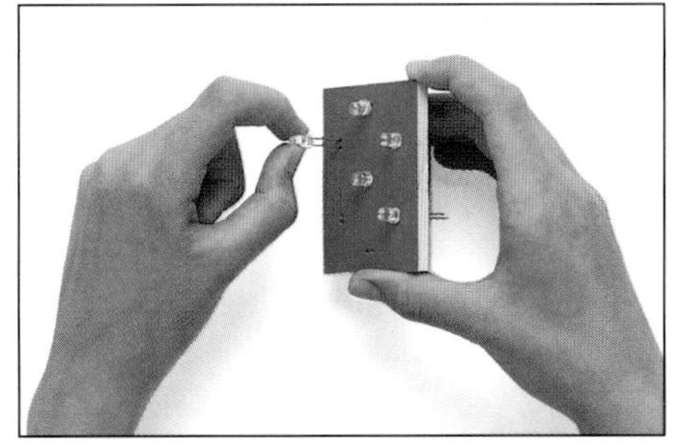

1 COPY THE BASE template (the smaller rectangle between the two red lines), and the positions of the red dots, onto one of the five foamcore pieces. Using a compass, make two adjacent holes through the foamcore where each pair of dots is marked. Carefully push the two pins of an LED through each pair of holes. The LED's should sit firmly on the foamcore base.

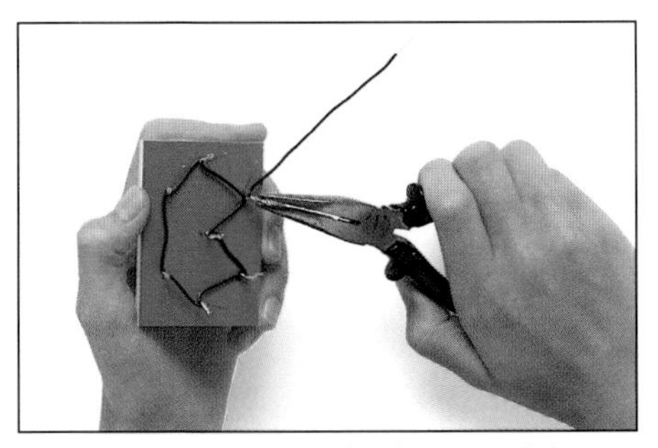

2 CONNECT THE cathode pins of the seven LED's with six short wires; wind the stripped ends of each wire around adjacent pins. Connect eight longer wires to the anode pins and one cathode pin, leaving one end of each wire free.

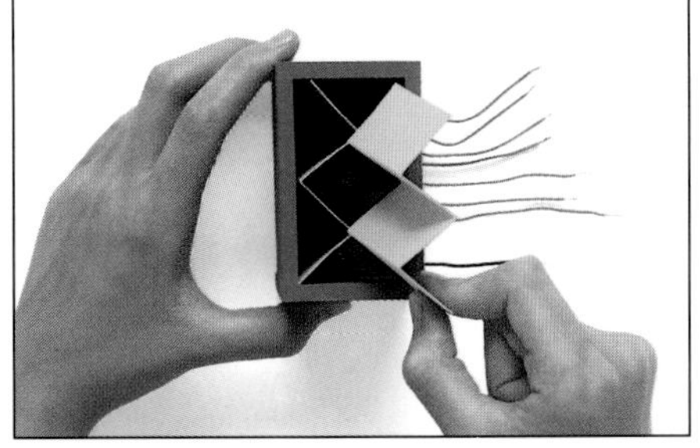

3 TAPE THE four foamcore sides to the display base as shown above. Fold the two cardboard pieces into quarters to make W-shaped partitions. Place them inside the display unit to separate the LED's from each other.

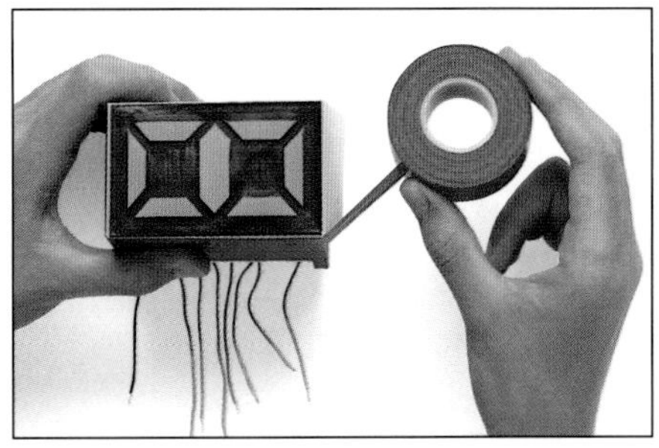

4 COPY THE display template onto tracing paper. Lay it over the display unit, and tape down the tabs. Connect the unit's wires to the upper breadboard, and trim to fit. The display should light up, showing 0, if the battery is connected.

Homebuilt computer 2

THE COMPUTER that you assembled using the instructions on the previous two pages is now ready to use. It displays numbers that you feed into it and can add two numbers together. However, since the display can show only one numeral, the total cannot exceed nine.

The first computer
The British engineer Charles Babbage (1792–1871) first had the idea of a computer in 1883. He designed a mechanical machine called the Analytical Engine that could be programmed to carry out different mathematical tasks, but it was never built. The forerunner of today's computers was the British Colossus (above), the first electronic computer. This machine was built in 1943, at the height of World War II, to crack enemy codes. Early computers were huge machines with masses of wiring connecting banks of electronic tubes and other components.

EXPERIMENT

Using your computer

Add two decimal numbers between zero and nine to make a sum not greater than nine, using your homebuilt computer. You must first convert the decimal numbers into four-bit binary numbers (p.164). Then input the two binary numbers into the computer. The computer stores the numbers in its memory and displays them in decimal form when you press the enable switches. From the memory the numbers are sent to the adder. The adder adds up the numbers and sends the result to the decoder, which converts the binary result into a decimal number that is shown on the display.

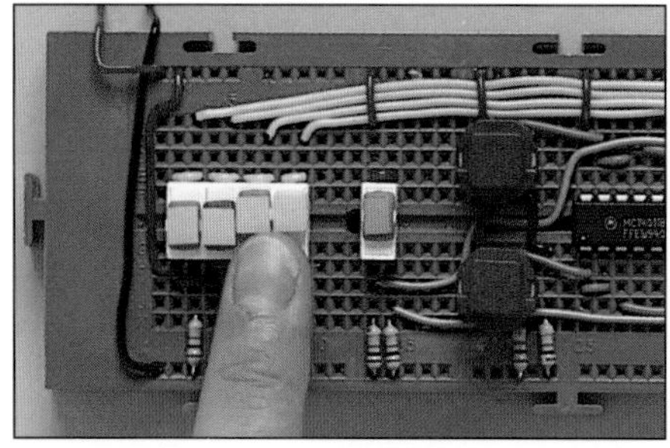

1 MOVE THE address switch to the upper position (**A**). Set the four-bit number to be input on the data switches, moving them up for each 1 and down for each 0. In the image above, the binary number 0011 (decimal 3) has been set.

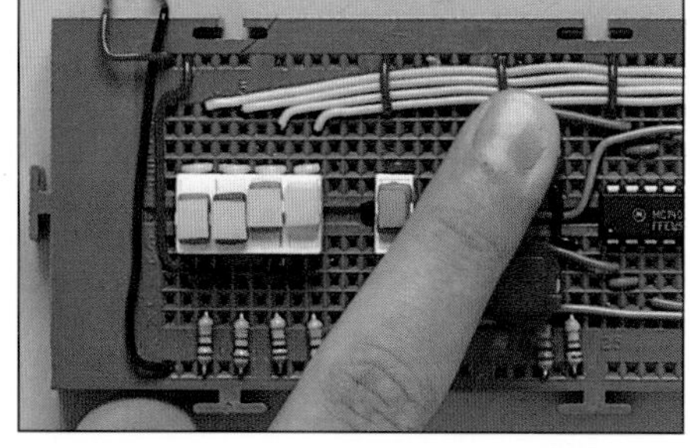

2 PRESS THE enable switch **A**. This sends the input number to the memory, then to the adder, which adds it to 0000. The result is decoded and sent to the display. The input number must not exceed 1001 (decimal 9).

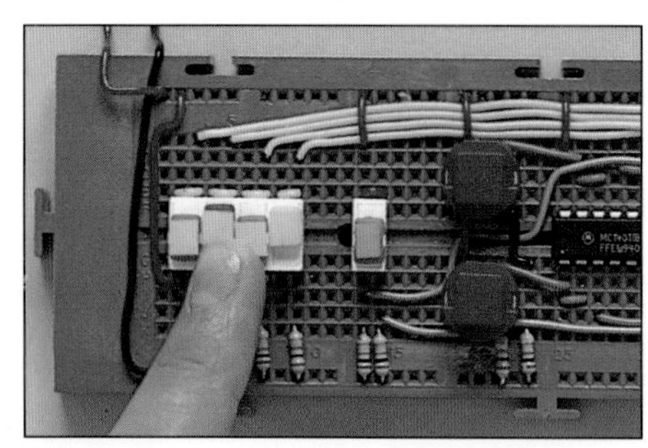

3 MOVE THE address switch to the lower position (**B**). Use the data switches again to set the second number that is to be input. In the image above, the setting for the binary number 0101 (decimal 5) is shown.

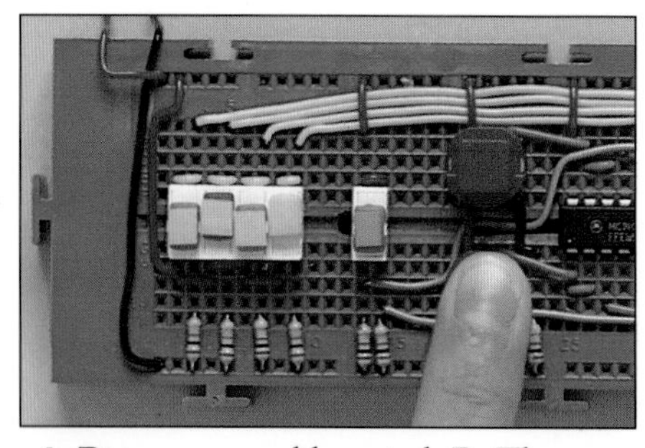

4 PRESS THE enable switch **B**. The input number is sent to the memory and on to the adder, which adds it to 0000. The result is decoded and sent to the display. The input number must not exceed 1001 (decimal 9).

How the computer works

The input, memory, processing, and display units in the computer are linked by data lines that carry binary numbers from one unit to the next. All of the data lines except the one going to the display unit consist of four wires, each of which carries one bit (a 0 or a 1) of a four-bit binary number. When a 1 is present, it is represented in the circuit by a flow of electric current along a wire. When a 0 is present, it is represented in the circuit by an absence of current.

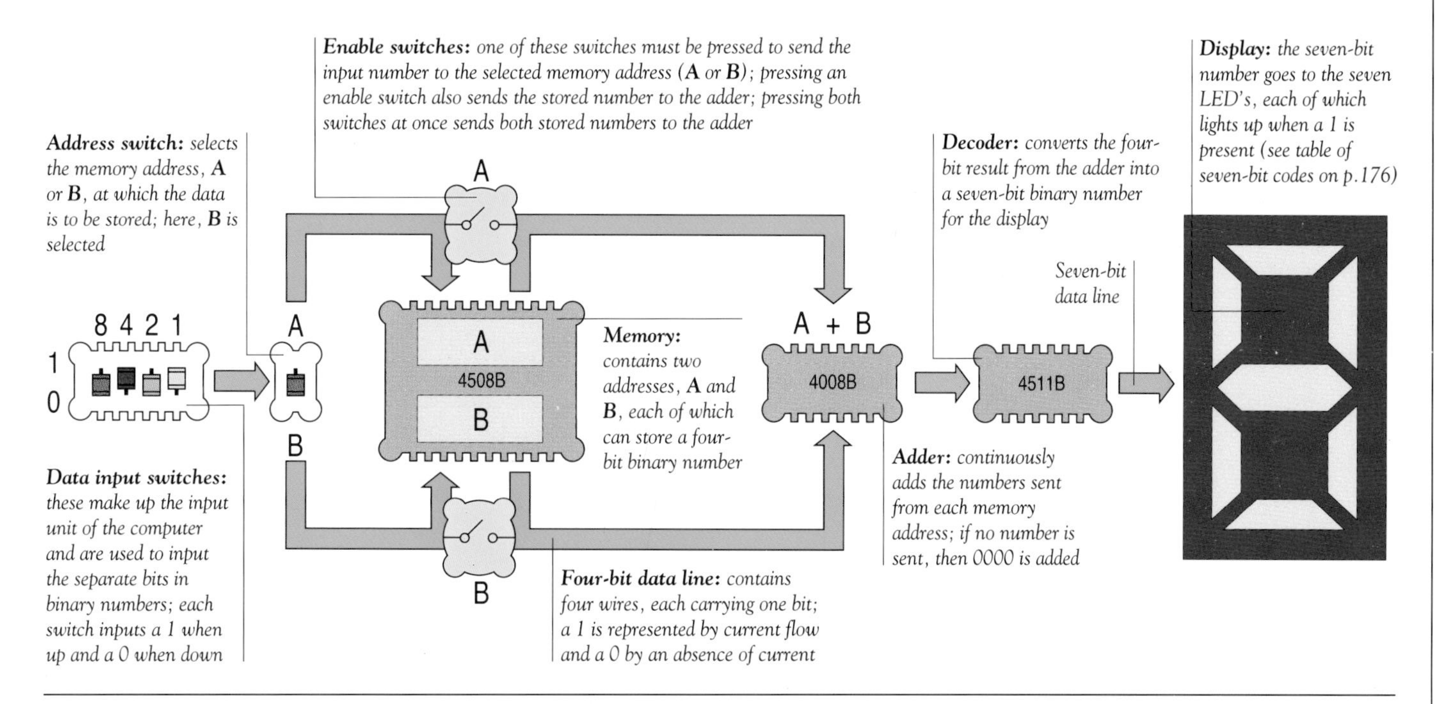

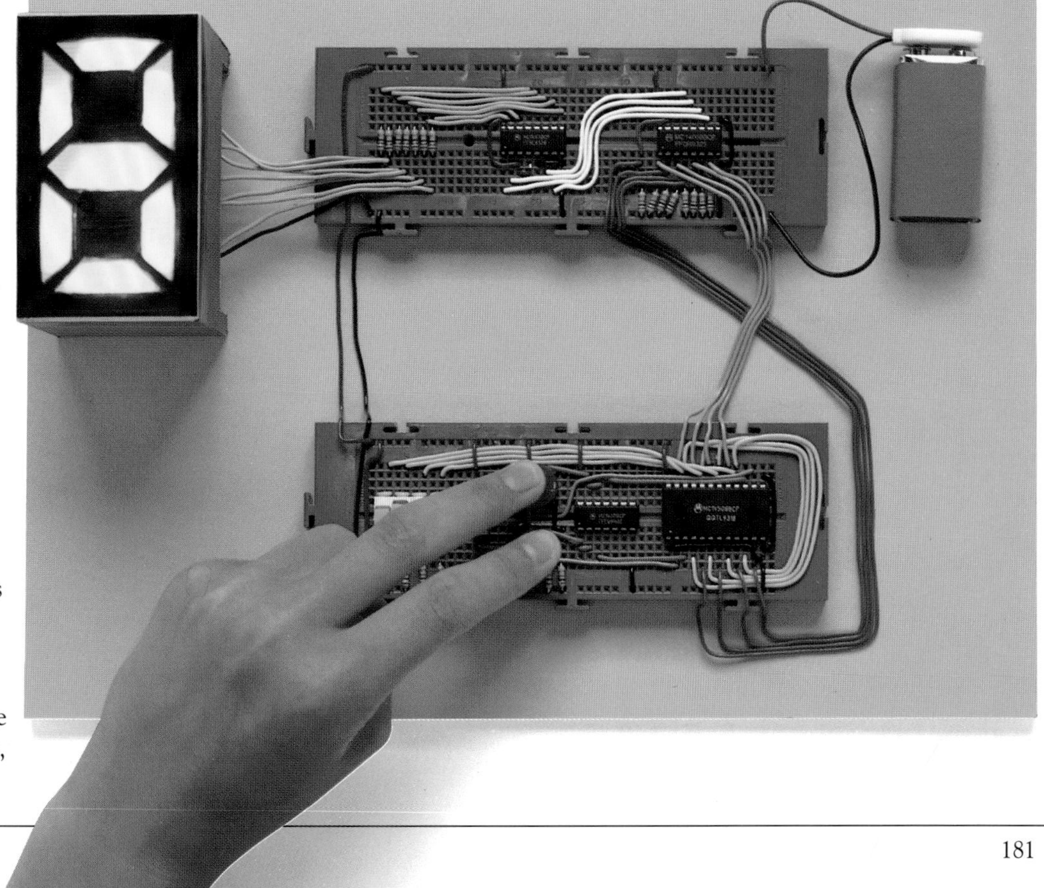

Seven-segment LED display
The computer's adder continuously sends a seven-bit binary number (a combination of seven 0's or 1's) to the seven LEDs in the display. When a 1 is present, an electric current flows along a wire and switches on an LED. The seven-bit binary code for the decimal number eight, shown here, is 1111111; so all seven LEDs are lit. For a list of other codes, see p.176

5 PRESS THE enable switches **A** and **B** together. The sum of the numbers from **A** and **B**, in this case 1000 (decimal 8), is displayed. If the sum exceeds 1001 (decimal 9), the display goes blank.

GLOSSARY

THE ENTRIES ON THESE PAGES explain many of the general terms used in this book. A term that appears in italics, either in the middle of the entry or following it, has its own entry elsewhere in the glossary. Many terms not defined here can be found in the index (pp.188-191).

ADDRESS A number that specifies where a particular piece of information or *data* is located in a computer's *memory*. Each address holds one or more *bytes* of data and is assigned a unique number.

ALTERNATING CURRENT (AC) An *electric current* in which the flow of electricity regularly reverses direction.

AMPLIFIER A device that increases the strength of an *electric signal.*

ANALOG In electronics, a term used to describe a continuously variable *electric signal* or a system that uses such signals. In timekeeping, analog refers to a watch or clock that has a traditional face and hands to show the time. *See Digital.*

ANTENNA A metal structure that picks up radio *waves*, such as radio and television signals.

APERTURE An opening of variable diameter in a camera *lens*. It controls the brightness of the image by varying the amount of light passing through the lens.

ATOM An extremely small particle. All physical matter is made up of these particles. There are about 100 different types of atoms, each of which is a particle of one of the different elements that make up all matter.

BIMETALLIC STRIP A strip made of two metals, such as brass and iron, that expand and contract by different amounts when heated and cooled, causing the strip to bend on heating and unbend on cooling.

BINARY CODE The code used by computers, which process and store *data* by converting it into *binary numbers* composed of 0's and 1's. Each number is represented by a *digital* code signal made up of on-off electric pulses in a microchip, on-off patterns of magnetism in a tape or *disk*, or sequences of microscopic pits and flat areas on a *compact disc* or *CD-ROM*.

BINARY NUMBER A number composed of the two numerals 0 and 1. In a binary number, the numeral in each place has twice the value of the numeral in the place to its right. *See Binary code.*

BIT Short for "binary digit." Each numeral (0 or 1) in a *binary number*, or each on-off pulse in a *binary code* signal, is a bit. An eight-bit binary number contains eight numerals, such as 10011011.

BREADBOARD A small plastic board containing a grid of holes connected together by metal strips. Electronic components and wires are inserted into the holes to build electric *circuits*. Also called a prototyping board.

BYTE A binary code signal containing eight *bits* or on-off pulses. A computer often breaks down binary code signals into groups of bytes. *See Megabyte.*

CAM A projection on a rotating shaft, which is in contact with another part of a machine and causes it to move up and down, or back and forth, while the shaft rotates.

CCD Short for "charge-coupled device." An array of small pieces of *semiconductor*, each producing an electric signal proportional to the amount of light falling on it.

CD-ROM Short for "compact disc read-only memory." A *compact disc* storing programs and data to be used on a computer.

CHARGE A quantity of *electricity*. Objects that receive *electrons* gain a negative charge; and objects that lose electrons gain a positive charge.

CHIP *See Microchip.*

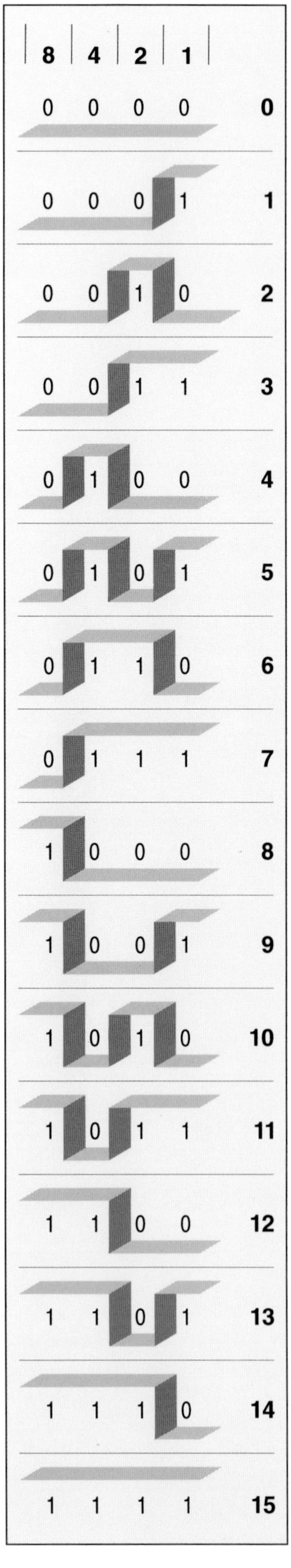

Binary codes and numbers
The decimal numbers 0 to 15 are shown at right with their binary equivalents. In a binary number, from right to left, a 1 signifies that each place has a value of 1, 2, 4, 8 and so on, doubling each time; a 0 signifies that a place has no value. So 1101 in binary means 8+4+0+1 (decimal 13). The pink band represents binary numbers in code form, such as on-off electric pulses.

8	4	2	1	
0	0	0	0	0
0	0	0	1	1
0	0	1	0	2
0	0	1	1	3
0	1	0	0	4
0	1	0	1	5
0	1	1	0	6
0	1	1	1	7
1	0	0	0	8
1	0	0	1	9
1	0	1	0	10
1	0	1	1	11
1	1	0	0	12
1	1	0	1	13
1	1	1	0	14
1	1	1	1	15

CIRCUIT A linked set of *components* powered by a source of electrical energy.

CIRCUIT BOARD A board on which electrical *components* are mounted to form a *circuit*.

CLUTCH A device situated between the engine and the *gearbox* in a motor vehicle. When the clutch is engaged, the engine is connected to the gearbox and power goes to the wheels.

COIL A wire wound into several turns. A coil produces a *magnetic field* when an *electric current* passes through it.

COMBUSTION ENGINE *See Fuel.*

COMMUNICATIONS SATELLITE An artificial satellite that relays telephone calls and television pictures. It receives radio signals from stations on Earth, and broadcasts the signals back to other stations or to dish antennas.

COMPACT DISC (CD) A disc on which music or information is recorded in *binary code* as a sequence of microscopic pits and flat areas. Conventional 5 in (12 cm) music CD's can hold up to 600 *megabytes* of data—equivalent to 75 minutes of music. *See CD-ROM*.

COMPONENT In mechanics, a part of a force. A force acting in one direction can split into smaller component parts that act in different directions. In electronics, the term component refers to a device, such as a *transistor* or *resistor*, that can be connected in an electric *circuit*.

COMPRESSION An action that squeezes something. Compressing a quantity of gas reduces its volume and raises its pressure. Squeezing a solid object exerts a compression force on the object that may strengthen it. *See Tension*.

CONDENSATION The change from *vapor* to liquid; the opposite of *evaporation*.

COUNTERWEIGHT A weight that balances a load in a machine so that the motor driving the machine does not have to raise the entire weight of the load.

DATA Factual information such as words, numbers, or pictures that describe, quantify, or identify things. Measurements, names, and addresses are examples of data.

DECIMAL NUMBER A number containing one or more of the ten numerals from 0 to 9, such as 385. The numeral in each place of a decimal number has ten times the value of the numeral in the place to its right. For example, in 385, 3 means 300, 8 means 80, and 5 means 5.

DIAPHRAGM A vibrating or moving plate or seal in a tube or chamber. An iris diaphragm is a special type of diaphragm with blades that can be adjusted to alter the size of a central aperture in the diaphragm.

DIESEL ENGINE An engine in which *compression* of the fuel/air mixture causes it to ignite.

DIFFERENTIAL A set of *gears* in the *transmission* of a car or other road vehicle that enables the wheels on either side to turn at different speeds as the vehicle turns a corner.

DIGITAL In electronics, a term used to describe a pulsed (on-off) electrical signal in which the pulses represent binary numbers, or a system that uses such signals. In timekeeping, a digital watch or clock is one with a *digital display*. *See Analog, Binary code*.

DIGITAL DISPLAY A display that shows a quantity, such as the time on a clock, as a *decimal number*.

DISK A rotating disk of metal or plastic such as a floppy disk or a hard disk used in computing to store *data*. It has a magnetic surface that stores *binary codes* as patterns of magnetism.

EFFORT The force applied to a device or machine to move a load.

ELECTRIC CURRENT A flow of electric *charge* from a source such as a battery or outlet. It consists of moving *electrons*.

ELECTRIC PULSE A short burst of electricity in an *electric signal* caused by switching an *electric current* on and off.

ELECTRIC SIGNAL A variable *electric current* that may represent *data*, *light*, or sound. The current may vary continuously, as in an analog signal, or in a series of on-off pulses, as in a *digital* signal. *See Electric pulse*.

ELECTRICITY A form of energy involving *charge*. There are two kinds: *electric current* and *static electricity*.

ELECTRODE A part in an electrical device that receives or gives off *electrons*.

ELECTROMAGNET A magnet consisting of a *coil* wound around an iron bar. The bar becomes magnetized when an *electric current* flows through the coil; the bar loses its magnetism when the electric current stops.

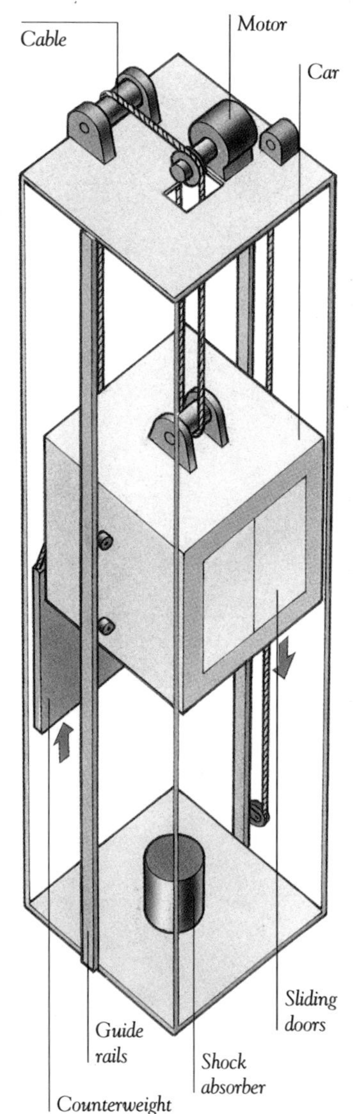

Elevator
A heavy counterweight at the other end of the cable carrying the car balances the weight of the car and an average load of passengers. The motor has only to move the car and raise little if any weight.

ELECTRON A tiny particle with a negative electric charge. Electrons are present in atoms. When they are freed from atoms, electrons may accumulate on an insulator, causing it to become negatively *charged* with static electricity, or they may flow through a conductor to produce an *electric current*.

ELEVATOR A flap on the horizontal tailplane of an aircraft. It hinges up or down to lower or raise the tail. The term elevator also refers to a car that carries passengers and freight up and down a shaft in a building.

ENERGY The capacity to cause work to be done, that is, to cause an action or process to happen. There are several forms of energy, including electrical, heat, light, sound, and *kinetic* (movement). *Fuel* is often thought of as energy; it is actually a source of energy rather than energy itself.

EVAPORATION The change of a liquid whose temperature is below boiling point to a *vapor*, caused by the escape of molecules from its surface.

EXPANSION An increase in size. Solids and liquids expand when heated, as do gases if not prevented from doing so by a container. On cooling, the solid, liquid, or *gas* contracts (gets smaller).

FLOPPY DISK *See Disk.*

FLUID A liquid or a gas.

FLYWHEEL A heavy wheel used to smooth out variations of speed in machines, or to store *energy* in the form of the motion of the flywheel.

FORCE A push or a pull. Force produces a change in motion, *pressure*, or *tension*.

FREQUENCY The rate at which something moves back and forth (as with a pendulum), or the rate at which the strength of an *electric signal*, a sound wave, or a radio wave varies. Frequency is measured in hertz (Hz), which is the number of complete cycles of movement or variation per second. *See Pitch.*

FRICTION The *force* produced whenever an object moves, which acts to slow down and eventually stop the object. It is caused by the object rubbing against other objects, or by the *resistance* of the surrounding air or water.

FUEL A substance, such as gasoline or coal, that is burned to provide heat. In a combustion engine, this heat *energy* is converted into *kinetic energy*, or motion.

GAS One of three states of matter, the others being liquid and solid. A gas flows and completely fills its container.

GEAR A toothed wheel. Two or more gears (or gear wheels) may connect together—either directly or by a chain or shaft—to change speed, *force*, or direction of movement.

GEARBOX A set of gears in an engine-powered vehicle.

HARD DISK *See Disk.*

HARDWARE In information technology and communications, machines (such as computers and printers) or their individual components (such as hard disks). *See Software.*

HARMONIC A musical note with a frequency (pitch) that is an exact number of times greater than another note (known as the fundamental). For example, a fundamental of 440 Hz has harmonics of 880 Hz (2 x 440), 1320 Hz (3 x 440), and so on.

HEAT A form of *energy* . It is the energy of the motion of the particles (*atoms* or molecules) that make up things. The particles move faster as something gains heat, and slower as it loses heat.

IMAGE A picture of a scene or object formed by a lens or mirror. The image may appear on a surface, such as a screen, or you may see an image by looking into a lens or mirror. The term is also used to refer to a picture on a television or computer screen.

INFORMATION Facts or *data*, in the form of words, numbers, pictures, or sounds. Information can be stored and processed by machines such as computers.

INFORMATION SUPERHIGHWAY A worldwide network that connects computers to bring information, entertainment, and services direct to the home on demand.

INPUT *Data* entered into a computer, or the process of entering data using a device such as a keyboard.

INSULATION A plastic coating around a wire that prevents *electricity* from escaping. Also, a material that reduces the amount of heat entering or leaving an object.

INTEGRATED CIRCUIT (IC) An electronic device manufactured in a single piece of semiconductor containing many components connected together to form a circuit.

KILOBYTE (K, KB, KBYTE) A measure of memory capacity in a computer equal to 1,024 bytes. *See Megabyte.*

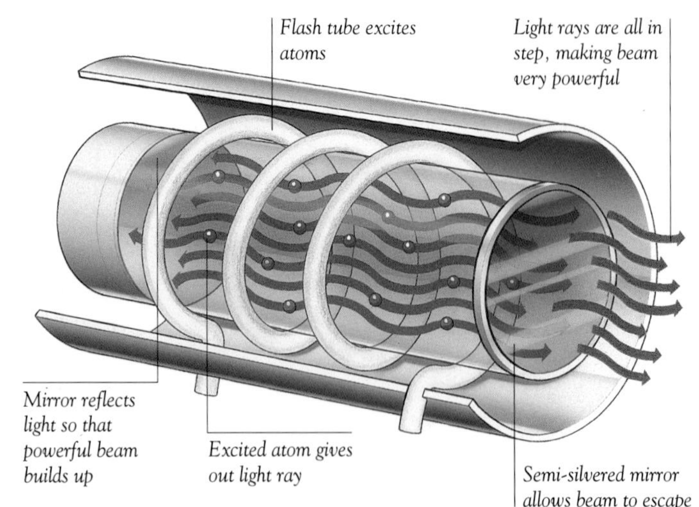

Laser
Atoms in a tube of laser material gain energy, for example from light produced by a flash tube. The excited atoms then suddenly lose this energy by giving off their own light rays, forming a powerful lasser beam that passes through a semi-silvered mirror to emerge from the laser material.

KINETIC ENERGY The energy an object has because it is moving. The faster an object is moving, the more kinetic energy it has.

LASER Short for "light amplification by stimulated emission of radiation." A device that produces a very powerful beam of light or invisible infrared rays.

LCD Short for "liquid-crystal display." *See Liquid crystal.*

LDR Short for "light-dependent resistor." An electronic component that can be used to detect light.

LED Short for "light-emitting diode." A special type of electronic component that produces light.

LENS A piece of transparent glass or plastic with curved sides that bends light rays passing through it to form an image.

LEVER A simple machine consisting of a rigid bar that tilts about a pivot (called the fulcrum). Levers are used in more complex machines to change the magnitude of and the distance moved by an applied force.

LIFT The force that acts on aircraft, helicopters, and balloons to raise them into the air. The term also refers to an elevator that carries people and freight up and down a shaft in a building.

LIGHT A form of energy that consists of visible electromagnetic *waves*. Other types of electromagnetic waves, such as *radio* waves and X-rays are invisible.

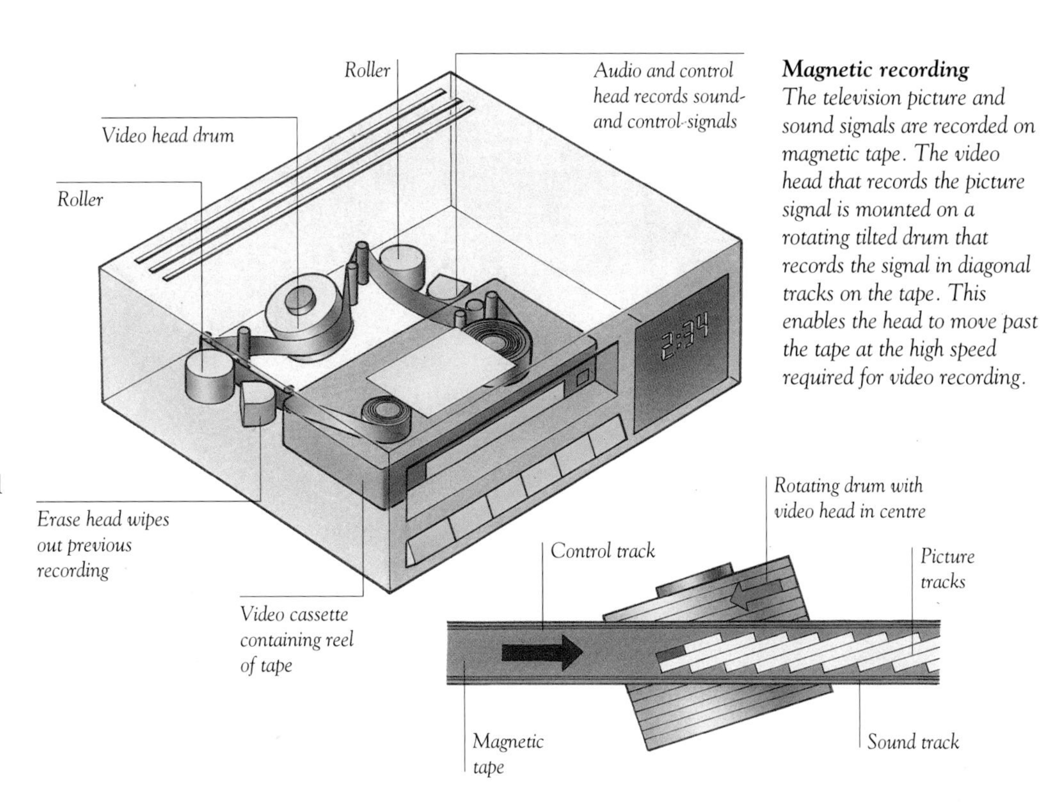

Magnetic recording
The television picture and sound signals are recorded on magnetic tape. The video head that records the picture signal is mounted on a rotating tilted drum that records the signal in diagonal tracks on the tape. This enables the head to move past the tape at the high speed required for video recording.

LIQUID CRYSTAL A special type of liquid that has an internal structure similar to that of a crystal. Applying an electric current to a liquid crystal affects the passage of light through it—this is the principle behind an *LCD.*

LOAD The force exerted by an object that is to be moved by a machine such as a lever or pulley, or the force that an object exerts on a structure such as a bridge. The object itself is also called a load.

MAGNETIC FIELD The region around a magnet in which a magnetic material, for example iron and other ferrous metals, is affected by the force of the magnet.

MAGNETIC POLE A center of magnetism in a magnet. Every magnet has two poles, called north and south, one at each end or on opposite sides. Like poles repel each other; opposite poles attract.

MECHANICAL A term applied to devices that rely solely on moving parts such as gears, levers, and pulleys to work, rather than functioning with electrical or electronic *components*.

MEGABYTE (MB, MBYTE) A measure of *memory* capacity in a computer equal to 1,048,576 bytes. *See Kilobyte.*

MEMORY Part of a computer or other electronic machine that stores data, information, or computer programs, as magnetic, electronic, or optical codes. *See Disk, RAM, ROM.*

MICROCHIP Also called simply a chip. An electronic *component* containing an *integrated circuit.*

MICROWAVES Invisible electromagnetic *waves* similar to *radio waves.*

OCTAVE *See Scale*

OUTPUT In computing, the results or *data* obtained from a computer, or the process of obtaining results or data using an output unit such as a screen or printer.

PHOSPHOR A luminescent substance used in television screens. It glows when struck by *electrons.*

PISTON A disk or short, solid cylinder that moves inside a tube or hollow cylinder. A piston moves a *fluid*, or is moved by the *pressure* of fluid on one side of it.

PITCH In music, the sensation of how high or low a note sounds. The pitch of a note is determined by the *frequency* of the sound *waves*: increasing the frequency raises the pitch, and vice-versa.

POLARIZED LIGHT *Light* in which all the *waves* vibrate in the same plane.

Power In strict scientific usage, power is the rate at which work is done. More generally, however, power is used to refer to the strength of a machine or the level of *energy*. Power is also used to mean electricity.

Pressure The *force* exerted over a given area.

Printed circuit A *circuit board* in which the links connecting *components* are imprinted with a material that conducts *electricity*.

Processing unit A computer unit that calculates results and controls other units, such as the display monitor.

Program A set of coded instructions for a computer that enables it to perform a particular task.

Pulley A grooved wheel, or set of wheels, around which a rope passes in order to raise a load. Two or more wheels reduce the effort needed to lift the load.

Radar Short for "radio detection and ranging." A device that can detect the position and speed of distant objects. It works by bouncing *radio* signals off such objects.

Radio A system of transmitting *information* (voices or music for example) using radio *waves*, which are electromagnetic *waves* similar to *light* rays but invisible. The term is also used to mean a radio receiver.

RAM Short for "random access memory". A part of a computer's *memory* whose contents can be changed. It holds *programs* or *data* temporarily while they are required by the computer to perform a specific task. *See ROM.*

Reaction In physics, a *force* (the action) is always accompanied by its opposite (the reaction). The reaction is in the opposite direction to the action and pushes on the object causing the action. Reaction pushes an aircraft or ship forward as its engines push air or water backward. In chemistry, a reaction is a process in which two or more substances meet and change.

Resistance The degree to which an object resists the flow of *electric current* through it, or the flow of air or water around it. In a lever, resistance is the *force* that has to be overcome for the lever to move.

Resistor An electrical *component* having a fixed or variable amount of electrical *resistance*.

ROM Short for "read-only memory." A part of a computer's *memory* that permanently holds programs or *data*. Unlike the contents of the *RAM*, the contents of the ROM cannot be changed.

Rotor A rotating part in a machine, such as the blades of a helicopter, the rotating coil in an electric motor or electric generator, or the moving part of a rotary pump.

Pneumatic drill or jackhammer
A pneumatic drill is driven by compressed air from a pump. The drill is used to break up concrete and tarmac, often on roads. The compressed air drives the blade or tool of the drill into the road surface in a four-stage cycle.

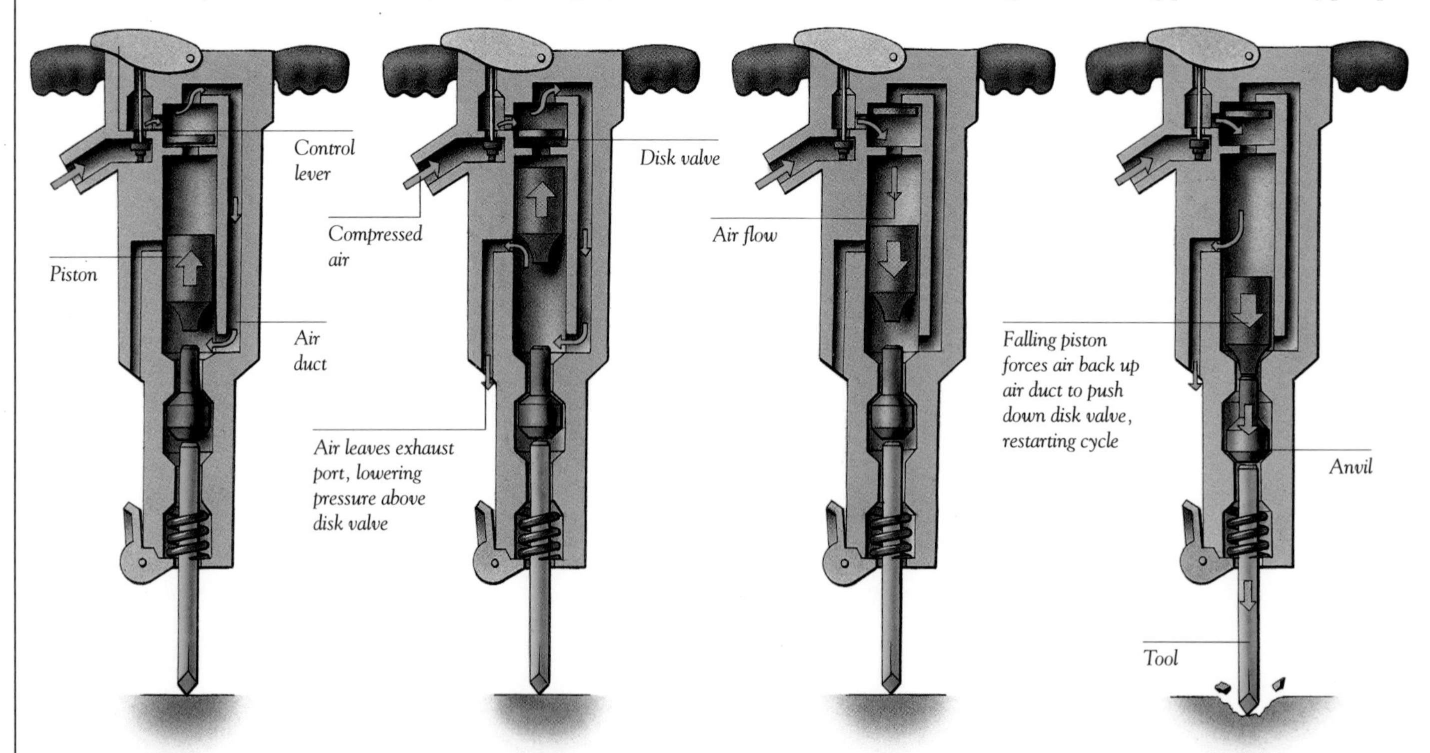

1 Piston rises
Pressing the control lever admits air to the base of the piston, which begins to rise.

2 Valve rises
The rising piston compresses the air above it, forcing up the disk valve that controls the air supply.

3 Piston drops
The disk valve admits the compressed air to the top of the piston, forcing it down.

4 Tool driven down
The piston moves down and strikes the anvil, driving the tool into the road surface, before rising again.

SCALE In music, a graduated sequence of notes in ascending or descending order of *pitch*. The sequence, which usually comprises eight notes, covers an octave. The scale of C major, for example, comprises the white notes of the piano between a given C and the C an octave higher or lower. The C an octave higher has a *frequency* twice that of the first C; the C an octave lower has a frequency half that of the first C.

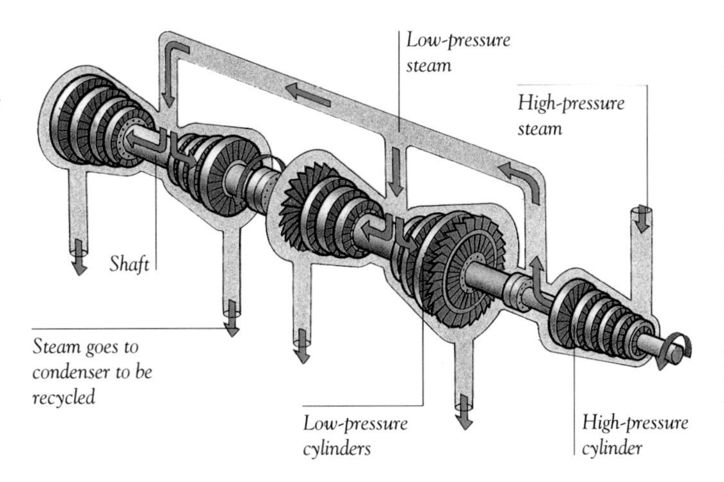

Steam turbine
At a power station, a steam turbine drives an electric generator. The turbine shaft is driven by steam flowing through several sets of blades in separate cylinders. The steam, created in a boiler, enters the first cylinder at high pressure, and leaves at low pressure. It then enters the low-pressure cylinders. The size of the blades increases as the pressure falls, enabling the turbine to extract as much energy as possible from the steam.

SEMICONDUCTOR A material, usually silicon, used in electronic *components* such as *transistors* and *integrated circuits*. The electrical *resistance* of a semiconductor can be altered by a control signal so that it blocks or passes an *electric current*. This property can be used to produce or process *electric signals* in *binary code*.

SENSOR A device that detects some property (such as *pressure*, movement, *light*, *heat*, smoke, or magnetism) and may also measure it.

SIGNAL In communications and information technology, a *radio* or *light* wave or an *electric current* that varies in strength to carry information. *See Electric signal*.

SMART MACHINE A machine that has a high degree of automatic control and can perform complex tasks unaided.

SOFTWARE The instructions or programs that enable computers to perform tasks. *See Hardware*.

SOLAR ENERGY *Electricity* or *heat* obtained by converting or collecting the Sun's *light* and heat.

SOLENOID A *coil* containing a central iron plunger. The *magnetic field* produced when an *electric current* flows through the coil pulls the plunger into the coil.

STATIC ELECTRICITY A form of *electricity* in which charge remains in one place rather than flowing as an *electric current*. When an electrical *insulator* loses or receives *electrons*, it gains a positive or negative charge of static electricity.

STEREO In sound reproduction, an abbreviation for "stereophonic." Stereo sound reproduction employs two loudspeakers so that the various individual sounds seem to come from different positions between the loudspeakers.

TELECOMMUNICATIONS An area of technology concerned with communications over long distances using machines such as the telephone, radio and satellites.

TENSION A *force* that acts to stretch or pull apart an object. *See Compression*.

THERMOSTAT A device that automatically regulates temperature.

TRANSFORMER A device that changes the *voltage* of an *alternating current*. It consists of two *coils* wound around an iron core. Alternating current fed to the input coil generates a varying *magnetic field*, which causes an alternating current to flow from the output coil. The change in voltage between the input and output currents depends on the number of turns in each coil.

TRANSISTOR An electronic *component* made of three pieces of *semiconductor* joined together. It may be used to amplify a weak *electric signal*, or to switch an electric current on or off.

TRANSMISSION In a vehicle, a set of shafts and other parts, such as gears, that transmits engine power to the wheels. In telecommunications, the sending of signals by *radio* or along wires or cables.

TURBINE An engine that basically consists of a set of blades connected to a shaft. The shaft rotates when a *gas* or liquid (such as steam or water) flows past the blades.

ULTRAVIOLET (UV) RAYS Electromagnetic *waves* that are similar to *light* rays but invisible.

VACUUM A totally empty space without air or any other substance.

VALVE A mechanical valve is a device that controls the flow or pressure of a fluid (a gas or liquid) in a pipe, container, or machine. A faucet, for example, is a valve that can be turned on or off to control the flow of water.

VAPOR A form of *gas* produced above the surface of a liquid by *evaporation*.

VIDEO A method of recording pictures and sound on *magnetic* tape; see illustration on p.185.

WAVE A way in which *energy* is transferred from one place to another. This transfer of energy always involves vibrations. Electromagnetic waves consist of vibrating electric and *magnetic fields*. Such *waves* include *radio* waves, *microwaves*, infrared rays, *light* rays, *ultraviolet* rays, and X-rays. Sound waves consist of vibrating air or other materials.

Index

F

G

H

I

J

K

L

S

T

U

V

X Y Z

Acknowledgments

NEIL ARDLEY wishes to thank the team at Dorling Kindersley for their energy and enthusiasm and for remaining undaunted by the demands of this project—especially Paul, Bryn, Mukul, and Phil for meticulous professionalism and inspiring further ideas when necessary; and Mukul and Phil in particular for transforming prototypes and sketches into the elegant, practical devices and clear diagrams that distinguish this book—and also Tim Kirk for his assistance and advice on electronic experiments.

DORLING KINDERSLEY would like to thank Will Hodgkinson, Edward Bunting, and Teresa Pritlove for editorial assistance; Claire Shedden for design assistance; Peter Pocock for his valuable advice and contribution; Joanna Thomas and Sharon Southren for picture research; Gary Ombler for photographic aid; Kay Wright for the index; all the staff and pupils at John Perryn Primary School, Acton, and the Soho Parish School; Prof. Roger M. Goodall, University of Technology, Loughborough; Mark Harding and Dave Woodcock of the Science Museum for information about exhibits in the Science Museum; Eric Kingdon, Sony Consumer Products; Jonathon Nicholson, CAA Press Office; Railway Technical Research Institute, Japan.

PICTURE CREDITS

t top; *c* center; *b* bottom; *r* right; *l* left; *a* above

The publisher would like to thank the following for their kind permission to reproduce photographs:

Arcaid/Richard Bryant: 44-45
BT Pictures/© British Telecommunications plc: 155*tl*
Bewator (UK) Ltd: 88*tr*
Central Office of Information: 21*tr*
Mary Evans Picture Library: 106 *tr*
Goddard Collection/Clark University Archives, Worcester, Massachusetts: 22 *tr*
Robert Harding Picture Library: 17*cr*, 23*t*, 47*br*, 65*bl*, *tr*, *cr*, 105*bc*; *G&P Corrigan* 65*cl*; *Walter Rawlings* 52*c*; *Adam Woolfitt* 49*tr*
Michael Holford: 74*cl*
Hoover European Appliance Group: 74*cr*
Hulton Deutsch Collection Ltd: 108*cr*, 147*br*
The Image Bank: K. Drinkwitz 68*tr*; *Dominique Sarraute* 152-153
Images Colour Library: 42*bl*
Institut Du Monde Arabe/Pessem: 43*tr*
E. Andrew McKinney: 66*c*
Otis Lift Company plc: 46*bc*
Panos Pictures/Bruce Paton: 46*tl*
Pictor International Ltd: 36*br*, 66*b*, 120*tr*, 125*br*
Planet Earth Pictures/William M. Smithey, Jr: 97*br*
Range/Bettmann: 149*t*
Railway Technical Research Institute, Tokyo: 105*tl*
Peter Sanders Photogaphy: 52*bc*
The Science Museum/Science & Society Picture Library: 81*tl*, 92*tr*, 108*t*
Science Photo Library: Alex Bartel 14*bl*; *Michael W. Davidson* 168*bl*; *Hank Morgan* 162*bl*; *John Mead* 52*bl*: *David Parker* 72-73, 163*tr*, *cr*; *Alfred Pasieka* 154*cla*; *David Scharf* 170*tl*; *Taheshi Takahara* 97*tr*
Sensory Systems: 163*tc*
Sharp Electronics (UK) Ltd: 162*br*
Shell UK Photographic Services: 71*cr*
Frank Spooner Pictures/Gamma/M. Wada: 163*bl*
Tony Stone Images: 97*tl*; *Glen Allison* 14*tl*; *Ken Biggs* 94-95; *John Callahan* 63*br*; *Paul Chesley* 123*tr*; *Lonnie Duka* 154*cr*; *John Edwards* 57*cr*; *Ambrose Greenway* 41*cr*; *Chris Kapolka* 105*br*; *Lester Lefkowitz* 54*bl*, 116-117; *Joe Ortner* 155*cr*; *Martin Rogers* 96*cl*; *David Ximeo Tajada* 97*cl*
The Van Wezel Performing Arts Hall, Sarasota, Florida/Stephen C. Traves: 52*c*
Wisconsin University/Professor H. Guckel: 12-13
Zefa: Mehlig 66*cl*; *Streichan* 66*clb*

Every effort has been made to trace the copyright holders. Dorling Kindersley apologizes for any unintentional omissions and would be pleased, in such cases, to add an acknowledgment in future editions.

ILLUSTRATIONS

Zirrinia Austin: 36*tr*, 115*br*
Rick Blakely: 16*br*, 54*tl*, 79*br*, 183*c*
Mick Gillah: 20*c*, 21*c*, 23*r*, 54*b*, 55*b*, 58*l*, 65, 68*b*, 70*b*, 76, 78*b*, 91*t*, 100*c*, 102*br*, 104*br*, 126*tr*, 128*tr*, 144, 145, 149*br*, 150*bl*, 166*bl*, 177, 183*cr*, 185*t*, 186*b*, 187*t*
Philip Ormerod: 11*cl*, 21*b*, 33*c*, 36*c*, 32*cr*, 33, 42*c*, 50*br*, 61*tr*, 74*br*, 86*b*, 88*b*, 89*bl*, 96*bl*, 98*cr*, 99*tr*, 101*br*, 110, 113*bl*, 124*bl*, 128, 129*t*, 142*bl*, 146*cr*, 158*b*, 161*tr*, 164*bl*, 167*t*, 173, 178*b*, 179*cl*, 182*c*
Bryn Walls: 26*tr*, 27*cl*, 30*b*, 40*cr*, 57*br*, 87*t*, 121*b*, 122*b*, 126*t*, 135*tr*, 157*br*
Alistair Wardle: 84, 96*cr*, 108*br*

MODEL MAKERS

Christina Betts: 137*c*, 143, 159
David Donkin: 2*tl*, 3, 4*tl*, 6, 7*tr*, *c*, 15, 17*br*, 18, 19, 27, 30*c*, 31*b*, 33, 47*t*, 57, 61*b*, 63, 64*b*, 75*tr*, 83, 92, 93, 101*bl*, 115*tr*, 118*cl*, 119*tr*, 123, 127, 129*br*, 131*br*, 133, 138*t*, 139, 151*b*, 155, 161, 165, 169, 170*b*, 176
Philip Ormerod: 7*br*, 37, 59, 125, 164*b*, 164*b*, 171, 172

MODELS

Fatima Anwar-Musah, Neil Ardley, Kim Armstrong, Laura Backhouse, Julien Baffour-Awuah, Christina Betts, Wayne and Sarah Black, Benjamin Bubb, Rebecca Bunting, Thomas Bunting, Lalya Camara, Marissa Campbell, Hayley Cherry, Rob Chong, Lacean Christie, Rebecca Cleveland, Priscilla Coburn, Jessica Coleman, Sarah Cummings, Marsha Denton, Shaun Dorkin, Bora Esen, Tariq el-Fiky, Lorna Franklin, Karley Gibbons, Sophie Goody, Briony Hassett, Elizabeth Hewitt, Cosima Hornak, Alison Hunzer, Kayode Josiah, Masuma Karim, Talia Keane, Damien Macintosh, Atif Malik, Geoff Manders, Michael Mangen, Isis Matrix, James Morcos, Wayne Murphy, Saira Nazeer, Hien N'Go, Chioma Obih, Sean O'Brien, Omoye Okoh, Danny O'Sullivan, Sharon Perrie, Toby Picket, Sheldon Pommell, Alima Rahman, Fahima Rahman, Hyatollah Ramatullah, Lorianne Rhoden, Adam Sales, Louise Sales, Kirsty Sheldon, Kirandeep Siddhu, Navjeet Siddhu, Sean Slattery, James Smith, Peggy Smith, Elliot Stewart, Sanjay Subarwal, Lisa-Rose Sutton, Sara-Jane Sutton, She Peng Tham, Kay Tsang, Vesna Vladusic, Aidan Walls, Vicky Watling

SPECIAL PHOTOGRAPHY

Brian Dowling: 149*b*
Mike Dunning: 158*cr*
Philip Gatward: 95, 157
Gary Kevin: 90
Dave King: 117
Clive Streeter: 118*cr*, 153, 166
Spike Walker (Microworld Services): 177*r*
Bryn Walls: 151, 158*b*